Cell Surface Carbohydrate Chemistry

ACADEMIC PRESS RAPID MANUSCRIPT REPRODUCTION

Cell Surface Carbohydrate Chemistry

edited by

ROBERT E. HARMON

Western Michigan University
Kalamazoo, Michigan

ACADEMIC PRESS, INC. New York San Francisco London 1978

A Subsidiary of Harcourt Brace Jovanovich, Publishers

ACADEMIC PRESS, INC.
111 Fifth Avenue, New York, New York 10003

United Kingdom Edition published by
ACADEMIC PRESS, INC. (LONDON) LTD.
24/28 Oval Road, London NW1

Library of Congress Cataloging in Publication Data

Cell Surface Carbohydrate Chemistry Symposium, San Francisco, 1976.
Cell surface carbohydrate chemistry.

Sponsored by the Carbohydrate Division of the American Chemical Society.

Includes index.
1. Plasma membranes–Congresses. 2. Cancer cells–Congresses. 3. Carbohydrates–Congresses. 4. Cytochemistry–Congresses. I. Harmon, Robert E. II. American Chemical Society. Division of Carbohydrate Chemistry. III. Title.

QH601.C398 1976 574.8'75 77-13638
ISBN 0-12-326150-3
PRINTED IN THE UNITED STATES OF AMERICA

Dedicated to my parents
Asa and Margauerite Crass
and to my siblings
Pat, Gerrie, and Cecil
and in memory of my father,
Everett Fulton Harmon

Contents

List of Contributors

Numbers in parentheses indicate the pages on which authors' contributions begin.

ALOJ, SALVATORE M. (103), Laboratory of Biomedical Pharmacology, National Institute of Arthritis, Metabolism, and Digestive Diseases, National Institutes of Health, Bethesda, Maryland 20014, and Centro di Endocrinologia ed Oncologia Sperimentale C.N.R., Naples, Italy

AMOS, HAROLD (181), Department of Microbiology and Molecular Genetics, Harvard Medical School, Boston, Massachusetts 02115

BECKER, JOSEPH W. (135), The Rockefeller University, York Avenue, New York, New York 10021

BEZKOROVAINY, ANATOLY (255), Department of Biochemistry, Rush-Presbyterian-St. Luke's Medical Center, Chicago, Illinois 60612

BRADY, ROSCOE O. (153), Developmental and Metabolic Neurology Branch, National Institute of Neurological and Communicative Disorders and Stroke, National Institutes of Health, Bethesda, Maryland 20014

BUNIEL, JOSEFINA F. (205), Department of Microbiology, School of Medicine, Case Western Reserve University, Cleveland, Ohio 44106

CHANG, C. (83), National Cancer Institute, National Institutes of Health, Bethesda, Maryland 20014

CHIEN, J. L. (241), Department of Biochemistry, Tulane University, New Orleans, Louisiana 60112

CODINGTON, JOHN F. (49), Laboratory for Carbohydrate Research, Departments of Biological Chemistry and Medicine, Harvard Medical School and Massachusetts General Hospital, Boston, Massachusetts 02114

COLL, J. (83), National Cancer Institute, National Institutes of Health, Bethesda, Maryland 20014

COLOFIORE, JOSEPH (181), Department of Microbiology and Molecular Genetics, Harvard Medical School, Boston, Massachusetts 02215

CULP, LLOYD A. (205), Department of Microbiology, School of Medicine, Case Western Reserve University, Cleveland, Ohio 44106

EDELMAN, GERALD M. (135), The Rockefeller University, York Avenue, New York, New York 10021

FISHMAN, PETER H. (153), Developmental and Metabolic Neurology Branch, National Institute of Neurological and Communicative Disorders and Stroke, National Institutes of Health, Bethesda, Maryland 20014

FLEISCHER, BECCA (27), Department of Molecular Biology, Vanderbilt University, Nashville, Tennessee 37235

FREESE, ERNST (153), Laboratory of Molecular Biology, National Institute of Neurological and Communicative Disorders and Stroke, National Institutes of Health, Bethesda, Maryland 20014

FRIEDMAN, ROBERT M. (103), Laboratory of Experimental Pathology, National Institute of Arthritis, Metabolism, and Digestive Diseases, National Institutes of Health, Bethesda, Maryland 20014

GAGNEJA, G. L. (269), Department of Biochemistry, University of Windsor, Windsor, Ontario, Canada

GLICK, MARY C. (13), Department of Pediatrics, University of Pennsylvania Medical School, Children's Hospital of Philadelphia, Philadelphia, Pennsylvania 19104

GONZALEZ, FEDERICO (181), Department of Microbiology and Molecular Genetics, Harvard Medical School, Boston, Massachusetts 02115

GROHLICH, DIETMAR (255), Department of Biochemistry, Rush-Presbyterian-St. Luke's Medical Center, Chicago, Illinois 60612

GROLLMAN, EVELYN F. (103), Laboratory of Biomedical Pharmacology, National Institute of Arthritis, Metabolism, and Digestive Diseases, National Institutes of Health, Bethesda, Maryland 20014

HENNEBERRY, RICHARD C. (153), Laboratory of Molecular Biology, National Institute of Neurological and Communicative Disorders and Stroke, National Institutes of Health, Bethesda, Maryland 20014

HUMPHREY, LINDA (181), Department of Microbiology and Molecular Genetics, Harvard Medical School, Boston, Massachusetts 02115

JAMIESON, G. A. (225), The American National Red Cross Blood Research Laboratory, Bethesda, Maryland 20014

JEANLOZ, ROGER W. (49), Laboratory for Carbohydrate Research, Departments of Biological Chemistry and Medicine, Harvard Medical School and Massachusetts General Hospital, Boston, Massachusetts 02114

KARNOVSKY, MORRIS J. (181), Department of Pathology, Harvard Medical School, Boston, Massachusetts 02115

KOHN, LEONARD D. (103), Laboratory of Biomedical Pharmacology, National Institute of Arthritis, Metabolism, and Digestive Diseases, National Institutes of Health, Bethesda, Maryland 20014

KORYTNYK, W. (311), Department of Experimental Therapeutics, Roswell Park Memorial Institute, Buffalo, New York 14263

LEDLEY, FRED D. (103), Laboratory of Biomedical Pharmacology, National Institute of Arthritis, Metabolism, and Digestive Diseases, National Institutes of Health, Bethesda, Maryland 20014

LEE, G. (103), Laboratory of Biomedical Pharmacology, National Institutes of Arthritis, Metabolism, and Digestive Diseases, National Institutes of Health, Bethesda, Maryland 20014

LI, S.-C. (241), Delta Regional Primate Research Center, Covington, Louisiana 70433

LI, Y.-T. (241), Department of Biochemistry, Tulane University, New Orleans, Louisiana 60112

LUBORSKY, S. W. (83), National Cancer Institute, National Institutes of Health, Bethesda, Maryland 20014

MELDOLESI, MARIA F. (103), Laboratory of Biomedical Pharmacology, National Institute of Arthritis, Metabolism, and Digestive Diseases, National Institutes of Health, Bethesda, Maryland 20014 and Centro di Endocrinologia ed Oncologia Sperimentale C.N.R., Naples, Italy

MORA, P. T. (83), National Cancer Institute, National Institutes of Health, Bethesda, Maryland 20014

MULLIN, BRIAN R. (103), Laboratory of Biomedical Pharmacology, National Institute of Arthritis, Metabolism, and Digestive Diseases, National Institutes of Health, Bethesda, Maryland 20014

PANCAKE, S. J. (83), National Cancer Institute, National Institutes of Health, Bethesda, Maryland 20014

PAOLINI, DOMENIC (181), Department of Microbiology and Molecular Genetics, Harvard Medical School, Boston, Massachusetts 02115

PAUL, B. (311), Department of Experimental Therapeutics, Roswell Park Memorial Institute, Buffalo, New York 14263

PIGMAN, WARD (1), Department of Biochemistry, New York Medical College, Valhalla, New York 10595 *

PODOLSKY, DANIEL K. (67), Gastrointestinal Unit, Massachusetts General Hospital, Boston, Massachusetts 02114

REEKE, GEORGE N., JR. (135), The Rockefeller University, York Avenue, New York, New York 10021

ROSEN, JONATHAN J. (205), Department of Biomedical Engineering, Case Western Reserve University, Cleveland, Ohio 44106

ROSSOW, PETER (181), Department of Microbiology and Molecular Genetics, Harvard Medical School, Boston, Massachusetts 02115

SANTER, URSULA (13), Department of Pediatrics, University of Pennsylvania Medical School, Children's Hospital of Philadelphia, Philadelphia, Pennsylvania 19104

SLY, DAYLE A. (255), Department of Biochemistry, Rush-Presbyterian-St. Luke's Medical Center, Chicago, Illinois 60612

TAYLOR, N. F. (269), Department of Biochemistry, University of Windsor, Windsor, Ontario, Canada

VAN DEN EIJNDEN, DIRK H. (49), Department for Medical Chemistry, Vrije Universiteit, Van der Boerchlorststraat 7, Amsterdam, The Netherlands

WEISER, MILTON M. (67), Gastrointestinal Unit, Massachusetts General Hospital, Boston, Massachusetts 02114

WIGLER, PAUL W. (291), Memorial Research Center, University of Tennessee Center for the Health Sciences, Knoxville, Tennessee 37920

Deceased

Preface

This volume is a record of the Cell Surface Carbohydrate Chemistry Symposium held on September 1 and 2, 1976 in conjunction with the American Chemical Society Centennial Meeting in San Francisco.

The objective was to bring together carbohydrate chemists and cell biologists who are involved in or interested in studying the oligosaccharides that make up the glycoproteins and glycolipids found in the cell membranes of normal or neoplastic cells.

Hopefully, by comparing the differences in oligosaccharide structure and/or composition of cancer cell membranes with those of normal cell membranes some insight may be gained into the process by which cells become malignant and how neoplastic cells could be more effectively eradicated by chemotherapy or through immunization techniques. The role of oligosaccharides in antigen recognition was discussed. It was stressed that scientists in this broad field must be innovative. They must take into account in their interpretations the example of numerous immune responses made by the body when attacked by foreign pathogenic organisms.

The "Central Dogma" in 1957 summarized principles for the flow of information from the fundamental deposit held in the DNA-codes of each cell nucleus to the process of protein biosynthesis by which distinctive enzymatic functions are implemented. Once the information has been passed into the protein it cannot get out again. The passage of information from protein to protein or from protein to nucleic acid as thus envisioned is forbidden. The theory as propounded is primarily concerned with proteins as enzymes and with the enzyme specific activities arising from unique amino acid sequences and conformations. The cell is thus genetically inviolate from its environment except for the potential information its nucleus store allows and is capable of implementation.

Biochemical research especially in immunology and cytology has shown since then that types of specified biological information exists other than DNA- and RNA-based codons. In particular, the mechanism of molecular recognition apparently implicit in antigen-antibody interactions and in a variety of cell recognition phenomena associated with the cell membranes indicates other forms of information being conveyed through macromolecules. Investigation of the determined groups conveying blood group specificity of the A, B, and H system has shown that both glycoproteins and glycolipids possess these functions. The specificity of the information is derived from a series of specific oligosaccharides attached to a protein at the time of or after its biosynthesis in an ordered sequence. An enzyme has been described that transfers ADP-ribose from nicotine adenine dinucleotide (NAD^+) to the amino groups of the peptides. The formation of some glycoproteins at least is accomplished by this mechanism. This attachment occurs in the following stages. The transfer RNA amino acid complex interacts with the ribosome to give a protein. The protein plus nicotine adenine dinucleotide plus an enzyme yields an ADP-ribosyl protein or (poly)ADP-ribosyl protein. The activated proteins are acted on by polymerase and transglycosidation enzymes to give glycoproteins. This mechanism is still being studied and it can be used to provide a plausible concept for antibody formation as well as for the formation of various blood group types. Previously this mechanism was given little credence in the belief that free peptides in the cell are rare or nonexistent. However, the literature is filled with papers showing their existence. At least some of the peptides have been shown to be attached to nucleotides. Each informational oligosaccharide is required to be complete in certain essential structural details if it is to perform its function and if bound to a membrane it is probably attached in some definitive topographical way. This volume is concerned with the exploration and explanation of membrane mediated information, its basis and regulation.

Though chemists and biologists should probably not concern themselves with the thought that some new class of biological molecules with comparative importance to the proteins remains to be discovered, there is now the very real likelihood that the informational content of proteins is not exhausted by the amino acid sequence. In addition to the postpolymerization glycosidation reaction mentioned previously, other modifications such as phosphorylation, sulfation, or possibly deamination offer significant and cognant means for adding distinctive information to proteins. It may be argued that such informational overtones must necessarily be of secondary significance since these can only be brought about by the action of specific enzymes themselves arriving from the sequence derived information content of a protein in that it appears to be susceptible to modification at the enzyme level (e.g., by kinetic and specificity effects) without recourse to the genome. A number of reports in this volume are concerned with these phenomena. The final molecular structure of a glycoprotein and possibly a glycolipid also (as far as the carbohydrate moiety is concerned) would thus appear to be dependent on the metabolic state of the cell and its environment. These possibilities are of great interest since they appear to imply further possible mechanisms by which biochemical individuality can be defined, as well as further degrees of flexibility, enabling cells to respond to limited changes required by their surroundings. Evidently, some critical control of growth, both normal and abnormal, must be mediated at the cell surface where the cells meet with each other and with the factors that influence their growth.

I would like to thank the participants at this symposium for their questions and discussions that created such a stimulating atmosphere. I am especially grateful to the speakers for being so prompt in submitting their manuscripts for this volume. I would like to acknowledge the administrative assistance of my wife, Marilyn K. Harmon, and Mr. A. L. Pence, Chief Accountant, Office of the Controller, American Chemical Society, who handled the on-site financial arrangements for the symposium. I would also express my appreciation to Mrs. Chris Harmon and Mrs. Barbara Hodsdon from the American Chemical Society Staff who selected the meeting room and made other physical arrangements for the symposium. I would like to thank Mrs. Marilyn K. Harmon for preparing the index for this volume. Lastly, I wish to thank Mrs. Kathy Kelley for doing the necessary typing.

This symposium was sponsored by the Carbohydrate Division of the American Chemical Society and was sponsored in part by Grant 1R13 CA20439-01 from the National Cancer Institute.

Executive Editor
Robert E. Harmon

THE INVOLVEMENT OF MEMBRANES IN THE BIOSYNTHESIS OF GLYCOPROTEINS*

WARD PIGMAN

New York Medical College
Biochemistry Department
New York, New York

University of Alabama
Medical Center
Birmingham, Alabama

INTRODUCTION

Many texts and books state that the free peptides are rare or nonexistent, yet the literature is replete with papers showing their existence. At least, some of these have been shown to be attached to nucleotides. An enzyme has been described by numerous workers that transfers AMP-ribose from NAD^+ to the amino groups of peptides.

The formation of some glycoproteins, at least, is described as occurring in the following stages

$$\text{t-RNA-amino acid} \xrightarrow{\text{ribosome}} \text{peptides}$$

$$\text{paptides} + NAD^+ \xrightarrow{\text{enzyme}} \text{AMP-ribose-peptides}$$

$$\text{activated peptides} \xrightarrow[\text{transglycosylases}]{\text{polymerases}} \text{glycoprotein}$$

*This work was supported by CA 17168 awarded by the National Cancer Institutes, DHEW.

Further evidence for this mechanism is in progress. It is used to provide a plausible concept for antibody formation.

PEPTIDES AND "ACTIVATED PEPTIDES" IN CELLS

Generally it is believed that previous workers have not been able to show the presence of small peptides in cells. Thus, Ramsay (1) states: "analyses show the presence in organisms of proteins and of free amino acids, but peptides of intermediate size are conspicuously absent." Spiegleman (2) stated: "No convincing evidence for the existence of peptide intermediates has been found under conditions where they might reasonably be expected to accumulate." Hendler (3) summarizes the work in the 1950s and early 1960s on the concept of peptide intermediates in protein synthesis and indicates that peptides are not generally present in cells, but could be present transiently.

Although the current concept of protein biosynthesis rejects the idea of peptides being intermediates in protein anabolism and catabolism, their occurrence in cells is not as uncommon as generally thought. Thus, in 1966, Waley (4) devoted 99 pages to a review on "Naturally Occurring Peptides" and reached the conclusion that peptides do exist widely in cells, but that early methodology was not suitable for detecting their presence. Medveder (5) also reviewed the subject and gives references to the existence of peptides activated by combination with nucleotides. It is our opinion that the early negative data on the subject created the general impression that peptides were absent from the cell, and that most later investigators made no attempt to look for them.

Actually, there are numerous references in the literature relating to the presence and isolation of peptides in both microorganisms and mammalian organs. Sorm and co-workers (6-8) and Wiseman and co-workers (9) showed that peptides mostly in the form of activated peptides are present in *Escherichia coli* and yeast cells, respectively. These activated peptides were found to be associated with nucleotides, the structure of which was established for several. Some of the components of these peptides were related to cell wall structures. Szafranski *et al.* (10) and Steinberg *et al.* (11) isolated similar compounds from guinea pig liver, and from rat kidney and liver, respectively.

In two papers, we have reported (12, 13) the occurrence of peptide materials in the ultrafiltrates of mildly homogenized cells from bovine and porcine submaxillary glands, rat liver, and *E. coli*. The yield can only be roughly estimated due to the presence of unknown amounts of other materials, but on the basis of the protein values (Lowry) for the ultrafiltrates, the peptide materials represented 0.1 to 2% of the dry weight of the original cells.

In our work (12, 13) the peptide materials obtained from *E. coli* cells and from animal sources exhibited strong absorption at 260 nm, which suggests that they are combined with nucleotides. The UV absorption was considered to be of particular significance in the case of bovine and porcine submaxillary glands, since the mucus glycoproteins, which constitute their major product, contain at most only small amounts of amino acids, which absorb at 280 nm. In the light of the earlier observations cited, it seems highly probable that a high proportion of these peptides are indeed linked to nucleotides.

We carried out the following analyses on the lyophilized ultrafilterable material from bovine and porcine submaxillary glands, rat liver, and *E. coli* cells:

Lowry protein

Ninhydrin value before and after hydrolysis

Amino acid composition

Ultraviolet and visible absorption measurements

Two-dimensional peptide mapping or high voltage electrophoresis

Carbohydrate analysis by the phenol-sulfuric acid procedure (in some cases) and hexosamine determinations

The increase in ninhydrin values after acid hydrolysis indicated the presence of peptides averaging from 2 to 20 amino acids in size. Only trace amounts of free amino acids were detected.

Both *E. coli* and rat liver fractions showed a total of 32 spots after peptide mapping. Resolution by high-voltage electrophoresis of the submaxillary gland fractions showed 16-18 spots.

The amino acid compositions of the hydrolyzed fractions were variable, and several unknown peaks were observed. One fraction from the bovine and porcine glands appeared to consist of a mixture of peptides having an average size of 20 amino acids; it showed a qualitative correspondence with the purified mucus glycoproteins derived from these glands.

Hexosamines were present only in traces, but the phenol-sulfuric acid method showed the presence of substantial amounts of "carbohydrates," presumably hexoses or pentoses (ribose?).

ORIGIN OF THE PEPTIDES AND "ACTIVATED PEPTIDES"

The possibility that the ultrafilterable material resulted from degradation of preformed proteins or glycoproteins during the isolation procedure was investigated in earlier work in this laboratory (12, 13) and was shown to be unlikely. Thus, when *E. coli* cells were placed in boiling water before sonication, the yield of ultrafilterable material was the same as that for unboiled cells and was independent of the degree of sonication. The same results for boiled and unboiled tissues were observed for rat liver and submaxillary glands. Furthermore, none of the ultrafiltrates showed an increase in ninhydrin value upon standing at room temperature for prolonged periods of time. Since bovine submaxillary glycoproteins contain large amounts of bound hexosamine, the degradation products resulting from endogenous proteolytic enzymes should also contain hexosamines, but only traces were found in the ultrafiltrate.

With yeast cells growing in the presence of ^{14}C-arginine, Cooper *et al.* (9) found that radioactivity was more rapidly incorporated in the peptide fractions than in the protein fractions. In unpublished preliminary work with *E. coli* cells, which had been grown in the presence of T-leucine, we were able to show that up to 10 min, T-leucine was incorporated in the ultrafilterable fraction more rapidly than in the nonfilterable (protein) fraction; after this time, the ratio of radioactivity for the two fractions became constant. These data indicate that the peptides are precursors.

Another possible source of peptides is in the normal catabolism of proteins. According to prevailing concepts, complete degradation to amino acids takes place. If instead of complete degradation to amino acids, cleavage to peptides occurred in the catabolic process, these peptides could be activated and returned to the pool of activated peptides. This process would conserve much peptide bond energy.

Using a cell-free system (yeast), Cooper *et al.* (9) produced some evidence for this concept and indicated that some of the peptides appeared to be formed from cellular proteins.

ACTIVATED PEPTIDES AS A PRECURSOR OF PROTEINS

Our original concept, that activated peptides may be the precursors of at least certain types of large secretory glycoproteins, was "accidental," as a result of our studies of the structures of mucus glycoproteins, especially bovine submaxillary glycoproteins (BSM). We have found that the core protein of BSM consists to a large extent of peptide repeating sequences containing about 28 amino acids or its trimer. Actually, these "repeating units" appear to represent a family of closely related peptides, probably differing in a few amino acids (14). The glycoprotein has a molecular weight of about 10^6 and on this basis may consist of as many as 500 homogeneous repetitive peptide sequences.

Shackleford and Wilborn (15) provided electron micrographs of bovine submaxillary glands. Light scattering data indicated that the BSM exists as an extended rod with dimensions about 4000 by 40 Å (diam.). Such molecules should be readily seen in the electron micrographs, which have about a 15 Å resolution. The particles in the cysternae of the endoplasmic reticulum were about 25 Å; larger molecules, however, could be seen in the Golgi sacs, as clumps in the case of the bovine glands, but apparently as individual molecules in the cat submaxillary sacs (16).

Using pulse labeling with amino acids, Jamieson and Palade (17) had found an accumulation of label to occur at the ribosomes of pancreatic tissues several minutes after exposure; several minutes later the label was at the Golgi membranes. These authors ascribed this observation to synthesis of proteins at the ribosomes, followed by "packaging" at the Golgi membranes.

The presence of small molecules in the cysternae of bovine submaxillary glands suggested the presence of small peptides and led to our successful attempts (12, 13) to isolate dialyzable peptides, as described earlier. Other tissues also yielded peptides. It may be fortuitous, but by two-dimensional peptide mapping we found 32 ninhydrin positive spots for *E. coli* and rat liver homogenates, almost the same number (30) reported by Sorm and associates (6-8).

We suggested that the "packaging" observed by Jamieson and Palade corresponds to the assembly of "activated" peptides by polymerases at the Golgi membrane.

For a large molecule composed of repeating sequences such as for BSM, one single polymerase might be sufficient. If this mechanism is of a more general nature and applicable to other large proteins and glycoproteins synthesized from a common pool of "activated" peptides, the number of such polymerases required could be small provided the specificity resides principally in the C- and NH_2-terminal amino acid regions of the peptides.

In the preliminary work, we found that after exposure of *E. coli* cells to T-leucine for short periods (3-10 min), the ultrafilterable material had a greater radioactivity than the nonultrafilterable fraction. After 10 min, the ratio became constant. This result suggests that the peptides are primarily precursors rather than degradation products. In further experiments, again using T-leucine, we separated the ultrafilterable and nonultrafilterable fractions and removed unreacted T-leucine on Sephadex G-10. "Hot" ultrafilterable material (in some cases boiled briefly) was incubated with cold nonultrafilterable material, and the mixture was placed on Sephadex G-150. High molecular weight "hot" protein was present in significant quantities. This result again suggested that the peptide fraction was converted to "protein" in the absence of amino acids. The use of boiled material excluded any possibility of degradation of this fraction to amino acids. This work is being continued and expanded with rat colonic cells.

OTHER EVIDENCE

The concept of protein biosynthesis via activated peptides appears to have been the concept of a considerable number of workers in the field until about fifteen years ago, when the current concepts of the role of DNA-RNA-ribosomal system was established. This, and the concept of "one gene, one protein" (18) became generally accepted despite the earlier evidence. As late as 1966, Waley (4), reviewing the literature, suggested that a two-stage mechanism is involved in the synthesis of some types of proteins, as has Fruton (19), the Sorm group (6-8), Wiseman, Harris, and associates (9), Szafranski and Bagdasarian (10), and many earlier workers. Hendler (20, 21) has published several papers showing that small "lipopeptides" are present in hen oviduct cells and that these are more active than free amino acids in protein synthesis. Since these were isolated by lipid solvents, their relation to the peptides, which we have obtained, is not clear. His "lipopeptides" have been only partially characterized.

Recently, amino acids have been shown to add covalently to plasma proteins both by an enzymatic and a nonenzymatic process. Fuller and Roush (22) found in many species of animals that tryptophan could add to plasma proteins. Leibowitz and Soffers (23) reported that leucyl, phenylalanyl t-RNA protein transferase will add leucine and phenylalanine to serum albumin, especially if the amino terminal group was arginine.

Erhan and Greller (24) have used a computer program for finding common peptide sequences in unrelated and related proteins and in the same molecule. They report numerous examples of common peptide sequences of statistical significance. For example, the hexapeptide active site region of trypsin, chymotrypsin, and *S. grieseus* matched a region in a Bence Jones immunoglobulin at a very high significance. Erhan and Greller are seeking the original "parent" proteins from which all subsequent ones were derived by mutations. They state: "Thus, it is conceivable that formation of present day proteins from a limited number of primordial peptides would not only circumvent the dilemma described by Salisbury but that such a primordial process is quite feasible." The dilemma mentioned is the inconceivable time required for evolution by single mutations.

The proposed two-stage synthesis may be a specialized mechanism for certain types of proteins. The Co-A system of Lipman (25) provides for some bacterial peptides a different mechanism not involving ribosomes. The two-stage mechanism still requires formation of activated peptides at the ribosomes in the conventional manner, but introduces a second assembly step.

Apte and Ziser (26) report that for certain pairs of mutants, *E. coli* β-galactosidase upon dissociation does not yield two independent polypeptides but only a single covalently bound protomer. Evidence was provided that two genes are required for the synthesis and that the monomers are combined by a "protein ligase."

The first issue of Trends in Biochemical Sciences contains a review by Hayaishi (27). Hayaishi reports that much work has been carried out on the formation of ADP-proteins. Three symposia were held in 1972, 1973, and 1974 on the subject. The reaction is

NAD^+ + "protein" → ADP-ribosyl-protein + nicotinamide

NAD^+ + "protein" → poly-(ADP-ribosyl)-protein + nicotinamide

The enzyme poly-ADP-ribose synthetase was found in the nucleus of both eucaryotic and procaryotic cells. Later, it was found in mitochondria. Because it was found in the nucleus initially, most attention was paid to the poly-ADP-ribose portion. However, the function of a poly-ADP-ribose has no precedent. We suggest, in light of the previous discussion, that this enzyme activates the peptides, which are formed at the ribosomes and are fed into the general pool of activated peptides.

DEFICIENCIES OF THE ORTHODOX CONCEPT OF PROTEIN BIOSYNTHESIS

The current concept of protein biosynthesis requires that the complete protein be formed at the ribosomal level controlled by one gene. The concept has been very useful, but it was developed mainly by work on microorganisms and applied to higher animals. As new data appear, it is becoming increasingly clear that it cannot explain important biological processes. Such a situation has been the fate of all important concepts. The orthodoxy of the unreactivity of the noble gases was changed easily when they were found to react under some conditions. The first law of thermodynamics is still useful, despite Einstein's theory and later proof that matter can be converted to energy. Several examples of the deficiencies of the orthodox mechanism of protein biosynthesis will be given. It will also be shown that they can be explained easily by the two-step mechanism.

By the current dogma, the DNA strands and m-RNA must be essentially infinitely long to explain antibody formation, the formation of large proteins, especially glycoproteins, and the differentiation of cells during tissue development. The first two of these will be discussed.

How does an antigen, which does not enter the plasma cell, for example, influence the formation of the antibody? The orthodox concepts are reviewed by Hood and Prahl (28). It has no contact with the nucleus. The current concepts have no explanation that satisfied many workers. The two-step mechanism explains it easily, with one more change in concept. This is the concept that the Golgi membranes are the origin or part of the plasma membrane. Biochemical evidence is accumulating for this concept, which really was implied in the Palade description of the removal of "zymogen granules" from secretory cells.

The new concept implies that the plasma cell has an antigen receptor. This attaches to the antigen which becomes a part of the plasma membrane. As shown in Fig. 1, components of the pool of activated peptides migrate to the membrane where polymerases form the immunoglobulin, which has "fixed regions" for all immunoglobulins and others complementary to the antigen for the variable regions. As long as the antigen remains in the membrane at the receptor site, the cell will continue to make only one type of immunoglobulin. The usual loss of protection after a period of time simply involves loss of the antigen and a freeing of the receptor site.

The current dogma requires that all daughter cells produce the same immunoglobulin. This concept has no experimental basis but is a result of the dogma that the DNA must have been altered. The two-stage concept does not make such demands.

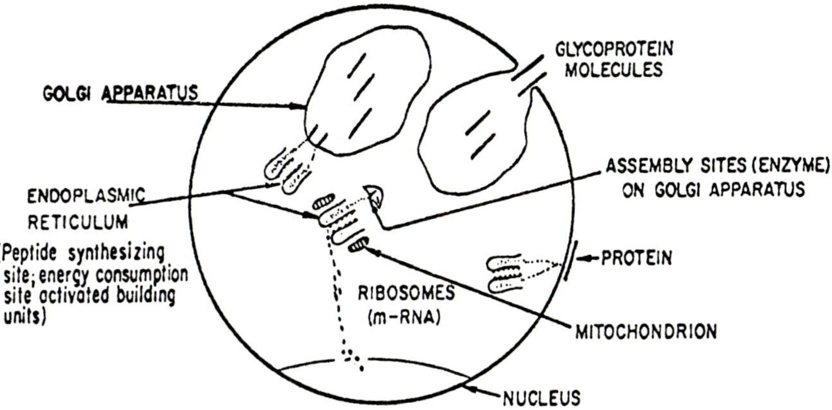

FIGURE 1 Schematic diagram of protein biosynthesis.

The dogma requires that after a heavy infection (exposure to antigens) no unaltered cells will be left. The animal will then have no protection against subsequent infections (exposure to new antigens). The new concept does not require changes at the DNA level and daughter cells will have the capacity to attract new antigens.

The orthodox concept assumes that the entire protein is formed at the ribosomal level under control of one gene. This concept was developed mostly from studies using microorganisms and relatively simple proteins. For the mucus glycoproteins, it becomes extremely complex. The mucus glycoproteins have a large protein core ($\sim 3 \times 10^5$) and hundreds of oligosaccharide side chains of variable size and complexity. The recent reports from the laboratory of Kabat (29) indicate that those from human ovarian cyst fluid have side chains of one carbohydrate unit to much longer than 30 component sugars. In the biosynthesis, the addition of each sugar to the growing chain appears to require a special transglycosylase. Since each enzyme requires a gene, the biosynthesis of the glycoprotein requires the interaction of many, many genes. Such interaction is difficult or impossible to explain by the current dogma.

The two-stage concept provides a conceivable mechanism. The pool of activated peptides formed at the ribosomes and contained in the endoplasmic reticulum are assembled at or in the Golgi membrane by an organized arrangement of peptide polymerases and glycosyltransferases. How they are organized is pure speculation at this time, but some type of organization obviously must exist.

CONCLUSIONS

The current concept of protein biosynthesis has been one of the greatest achievements of biological science. However, an increasing number of phenomena are not easily explained by the dogma. As with most past great advances, modifications are required, which will not alter most of its postulates, but which will modify them to explain its deficiencies. By tradition, scientists must be prepared to constantly question all dogmas and to alter them as they fail to explain basic phenomena.

REFERENCES

1. Ramsay, J. A. (1966). "The Experimental Basis of Modern Biology," pp. 259-260. Cambridge Univ. Press, Cambridge, Massachusetts.
2. Spiegleman, S. (1957). "The Chemical Basis of Hereditary" (eds. W. D. McElroy and B. Glass). The Johns Hopkins Press, Baltimore, Maryland.
3. Hendler, R. W. (1968). "Protein Biosynthesis and Membrane Biochemistry." Wiley, New York.
4. Waley, S. G. (1966). *Advances in Protein Chemistry 21*: 1-112.
5. Medvedev, Zh. A. (1966). "Protein Biosynthesis and Problems of Hereditary Development and Aging." Plenum Press, New York, Chapter VI.
6. Grünberger, D., Cerná, J., and Sorm, F. (1960). *Experientia 16*: 54.
7. Cerná, J., Rychlik, I., Grünberger, D., and Sorm, F. (1961) *Coll. Czechoslov. Chem. Commun. 28*: 1215.
8. Cerná, J., Grünberger, D., and Sorm, F. (1962). *Ibid., 27*: 1422.
9. Cooper, A. H., Harris, G., Neal, G. E., and Wiseman, A. (1963). *Biochem. Biophys. Acta 68*: 68.
10. Szafranski, P., and Bagdasarian, M. (1961). *Nature 190*: 719.
11. Steinberg, D., Vaughn, M., Sherman, M., and O'Dell, B. L. (1960). *Biochim. Biophys. Acta 40*: 225.
12. Moschera, J., Mound, R., Payza, N., Pigman, W., and Weiss Weiss, M. (1970). *FEBS Letters 6*: 326.
13. Herp, A., Liska, M., Pigman, W., and Vittek, J. (1970). *FEBS Letters 6*: 321.
14. Pigman, W., Moschera, J., Weiss, M., and Tettamanti, G. (1973). *Eur. J. Biochem. 32*: 148.

15. Shackleford, J. M., and Wilborn, W. H. (1970). *In* "Blood and Tissue Antigens" (ed. D. Aminoff), pp. 223-227. Academic Press, New York, New York.
16. Shackleford, J. M., and Wilborn, W. H. (1970). *J. Morphology 131*: 253.
17. Jamieson, J. D., and Palade, G. E. (1967). *J. Cell. Biol. 34*: 577.
18. Beagle, G. W. (1946). *In* "Currents in Biochemical Research" (ed. D. E. Green), p. 6. Interscience, New York.
19. Fruton, J. S. (1963). *In* "The Proteins" (ed. H. Neurath), 2nd ed., Vol. 1, p. 190. Academic Press, New York, New York.
20. Hendler, R. W. (1965). *Proc. Natl. Acad. Sci. 54*: 1233.
21. Hendler, R. W. (1963). *Biochim. Biophys. Acta 74*: 667.
22. Fuller, R. W., and Roush, B. W. (1973). *Comp. Biochem. Physiol. (B) 46 (2)*: 273.
23. Leibowitz, M. J., and Soffers, R. L. (1971). *J. Biol. Chem. 246*: 4431.
24. Erhan, S., and Greller, L. D. (1974). *Int. J. Peptide Protein Res. 6*: 175; *Nature 251*: 353; Erhan, W. (1976). Medical Hypotheses, in press.
25. Lipman, F. (1971). *Science 173*: 875.
26. Apte, B. N., and Zioser, D. (1973). *Proc. Natl. Acad. Sci. 70*: 2969.
27. Hayaishi, O. T. (1976). *Trends Biol. Sci. 1*: 9-10, Jan.
28. Hood, L., and Prahl, J. (1971). *Adv. Immunology 14*: 291-351.
29. Rovis, L., Anderson, B., Kabat, E. A., Greuzo, F., and Liao, J. (1973). *Biochemistry 12*, 1955, 5340.

Note added in proof: We have found with rat colonic cells that ADP- and GDP-peptides are present to the extent of 11% of the dry weight of the cells.

The following pertinent references have been found recently:

Elzinga, M., Maron, B. J., and Adelstein, R. V. (1976). *Science 191*, 94.

Walker, D., and Bridgen, J. (1976). *Eur. J. Biochem. 67*, 283.

Cantore, M. L., Fernandéz de Recando, M. E., and Recondo, E. (1972). *In* "Biochemistry of the Glycosidic Linkage" (R. Piras and H. G. Pontis, eds.). Academic Press, New York.

SURFACE MEMBRANES OF TUMOR CELLS

MARY CATHERINE GLICK
URSULA SANTER

Department of Pediatrics
University of Pennsylvania Medical School
Children's Hospital of Philadelphia
Philadelphia, Pennsylvania

Surface membrane glycoproteins of tumor cells contain glycopeptides characteristic of transformed fibroblasts and epithelial cells. Loosely associated and more intrinsic membrane glycoproteins of transformed cells, metabolically labeled, show striking differences in the specific activity of the hydrolyzable fucose residues. Pronase digestion and chromatographic purification separate four fucose-containing glycopeptides unique for transformed cells from three common to normal and transformed cells. One transformation-specific glycopeptide, which has been sequenced by treatment with exoglycosidases and quantitation of the released monosaccharides, contains three oligosaccharide branches, each terminating in sialic acid. The core contains β-D-mannose and α-L-fucose. The putative structure is given and reinforces previous suggestions that transformation specific glycopeptides are more complex than those from normal cells.

INTRODUCTION

Glycoproteins from the surface of primary tumor cells appear to have complex saccharide structures similar to those described from virus transformed fibroblasts (1, 2) or transformed epithelial cells (3) in culture. This has been shown in human neuroblastoma (4) and leukemic cells (5, 6) and in hamster and mouse tumors (2, 7, 8). The normal counterparts appear to be less complex. The latter include nontransformed hamster or human fibroblasts and epithelial and brain cells. The complexity is expressed in the carbohydrate structures of the membrane glycopeptides (9).

The altered glycoproteins found in the surface membrane of the malignant cell could be the result of the expression of single or multiple events. Defining the causal event is complicated because of the variety of factors involved in glycoprotein metabolism and membrane assembly. Any number of areas can be examined such as the synthetic and degradative enzymes, dolichol metabolism, and the attachment of the carbohydrates to the polypeptides. However, a more direct approach to this problem is to analyze in detail various oligosaccharide units derived from the membrane glycoproteins of malignant and nonmalignant cells. After the structures are known, the enzymes responsible for the phenotypic expression of the key differences in the structures can then be explored as causal events. Using this approach, we have defined alterations in the glycoproteins from membranes of malignant cells, and from the detailed analyses we have shown that more complex and highly branched oligosaccharides are more representative of the virus transformed cell surfaces than their normal counterparts. We will first review the data showing the membrane glycopeptide alterations in tumor and virus transformed cells, then describe briefly the methods and results of oligosaccharide analyses, and finally give the putative structure of a membrane glycopeptide and discuss the implications of the structure to tumorigenesis.

GLYCOPEPTIDE ALTERATIONS IN TUMOR AND VIRUS TRANSFORMED CELLS

A number of methods are available for removing glycopeptides from the cell surface. In our laboratory, trypsin under controlled conditions has been the most used; the procedures for these trypsinizations are shown in Fig. 1.

It was proposed that trypsinate A, the result of the mildest procedure, contains the more loosely associated glycoproteins, while trypsinate C contains the loosely associated glycoproteins as well as those more intrinsic to the membrane (9). The B trypsinization, which follows trypsinization A, removes mainly the more intrinsic glycoproteins. Surface membranes containing the remaining glycoproteins can be prepared after any of the trypsinization procedures. For detection, the glycoproteins were labeled metabolically with radioactive precursors by growing the cells in the presence of L-[^{3}H]- or [^{14}C]fucose. The glycopeptides, obtained by any of the three trypsinizations or preparation of surface membranes, were further digested with Pronase to remove most of the remaining amino acids and examined by chromatography on Sephadex G-50. Examples of the glycopeptides removed by each of these proce-

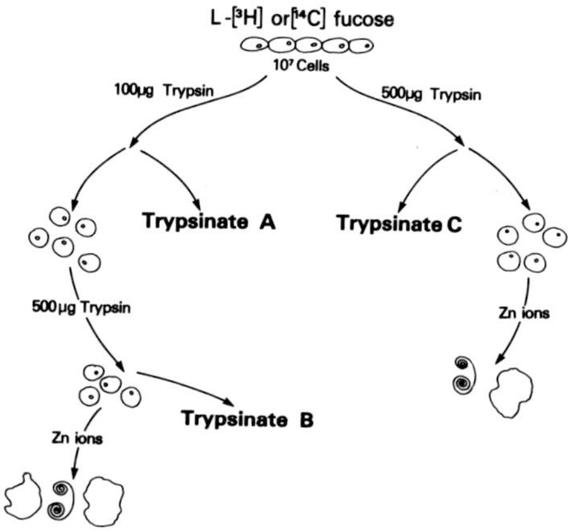

FIGURE 1 Schematic outline of the protocol for the preparation of trypsinates and surface membranes.

dures from baby hamster kidney fibroblasts transformed by the Bryan strain of Rous sarcoma virus ($BHK_{21}C13/B4$) are shown by the profiles of the Sephadex G-50 columns (Fig. 2). In all of these trypsinization procedures, two groups of glycopeptides (fractions 28-35 and 36-48) were obtained from all virus transformed fibroblasts thus far examined although different proportions of these groups were obtained from the different cell lines (compare Figs. 2 and 3) and as a result of the trypsinization procedure used (Fig. 2). Fractions 63-80 (Figs. 2 and 3) contained dialyzable material including free fucose, and will not be discussed here. Note that this material was not present in surface membranes (Fig. 2c).

An example of the glycopeptides obtained in trypsinate C from transformed epithelial cells is shown in Fig. 3, where a comparison is made to the nontransformed counterpart. The nontransformed cell surface material contained only one group of glycopeptides (fractions 35-50). Further evidence that the early eluting group of glycopeptides (fractions 28-34) represents the transformed state was obtained with a mutant epithelial cell line, temperature sensitive for the transformed phenotype. The trypsinates from these mutant cells showed an increased proportion of the early eluting group of glycopeptides at permissive temperature, and at nonpermissive temperature showed a shift in the Sephadex G-50 profile to a pattern resembling that of the normal epithelial cells (3).

Examples of glycopeptides obtained from primary tumor cells are shown in Fig. 4. The membrane glycopeptides from tumor cells of a human neuroblastoma (4) and from the leukemic

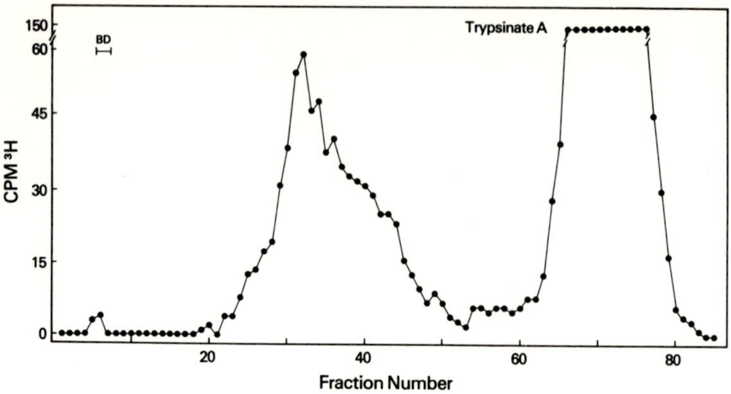
Trypsinate A
BD
150
60
45
30
15
0
CPM ³H
20
40
60
80
Fraction Number

A

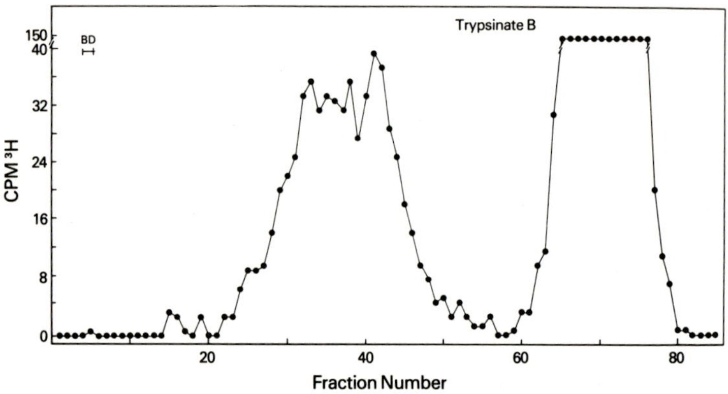
Trypsinate B
BD
150
40
32
24
16
8
0
CPM ³H
20
40
60
80
Fraction Number

B

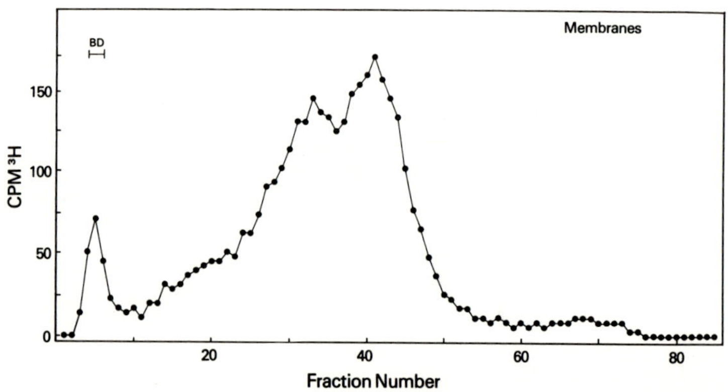
Membranes
BD
150
100
50
0
CPM ³H
20
40
60
80
Fraction Number

C

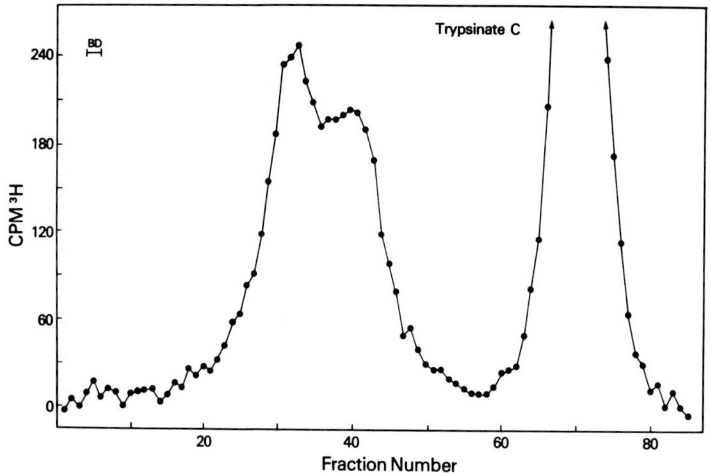

D

FIGURE 2 Sephadex G-50 chromatography of Pronase digested trypsinates and surface membranes of transformed baby hamster kidney fibroblasts C_{13}/B_4. (A) Trypsinate A (B) trypsinate B (C) surface membranes and (D) trypsinate C were prepared as shown in Fig. 1 from cells grown in the presence of L-[^{3}H]-fucose for 72 hr. BD represents fractions in which Blue Dextran 2000 eluted.

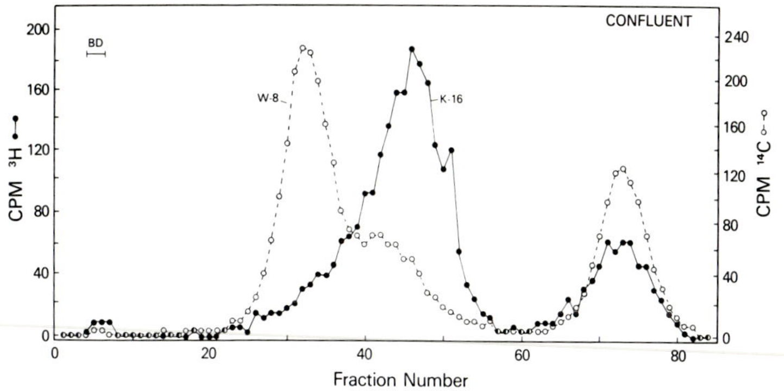

FIGURE 3 Sephadex G-50 chromatography of Pronase digested trypsinate C of normal epithelial cells K-16 (●) and transformed epithelial cells W-8 (o) labeled with L-[^{3}H]- or [^{14}C]-fucose, respectively (3).

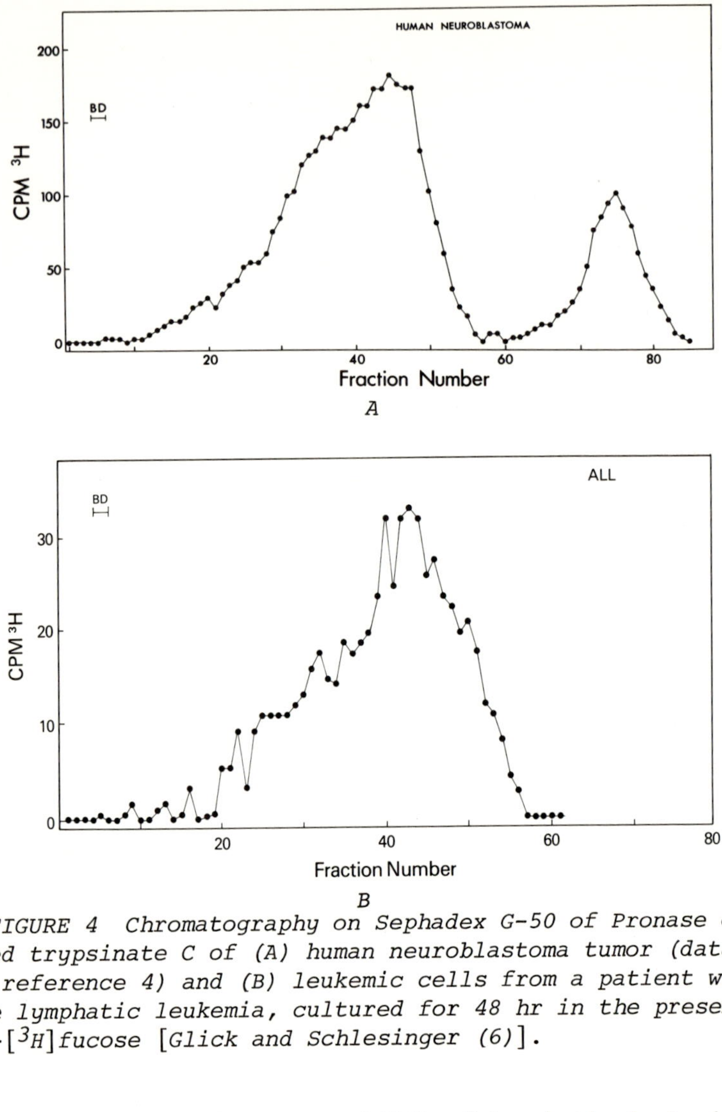

FIGURE 4 Chromatography on Sephadex G-50 of Pronase digested trypsinate C of (A) human neuroblastoma tumor (data from reference 4) and (B) leukemic cells from a patient with acute lymphatic leukemia, cultured for 48 hr in the presence of L-[^{3}H]fucose [Glick and Schlesinger (6)].

cells of a patient with acute childhood lymphatic leukemia also had glycopeptide profiles suggesting more complex oligosaccharides (fractions 20-35, Fig. 4).

COMPARISON OF LOOSELY ASSOCIATED AND INTRINSIC GLYCOPEPTIDES

In addition to metabolic labeling of the glycoproteins with radioactivity, a number of other techniques are available that can be used to directly label the cells on the monolayer. One of these is the periodate oxidation of the terminal sialic acid moieties followed by reduction with borotritide (10). This method is used to distinguish glycoproteins on the external side of the cell membranes from the more internal ones. The glycoproteins, labeled with [^{3}H] borohydride on the monolayer and removed from the cells as trypsinate A, showed profiles on Sephadex G-50 somewhat similar to those of the metabolically labeled cells (Figs. 5 and 2a). However, trypsinate A radioactively labeled in the sialic acid moiety contained an additional group of glycopeptides (Fig. 5, fractions 46-55) not seen to the same extent when the cells were labeled metabolically with radioactive L-fucose (Fig. 2a). Since labeling of sialic acid was done on the monolayer, it remains to be demonstrated whether this group of glycopeptides (Fig. 5, fractions 46-55) represents a significant portion of membrane glycoproteins or contaminating glycoproteins from the serum in which these cells were grown (11). These results provide further evidence that some of the glycopeptides in trypsinate A are on the external side of the membrane, while removal by mild trypsinization suggests that they are loosely associated.

Further comparisons of the more loosely associated glycopeptides (trypsinate A) and the more intrinsic glycopeptides (trypsinate B) were made by examining the specific activity of radioactive L-[^{3}H]fucose incorporated into the glycopeptides removed by each of these procedures from the surface of the transformed hamster fibroblasts. If we assume that all of the glycoproteins assembled within the cell incorporate sugars from similar pools while those made for export or assembled externally have alternate pools, then this should be reflected in different specific activities. The larger glycopeptide units, removed by trypsinization, were purified several times over Biogel P2, until they contained no low molecular weight material. Approximately 40% of the fucose from the loosely associated glycoproteins (trypsinate A) was hydrolyzed by acid (Table I) and of a high specific activity (11,400 cpm/nmol). The remaining fucose was not readily hydrolyzed by acid but subsequent treatment with α-L-fucosidase, purified from rat testes, removed fucose with a lower specific activity (530 cpm/nmol) from the glycopeptide units. Fucose released from trypsinate B by α-fucosidase had a specific activity three times greater than that released from trypsinate A by the enzyme (1850 cpm/nmol). In all cases, fucose was characterized by gas liquid chromatography of the alditol acetate derivative

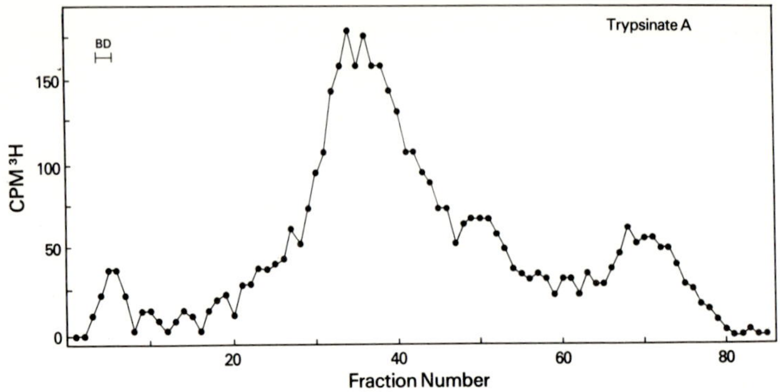

FIGURE 5 Chromatography on Sephadex G-50 of Pronase digested trypsinate A from C_{13}/B_4. The cells were made radioactive by externally labeling while on the monolayer with [^3H]borohydride after periodate oxidation [Flowers and Glick (11)].

TABLE I Specific Activity of L-[^3H]fucose in Trypsinates of Transformed Hamster Fibroblasts (C_{13}/B_4)[a]

Trypsinate	Stepwise hydrolysis	Radioactivity released (%)	Specific activity (cpm/nmol)
A	1. 0.1N H_2SO_4/100°/ 2 hr	43	11,400
	2. 0.1N H_2SO_4/100°/ 1 hr	5	b
	3. α-fucosidase[c]	20	530
B	1. α-fucosidase[c]	61	1850

[a]Data of Fischer and Glick (12).
[b]Quantity not sufficient to analyze.
[c]Free of other glycosidases.

and radioactive fucose was recovered by split stream from the gas liquid chromatograph, representing a recovery of 90% of the total radioactivity (12).

These results suggest further differences between the loosely associated (trypsinate A) and the more intrinsic glycopeptides (trypsinate B). The high specific activity of the trypsinate A, which was acid hydrolyzed, suggests that approximately 40% of the loosely associated glycopeptides contain fucose derived from pools different than those for the intrinsic glycopeptides. Alternatively, the loosely associated glycopeptides could have a different rate of turnover than those of trypsinate B or perhaps be fucosylated by an extracellular fucosyltransferase. The reasons that approximately 50% of the fucose was more stable to acid hydrolysis are not apparent at the moment and are under investigation. The release of fucose by α-L-fucosidase, an exoglycosidase, argues against the possibility that the fucose is in a penultimate position in the oligosaccharide chain.

PURIFICATION OF THE MEMBRANE GLYCOPEPTIDES

The radioactive fucose-containing glycopeptides from the virus transformed baby hamster kidney fibroblasts (C_{13}/B_4) and their normal counterpart (BHK_{21}/C_{13}) were removed from the cell surface as trypsinate C (Fig. 1). The glycopeptides, further digested with Pronase, were subsequently purified over DEAE cellulose. Figure 6 shows a comparison of L-[^{3}H]- and [^{14}C]fucose-containing glycopeptides, metabolically labeled, from the normal and virus transformed cells (9). Glycopeptides IV-VII were found in the membranes of the transformed fibroblasts to a greater extent than in the nontransformed cells. Glycopeptides I-III were found in similar amounts in both cell types. Glycopeptides IV-VII represented fractions 25-35 from the Sephadex G-50 column and glycopeptides I-III represented fractions 36-45 (see Fig. 2d). Thus the two groups of fucose-containing glycopeptides separated on Sephadex G-50 resulted in seven or more groups of glycopeptides when chromatographed on DEAE cellulose. All of these glycopeptides will be used for enzymatic sequencing of the oligosaccharide units.

SEQUENCING OF GLYCOPEPTIDE V

Glycopeptide V was found in significant amounts only in the membranes of the transformed cells. Table II summarizes the exoglycosidases, the sequence used, and the residues of

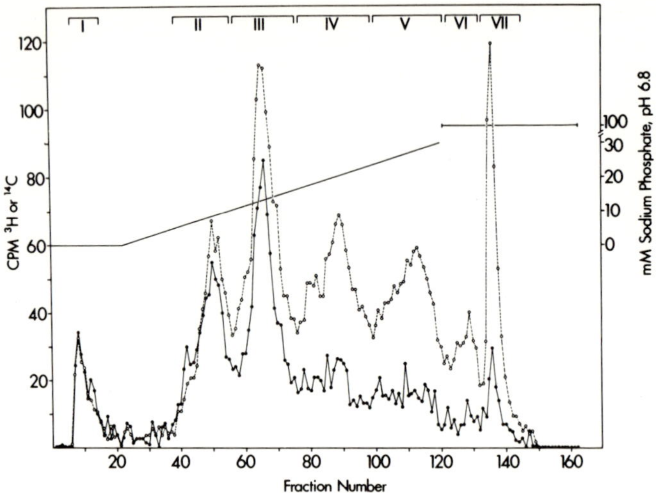

FIGURE 6 Chromatography on DEAE cellulose of Pronase digested trypsinates of hamster fibroblasts BHK_{21}/C_{13} (●) and the transformed counterpart C_{13}/B_4 (o) labeled with L-[^3H]- or [^{14}C]fucose, respectively. The numbers represent the fractions combined for sequencing (9) (with permission, Academic Press).

monosaccharides released after each treatment of glycopeptide V. The assignment of a particular sequence to the glycopeptide was based on the recovery and quantitation of the released monosaccharides as alditol acetate derivatives by gas liquid chromatography, correcting for the losses in each step with radioactivity, [^{14}C]fucose. Sialic acid was measured after acid hydrolysis by the thiobarbituric acid assay modified to a microscale (13). All enzymes were purified until free of other glycosidases (14). Aliquots of glycopeptide V were chemically hydrolyzed and the amounts of monosaccharides in the hydrolysate were considered 100%.

TENTATIVE STRUCTURE FOR GLYCOPEPTIDE V

Using the data presented in Table II, glycopeptide V can be depicted as shown in Fig. 7. It is composed of three branches, each terminating in sialic acid and assembled around the β-Man core. The sequence of the monosaccharides enclosed within the dotted lines has not been determined. The core

TABLE II Sequential Removal of Monosaccharides from Glycopeptide V Isolated from the Cell Surface of Transformed Hamster Fibroblasts (C_{13}/B_4)

	Residues released				
Treatment	SA[a]	Gal	GlcNAc	Man	Fuc
Intact glycopeptide	3	3	6	3	1
1. β-Galactosidase and β-N-acetyl-hexosaminidase		0	0		
2. Desialization	3				
3. Repeat 1		3	3		
4. α-Mannosidase				2	
5. Repeat 3 and 4		0	0.6	0	
6. β-Mannosidase				0.6	
7. Repeat 6 and β-N-acetylhexosaminidase			0	0.4	
8. α-Fucosidase					1

[a]Sialic acid.

region is depicted as a chitobiose structure as reported by Ogata *et al.* (15) using a heterogeneous mixture of glycopeptides. However, the lack of removal of GlcNAc in step 7 after β-mannosidase treatment suggests that α-Fuc may be positioned on the second GlcNAc rather than the GlcNAc attached to the amino acid as reported with the use of endo-β-N-acetylglucosaminidase D (15). Further studies are needed to clarify this point and the location of the sixth residue of GlcNAc.

IMPLICATIONS OF THE SEQUENCING

Preliminary results, obtained by enzymatic sequencing of the oligosaccharides, show that the glycopeptides that appear to be characteristic of the surface of the virus transformed and tumor cells are highly branched. In contrast, the linear structure of the glycopeptides most predominant on the surface of the normal hamster fibroblasts reveal that these oligosac-

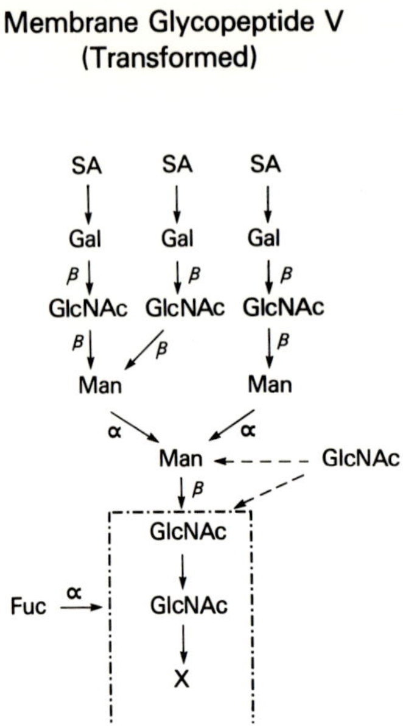

FIGURE 7 Linear structure of glycopeptide V from virus transformed cells (C_{13}/B_4) assembled from the data in Table II.

charide units are less branched (unpublished observations). These results support our original observations that the transformed glycopeptides are more complex than the normal glycopeptides (9). The original proposal was based on the molar ratios of the monosaccharides in the individual glycopeptides as well as the amounts present per cell. The verification is based on sequencing of the oliogosaccharides with the use of enzymes and quantitation of the released monosaccharides. The small amounts of glycopeptides available from the cell surfaces make additional analyses by chemical methods, such as Smith degradation, not feasible.

In all of the transformed glycopeptides thus far examined, fucose is present near the carbohydrate-amino acid bond, in contrast to the position of fucose on blood group substances. However, the release of only 40% of the fucose by acid hydrolysis (Table I) shows a heterogeneity suggesting more than one position. A number of hypotheses are under investigation to integrate the results of the two approaches. Sialic acid, on the other hand, is invariably in a terminal position since prior to desialization there is no release of Gal or GlcNAc. The structures of the branches are in keeping with several other complex asparagine-linked oligosaccharides (16).

Attempts to determine the monosaccharide to amino acid linkage have been made using aspartyl-glucosaminidase purified from rat liver. This enzyme hydrolyzes the synthetic substrate, 2-acetamido-1(L-β-aspartamido)-1,2-deoxy-β-D-glucose, but will not release N-acetylglucosamine from the membrane glycopeptide core, suggesting that amino acids remain after the exhaustive Pronase digestion.

These studies proceed slowly but we can see already that the linear structures will give enough information to aid in a rational approach to find the defect of the transformed phenotype. To date, we see several possible implications of the sequencing: (1) virus transformation brings about an alteration leading to branching of the glycoproteins; (2) transformation could be the acquiring of a specific N-acetyl glucosamine transferase making possible this branching; (3) the normal glycoprotein may have mannose substituted by a monosaccharide, which prevents the continuation of the branch GlcNAc-Gal-SA.

Whatever the ultimate defect will be that leads to the altered glycopeptides in the membranes of tumor cells, it is certain that the structural studies will aid in revealing the alteration. The highly branched oligosaccharide structure that appears to make distinct the membrane glycoproteins of virus transformed cells can now be used to postulate that the growth of the oligosaccharide chain may be the area of translation to decipher the enigma of the altered cancer cell surface.

ACKNOWLEDGMENTS

Excellent technical assistance of Mrs. Ray-Jen Chang for the gas liquid chromatographic analyses is gratefully acknowledged.

Supported by USPHS Grant CA 14037 and CA 14489 and American Cancer Society BC 109 and PRA 68.

REFERENCES

1. Buck, C. A., Glick, M. C., and Warren, L., *Biochemistry* (1970), *9*: 4567-4576.
2. Glick, M. C., Rabinowitz, Z., and Sachs, L., *J. Virol.* (1974), *13*: 967-974.

3. Pietropaolo, C., Yamaguchi, N., Weinstein, I. B., and Glick, M. C., submitted for publication.
4. Glick, M. C., Schlesinger, H., and Hummeler, K., *Cancer Res.* (1976), *36*, 4520-4524.
5. Van Beek, W. P., Smets, L. A., and Emmelot, P., *Nature* (1975), *253*: 457-460.
6. Glick, M. C., and Schlesinger, H., unpublished observations.
7. Glick, M. C., Rabinowitz, Z., and Sachs, L., *Biochemistry* (1973), *12*: 4864-4868.
8. Warren, L., Zeidman, I., and Buck, C. A., *Cancer Res.* (1975), *35*: 2186-2190.
9. Glick, M. C., *in* "Biology and Chemistry of Eucaryotic Cell Surfaces, Miami Winter Symposia" (1974), Vol. 7, E. Y. C. Lee and E. E. Smith, eds., Academic Press, N.Y., pp. 213-240.
10. Van Lenten, L., and Ashwell, G., "Methods in Enzymology" (1972), Vol. XXVIII, Academic Press, N.Y., pp. 209-211.
11. Flowers, H., and Glick, M. C., unpublished observations.
12. Fischer, A., and Glick, M. C., manuscript in preparation.
13. Glick, M. C., *in* "Methods in Membrane Biology" (1974), Vol. 2, E. D. Korn, ed., Plenum Press, N.Y., pp. 187-204.
14. Santer, U., and Glick, M. C., manuscript in preparation.
15. Ogata, S., Muramatsu, T., and Kobata, A., *Nature,* (1976), *259*: 580-582.
16. Kornfeld, R., and Kornfeld, S., *Ann. Rev. Biochem.* (1976), *45*: 217-237.

THE ROLE OF THE GOLGI APPARATUS IN THE SYNTHESIS OF GLYCOPROTEINS AND GLYCOLIPIDS

BECCA FLEISCHER

Department of Molecular Biology
Vanderbilt University
Nashville, Tennessee

Golgi is the main locus in mammalian cells for a number of enzymes involved in glycoprotein and glycolipid synthesis. Galactosyl- and sialyltransferases, which add terminal sugars to glycoproteins, are localized in Golgi in both liver and kidney. Galactosyltransferase can be solubilized using Triton X-100 and is probably an intrinsic membrane protein. Sulfo- and galactosyltransferases involved in the terminal steps of sulfatide and lactosylceramide synthesis are localized in kidney Golgi, while a sialyltransferase involved in ganglioside formation is localized in Golgi of both liver and kidney. In vivo, incorporation of ^{35}*S-sulfate into sulfatide in the Golgi precedes that into sulfatide of plasma membranes in rat kidney, consistent with the view that sulfatide is synthesized first in the Golgi apparatus and then transferred to the plasma membrane.*

INTRODUCTION

Carbohydrates occur on the cell surface primarily in the form of glycoproteins and glycolipids. Those accessible on the outer surface of the cell have biological importance as immunological determinants, cell recognition and adhesion sites, viral receptors, and possibly hormone receptors as well. The biosynthesis of these compounds is not completely understood, but a good deal of progress has been made in the last ten years.

Two distinct pathways are now known for the addition of carbohydrate units to glycoproteins. "Core" sugars, i.e., sugars most proximal to the protein backbone such as glucosamine and mannose, appear to be assembled stepwise on a carrier lipid and then transferred as a unit to the polypeptide backbone via an N-glycosidic bond between the first N-acetylglucosamine residue of the oligosaccharide chain and an asparagine residue of the polypeptide chain (for a comprehensive review of this pathway see reference 1). In mammalian cells the carrier lipid for the oligosaccharide unit appears to be dolichol pyrophosphate, while dolichol phosphate is the carrier for the outer mannose units added as monosaccharides. The more terminal sugars of the glycoproteins, such as N-acetylglucosamine, galactose, and sialic acid, are added by a second mechanism. They are attached stepwise to the nonreducing end of the carbohydrate side chain of the incomplete glycoprotein by the action of specific glycosyltransferases using nucleotide sugars as donors (2, 3). Some glycoproteins such as collagen may be glycosylated entirely by this pathway (3).

In liver and kidney, the bulk of the glycoproteins synthesized are destined for secretion. The subcellular pathway for the secretion of glycoproteins is believed to begin in the rough endoplasmic reticulum by formation of the polypeptide backbone of the protein on the ribosome followed by transfer across the membrane into the lumen of the endoplasmic reticulum (4, 5). Glycosylation of the protein begins in the endoplasmic reticulum probably by attachment of the core sugars via glycolipid intermediates. The more terminal sugars, however, are added directly to the incomplete glycoprotein as it passes through the Golgi apparatus (6). Finally, the completely glycosylated protein is secreted by fusion of Golgi-derived secretory vesicles with the plasma membrane (7). It is not clear at present whether membrane-bound glycoproteins follow a similar intracellular pathway or are formed on the membranes in which they occur.

A second major class of carbohydrate components present in mammalian cells is the glycosphingolipids. The level of these components varies widely from tissue to tissue being very high in brain, moderately high in kidney, and low in liver. Like many of the glycoproteins, the glycosphingolipids appear to be formed by stepwise addition of carbohydrates mediated by specific glycosyltransferases (2). The glycolipids are generally not secreted but remain in the cell as membrane components.

SUBCELLULAR LOCALIZATION OF SOME PROTEIN GLYCOSYLTRANSFERASES

Our approach to understanding the role of the Golgi apparatus in the cell has been to isolate fractions from mammalian tissues rich in Golgi apparatus and to assay directly what activities are present in these fractions compared to other purified cell components such as plasma membranes, mitochondria, nuclei, and endoplasmic reticulum. In order to accomplish this, we were obliged to develop methods for the isolation of Golgi apparatus from liver since methods for the isolation of the other organelles were already well developed. We then applied these techniques to isolated Golgi apparatus and other purified cell organelles from kidney, which we believed was a better source for study of the biosynthesis of glycolipids.

The Golgi apparatus consists of a series of flattened, smooth membrane-bound cisternae surrounded by what appear in sections to be small vesicles and often by large secretory vesicles (Fig. 1). This morphology is generally lost during the usual homogenization and fractionation techniques employed to isolate other cell organelles. By a combination of gentle homogenization coupled with a single flotation in a sucrose step gradient (8, 9), we have isolated largely intact, recognizable, purified Golgi apparatus from rat liver (Fig. 2). It was known from the work of Neutra and LeBlond (10) that galactose injected into animals is incorporated first into the Golgi region of most cells. We therefore looked in our purified cell fractions for an enzymic activity responsible for the incorporation of galactose into proteins. We were successful in demonstrating directly that the Golgi apparatus fraction had the highest concentration of UDP-galactose:N-acetylglucosamine galactosyltransferase activity [Eq. (1)], the enzyme probably responsible for the addition of the penultimate galactose unit to most serum glycoproteins synthesized by the liver [Eq. (2)] (2, 11, 12):

$$\text{UDP-}{}^{14}\text{C-Gal} + \text{GlcNAc} \xrightarrow[\text{Mn}^{++}]{\text{galactosyl transferase}} \text{UDP} + \text{GlcNAc-Gal} \qquad (1)$$

$$\text{UDP-Gal} + \text{protein--GlcNAc} \xrightarrow[\text{Mn}^{++}]{\text{galactosyl transferase}} \text{UDP} + \text{protein--GlcNAc-Gal} \qquad (2)$$

About 20 to 25% of the galactosyltransferase activity of the homogenate is recovered in the purified Golgi fraction

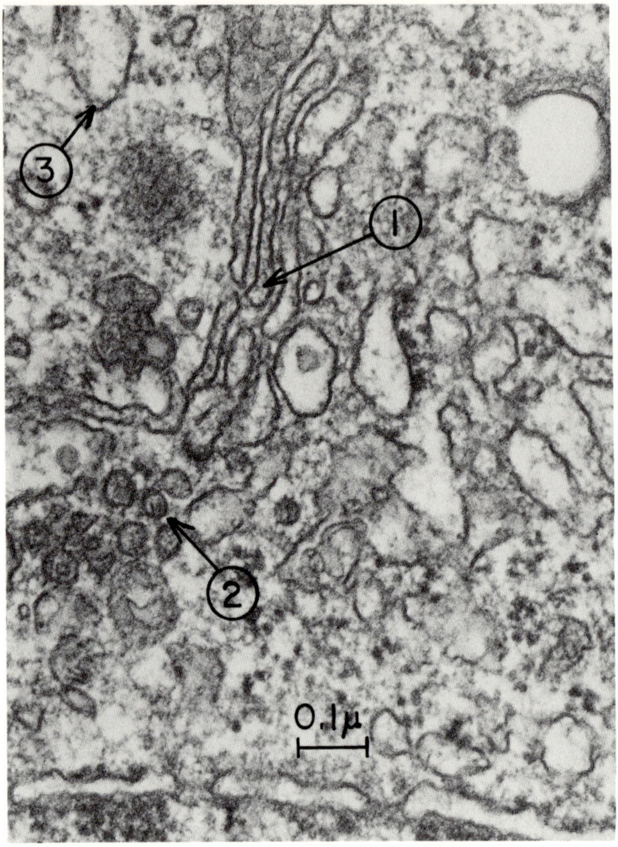

FIGURE 1 Electron micrograph of a section of a rat liver cell showing the characteristic morphology of the Golgi apparatus. (1) Cisternae arranged in parallel arrays; (2) peripheral vesicles or anastomizing tubules; (3) large secretory vesicles. The bar is equal to 0.1 μm.

with about a 100-fold concentration over the activity of the homogenate (Table I). As judged by its content of various marker enzymes for other organelles, the fraction is about 80 to 90% pure (9, 13). When fixed immediately upon isolation from the sucrose gradient, one finds a striking preservation of the structure of the Golgi apparatus during the isolation procedure (Fig. 2A). A higher magnification illustrates the characteristic features of the Golgi apparatus from liver that are found in this preparation (Fig. 2B). It consists of three to four flattened cisternae arranged in parallel rows, a num-

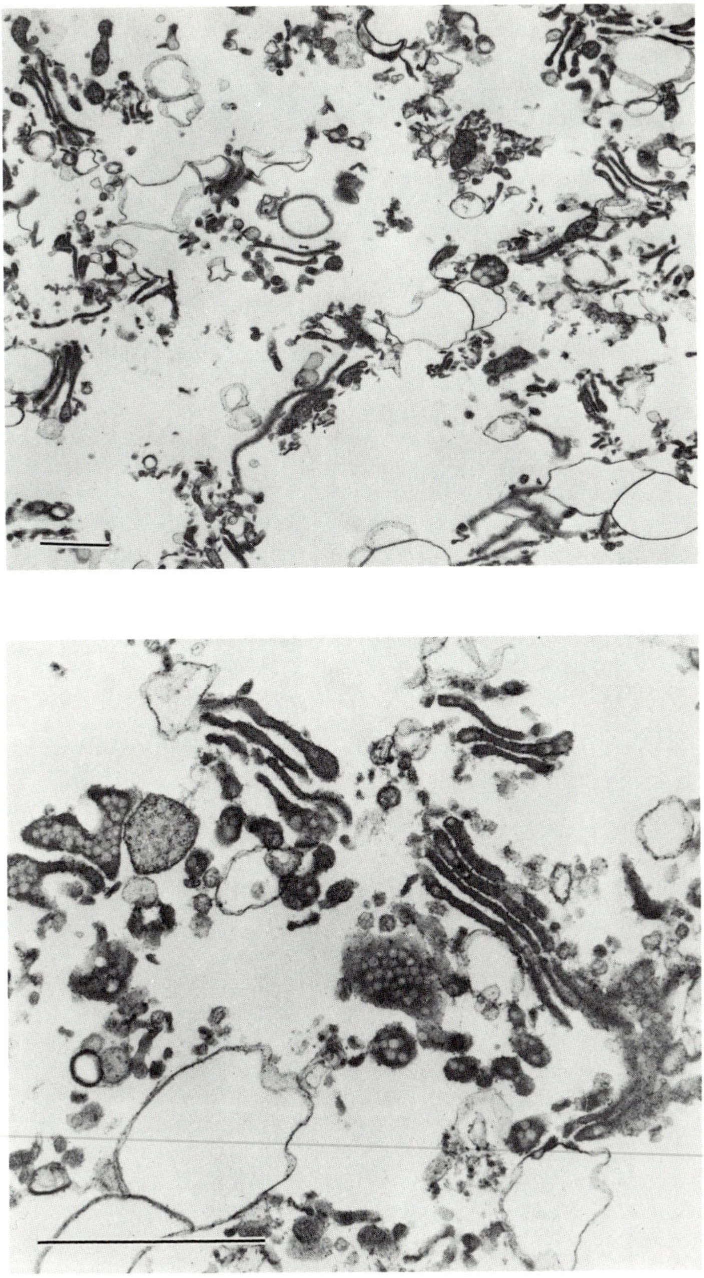

FIGURE 2 (A) Electron micrographs of Golgi-rich fraction isolated from rat liver homogenate by centrifugation in a sucrose step-gradient. (B) Higher magnification of (A). The bars are equal to 0.5 μm.

TABLE I Specific Activity of Some Marker Enzymes in Subcellular Fractions of Rat Liver[a]

Fraction	Bound P (μg/mg prot.)	Glu-6-P'ase[b]	RI NADH-Cyt.c red.	Succinate Cyt.c red.	ATP'ase	5'-AMP'ase	Gal. transf.[b]
Homogenate	18.2	0.096	0.317	0.132	0.177	0.036	6
Mitochondria	10.4	0.003	0.185	0.457	0.420	0.000	0
Golgi	23.1	0.035	0.198	0.005	0.132	0.089	764
Rough microsomes	40.1	0.244	2.27	0.012	0.065	0.046	12
Smooth microsomes	33.6	0.197	2.53	0.005	0.088	0.192	41
Plasma membranes	23.1	0.040	0.139	0.020	1.24	0.990	5
Nuclei	40.8	0.036	0.075	0.002	-	-	2
Supernatant	14.9	0.003	0.004	0.002	-	-	0

[a]Specific activity expressed as μmol/min/mg protein except galactosyltransferase, which is in nmol/hr/mg protein.

[b]At 37° C, all others at 30° C.

TABLE II Relative Specific Activities of Lipid Synthesizing Enzymes in Golgi Apparatus from Rat Liver

	Microsomes			
Formation of lecithin	*Smooth*	*Rough*	*Golgi*	*PM*
CDP-choline + diglyceride	*100*	*64*	*9.6*	*6.1*
Acyl-CoA + lysolec	*100*	*48*	*5.6*	-
N-methylation	*100*	*149*	*5.3*	*10.0*
Formation of triglycerides				
Acyl-CoA + diglyceride	*100*	*186*	*3.4*	-
Formation of sphingomyelin				
CDP-choline + ceramide	*100*	*110*	*4.1*	*9.9*
% Contamination with smooth microsomes	-	-	*8*	*9*

ber of small vesicles surrounding the cisternae that are probably cross sections of tubules attached to the cisternae, and large vesicles full of particles that are probably precursors of low density lipoproteins secreted by the Golgi apparatus of liver (14).

We have carried out extensive chemical and enzymatic characterization of the Golgi apparatus of liver as compared to endoplasmic reticulum and plasma membranes, two membranes with which it must interact to mediate the transport of secretory products out of the cell. Golgi apparatus differs from these membranes in its protein profile (8) as well as its lipid composition (15). The major pathways of lecithin and sphingomyelin biosynthesis present in endoplasmic reticulum do not appear to be present in rat liver Golgi or in plasma membranes (Table II), indicating that both the protein and lipid portions of these membranes are synthesized in the endoplasmic reticulum (16, 17).

In addition to serum glycoproteins and lipoproteins, liver Golgi apparatus is involved in the secretion of serum albumin, a simple protein (18). The conversion of proalbumin to albumin takes place at least in part in the Golgi apparatus (19).

We have applied our procedure for the isolation of liver Golgi with some modification to rat kidney and have succeeded in obtaining Golgi apparatus that is about 70% pure from this tissue (20). As illustrated in Fig. 3, the fraction is rich in morphologically identifiable Golgi, that is, stacked cis-

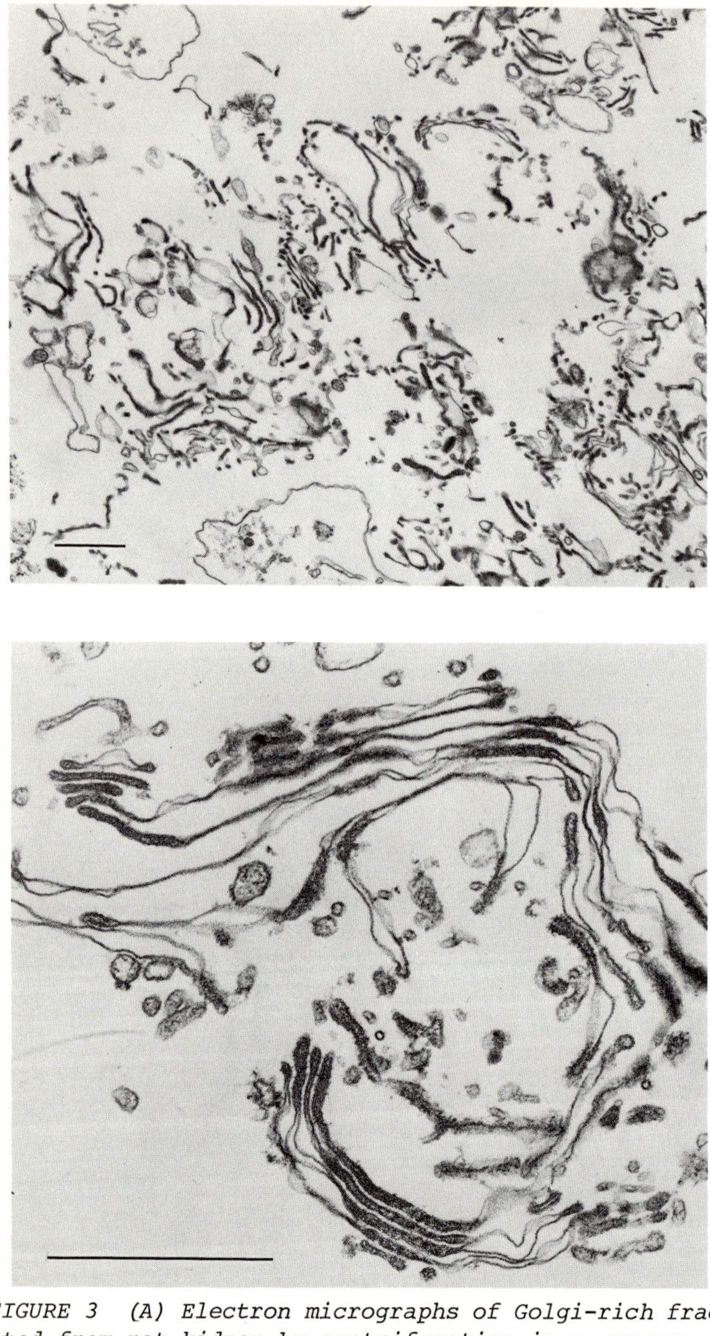

FIGURE 3 (A) Electron micrographs of Golgi-rich fraction isolated from rat kidney by centrifugation in a sucrose step-gradient. (B) Higher magnification of (A). The bars are equal to 0.5 μm.

TABLE III Distribution of Galactosyltransferase and Cerebroside Sulfotransferase in Purified Subcellular Fractions of Rat Kidney[a]

Fraction	*Galactosyl transferase*	*Cereboside sulfotransferase*
Homog.	7.2	0.3
Nuclei	3.9	0.2
Rough micr.	3.6	0.4
Smooth micr.	38	2.6
Golgi	650	25.6
Mitochondria	1.7	0.2
Plasma membranes	1.2	0.3
Super.	0.0	0.0

[a]*Specific activities expressed as nmol/hr/mg protein at* 37°.

ternae with attached tubules. When compared to other purified kidney cell fractions (Table III) the Golgi fraction contains the highest specific activity of galactosyltransferase activity, about 90-fold concentrated over the homogenate. We have also found that cerebroside sulfotransferase activity, that is, the ability to transfer sulfate from 3'-phosphoadenosine 5'-phosphosulfate (PAPS) to cerebrosides to form sulfatides, is also localized almost entirely in the Golgi fraction from kidney:

$$\text{ceramide-Gal} + \text{PAPS} \xrightarrow{\text{sulfotransferase}} \text{ceramide-Gal-SO}_4 + \text{PAP} \qquad (3)$$

Golgi apparatus from liver is devoid of the sulfotransferase activity. This is compatible with the fact that practically no sulfatides occur in liver. In testes, a novel sulfolipid, sulfoglycerogalactolipid, is the predominant glycolipid present. Golgi apparatus from testes is capable of sulfating galactosyl diglyceride prepared from the sulfolipid (21). The finding that the enzyme responsible for the synthesis of sulfatide is localized in Golgi brings to light a new functional aspect of the Golgi apparatus in mammalian cells. It demonstrates that Golgi can modify glycolipids that are membrane components as well as mucopolysaccharides and glycoproteins destined for secretion.

TABLE IV Distribution of Glycoprotein Sialyltransferase Activity in Rat Kidney and Liver Subcellular Fractions[a]

Fraction	Kidney	Liver
Homog.	1.4	7.2
Rough micr.	2.1	7.8
Smooth micr.	8.7	33.1
Golgi	135	550
Mito	0.8	0.3
Plasma memb.	2.7	22.5
Nuclei	0.6	1.9
Super.	0.0	0.1

[a]*Specific activity expressed as nmoles N-acetylneuraminic acid transferred to desialylated human transferrin per hour per mg protein.*

The distribution of glycoprotein sialyltransferase in our purified fractions from liver and kidney is shown in Table IV. This activity appears to be mainly localized in the Golgi apparatus in both tissues. This confirms and extends the work of Schacter *et al.* (22), who first showed that glycoprotein galactosyltransferase and sialyltransferase activities were present in isolated Golgi from rat liver.

We are currently carrying out experiments aimed at characterizing glycoprotein galactosyltransferase (GT) and cerebroside sulfotransferase (ST) of Golgi. These enzymes are not released by washing with salt or by disruption of the Golgi saccules using nitrogen decompression (9). The enzymes are released from the membranes by treatment with Triton X-100.

We have determined the sedimentation constant and the partial specific volume of the galactosyltransferase-Triton-X-100 complex by sedimentation of the solubilized rat liver Golgi membrane proteins in sucrose gradients made in water or in D_2O containing Triton X-100 (23). A representative sedimentation experiment is illustrated in Fig. 4. Using the sedimentation behavior observed for the active enzyme in the two systems, it is possible to calculate the sedimentation constant and the partial specific volume of the complex graphically, as illustrated in Fig. 5 (23, 24). The rather large partial specific volume found for the active enzyme in the presence of the detergent is probably due to a significant amount of Triton-X-100 bound to the enzyme. If detergent is

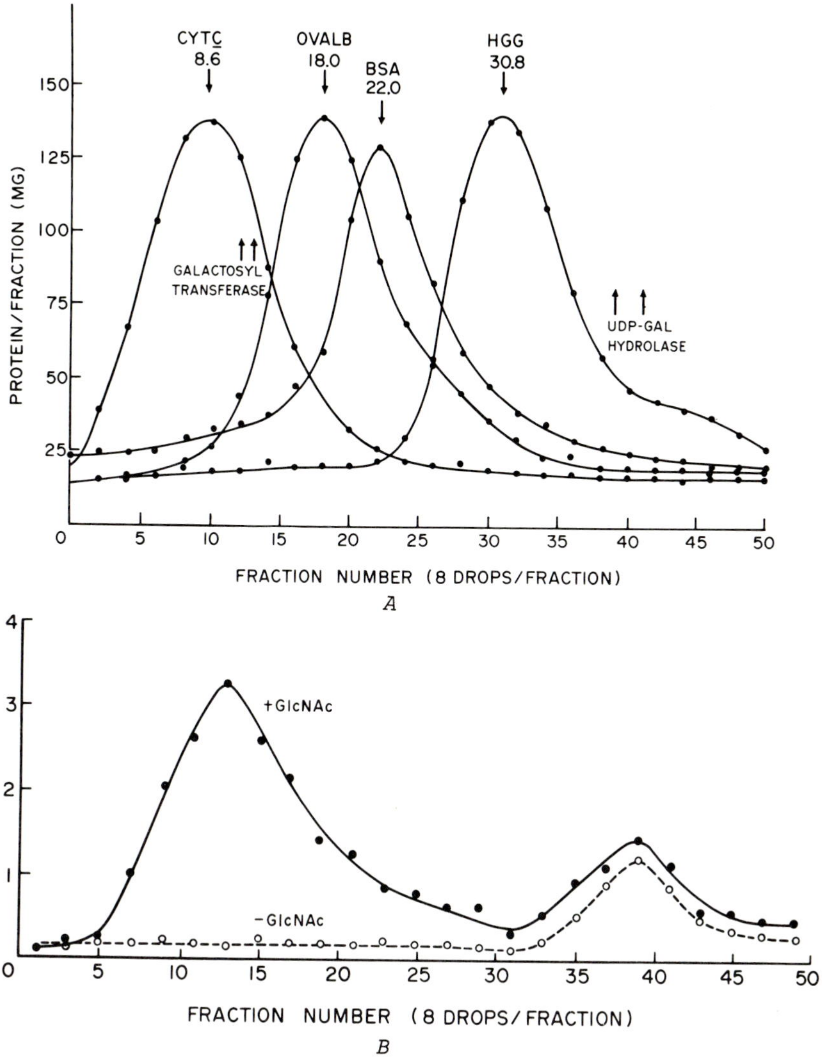

FIGURE 4 Sedimentation of standards and solubilized rat liver Golgi membranes centrifuged for 20 hr, 45,000 rpm, into sucrose-H_2O density fradients in a Spinco SW 50.1 rotor. The gradient contained 2 mg/ml Triton X-100, 0.15 M NaCl, and 10 mM HEPES pH 7.5 throughout. (A) Protein standards: cyt. c, horse heart cytochrome c; BSA, bovine serum albumin; OVALB, ovalbumin; HuGG, human gamma globulin. Numbers refer to positions of peaks. Arrows show positions of peaks of standards and of duplicate enzymic activities determined in a single experiment. (B) Galactosyltransferase activity: (●) in presence of GlcNAc (transfer and hydrolysis); (o) in absence of GlcNAc (hydrolysis only).

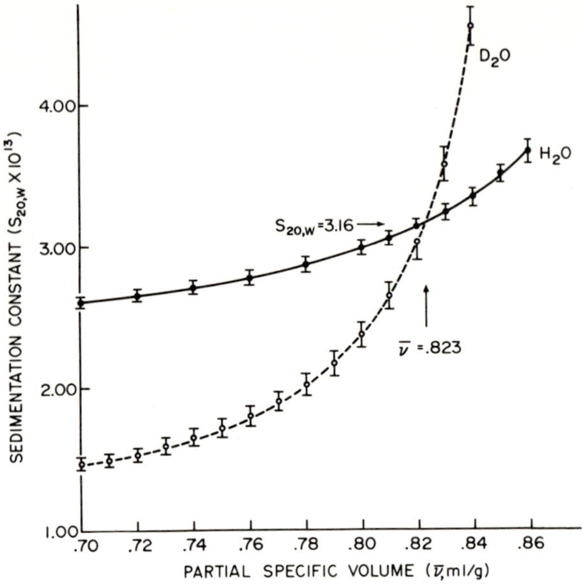

FIGURE 5 Determination of $S_{20,w}$ *and* $\bar{\nu}$ *of UDP-Gal:GlcNAc galactosyltransferase of rat liver Golgi membranes solubilized in Triton X-100, by centrifugation in* H_2O*- and* D_2O*-sucrose density gradients.*

omitted from the gradients, the enzyme aggregates and sediments to the bottom of the tube during the sedimentation. If we assume that the partial specific volume of the protein is 0.735 and that that of the detergent bound to the complex is 0.908 (25), the amount of detergent bound to the protein can be estimated to be 52.2% of the weight of the complex.

The molecular weight of the complex was determined from the sedimentation constant obtained as discussed above and from the Stokes' radius of the complex obtained by chromatography of the solubilized Golgi membrane proteins on Sepharose 6B in the presence of Triton X-100 (Fig. 6). The molecular weight of the complex was found to be 97,200 of which 52.2% is due to bound Triton X-100. The molecular weight of the enzyme is therefore estimated to be 46,500 (Table V). This value is remarkably similar to values of 43,000 to 54,000 determined by polyacrylamide gel electrophoresis in sodium dodecyl sulfate for the three soluble galactosyltransferase isoenzymes purified from bovine milk by Barker *et al.* (26). In contrast to the enzymes found in milk, the enzyme present in rat liver

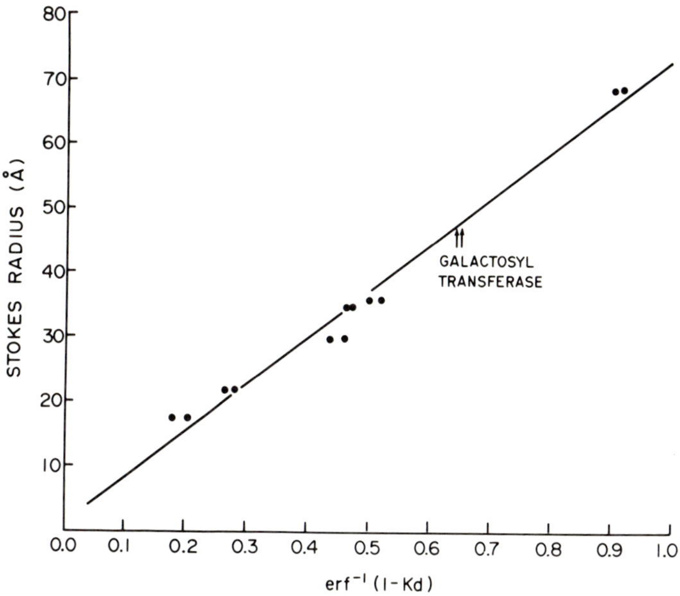

FIGURE 6 Determination of the Stokes' radius of galactosyltransferase by chromatography of solubilized membranes and standards on Sepharose 6B in 2 mg/ml Triton X-100, 0.15 M NaCl and 10 mM HEPES, pH 7.5. Standards used were run in duplicate and were, in order of increasing Stokes' radius, RNase, chymotrypsinogen, ovalbumin, bovine serum albumin, human serum transferrin, and β-galactosidase. Arrows show peak position of galactosyltransferase activity of samples run in duplicate.

Golgi is highly lipophilic and therefore can be classified as an intrinsic membrane protein.

SUBCELLULAR LOCALIZATION OF SOME GLYCOLIPID GLYCOSYLTRANSFERASES

The finding that sulfation of galactosylceramide to form sulfatide takes place in the Golgi apparatus of kidney led us to study the distribution of some glycolipid glycosyltransferase activities that could be in the pathway of formation of sulfatides and gangliosides as illustrated in Fig. 7. Some of our findings are summarized in Tables VI and VII. Enzyme (1)

TABLE V Properties of Triton-Solubilized UDP-Gal:N-acetylglucosamine Galactosyltransferase of Rat Liver Golgi Apparatus

$s_{20,w}$	3.2×10^{-13} sec
$\bar{\nu}$	0.82
R_S	48 Å
M^*	97,200 g complex/mole
f/f_o	1.41
X_p	0.48 g protein/g complex
X_d	0.52 g Triton X-100/g complex
$M_{protein}$	46,500 g protein/mole

(Fig. 7 and Table VI) the first step in sulfatide formation, is membrane bound in kidney but does not appear to be localized in any single cell organelle. This is in sharp contrast to the sulfotransferase enzyme (2) which is localized in Golgi apparatus in kidney (Table III). Enzyme (4), which converts glucosylceramide to lactosyl ceramide, a precursor of gangliosides, appears to be a Golgi enzyme in kidney. Kidney Golgi also converts galactosyl ceramide to digalactosyl ceramide. This activity and enzyme (1) are very low or absent in liver, which indicates that these enzymes are distinct from glycoprotein galactosyltransferase. Enzyme (5), which converts lactosyl ceramide to sialyllactosylceramide (Table VII), is present in both liver and kidney and appears to be localized predominantly in the Golgi apparatus in both tissues. Enzymes that convert sialyllactosylceramide to more complex gangliosides have also been reported to be largely localized in the Golgi apparatus of liver by Keenan *et al.* (28).

SYNTHESIS OF SULFATIDES IN VIVO

If cerebroside sulfotransferase is required for the synthesis of sulfatide in the kidney and the enzyme is localized in Golgi, it follows that, *in vivo*, sulfatide should be formed first in the Golgi apparatus and appear later in other membranes such as plasma membranes. We have investigated the rate of appearance of radioactivity into sulfatide of Golgi apparatus and of plasma membranes after a single iv injection of $[^{35}S]$-sulfate into rats. At given times after injection, the kidneys were removed and Golgi apparatus and plasma membrane fractions isolated from the same homogenate. The lipids

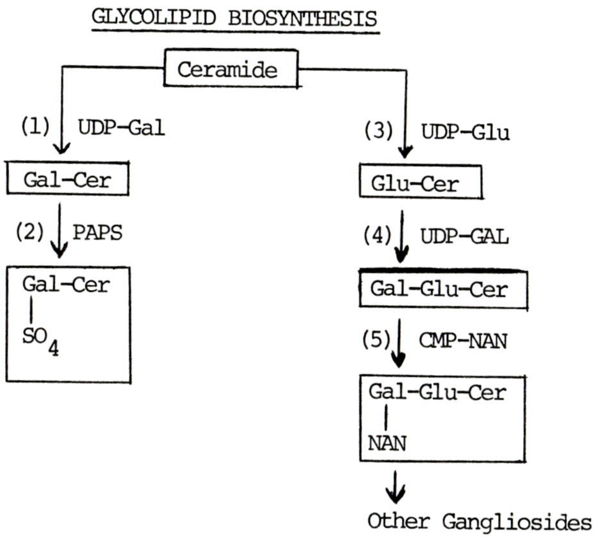

FIGURE 7 Two possible pathways in the formation of sulfatides and gangliosides in kidney.

were extracted from each cell fraction and the glycolipid fractions isolated (29). The radioactivity and the sulfatide content (30) of the glycolipid fractions were determined.

Table VIII summarizes some of the chemical and enzymic characteristics of the Golgi-rich and plasma membrane-rich fractions used in this study. The sulfatide content of the Golgi apparatus fraction is the highest of all the membranes we have studied in kidney, including plasma membranes (30), which contradicts the commonly held belief that glycolipids are concentrated largely in the plasmalemma of cells in which they occur.

Figure 8 summarizes our results on the incorporation of radioactivity into the sulfatide of the two membranes (31). Sulfatide in the Golgi is labeled before that in plasma membranes, reaching a peak value about 30 min after injection. The specific activity of sulfatide in plasma membrane, on the other hand, rises more slowly and reaches a plateau while the activity of the sulfatide in the Golgi is declining. This pattern of labeling is consistent with a precursor product relationship between sulfatide of the Golgi and that of the plasma membrane.

TABLE VI Distribution of Some Glycolipid Galactosyltransferases in Rat Kidney Subcellular Fractions Specific Activity[a]

	Ceramide[b] (1) ↓UDP-Gal Gal-Cer		Gal-Cer ↓UDP-Gal $(Gal)_2Cer$		Glu-Cer (4) ↓UDP-Gal Gal-Glu-Cer
Fraction	*Kidney*	*Liver*	*Kidney*	*Liver*	*Kidney*
Homog.	0.110	0.015	0.2	0.03	0.1
R. micr.	0.048	0.01	0.2	0.03	0.2
S. micr.	0.087	0.03	0.2	0.13	0.4
Golgi	0.113	0.03	18.0	0.14	8.0
Mito	0.030	0.01	0.1	0.01	0.1
PM	0.057	0.01	0.3	0.09	0.2
Supernatant	0.001	-	0.0	-	0.0

[a]*Expressed as nmoles galactose transferred per hour per mg protein at* 37°.
[b]*Ceramides containing hydroxyfatty acids only were used as substrate (27).*

TABLE VII Distribution of Glycolipid Sialyltransferase Activity in Rat Kidney and Liver Subcellular Fractions[a]

	Gal-Glu-Cer (5) ↓CMP-NAN NAN-Gal-Glu-Cer	
Fraction	*Kidney*	*Liver*
Homog.	*0.5*	*0.07*
Rough micr.	*0.9*	*0.07*
Smooth micr.	*2.4*	*0.70*
Golgi	*61.4*	*12.5*
Mito	*1.2*	*0.0*
Plasma memb.	*1.2*	*0.2*
Nuclei	*0.1*	*0.06*
Super	*0.0*	*0.0*

[a]*Specific activity expressed as nmoles N-acetylneuramic acid transferred per hour per mg protein at* 37°.

CONCLUSIONS

Our approach to understanding the role of the Golgi apparatus in cellular function has been to develop methods to isolate this organelle as well as the other main organelles of the cells and to assess the purity of the preparations by the use of marker enzymes. In this manner, it has been possible to show that the Golgi apparatus is the main locus in liver and kidney of galactosyltransferase and sialyltransferase activities involved in the terminal glycosylation of glycoproteins destined for secretion. In addition, we have shown that a number of glycosyltransferases and a sulfotransferase involved in the formation of glycolipids are localized in the Golgi apparatus from kidney. Glycolipids formed in the Golgi apparatus are probably not made for secretion but are membrane components and transported to other membranes such as the plasma membrane. The pathway for the transfer of glycolipids from the Golgi to the plasma membrane remains to be eluci-

TABLE VIII Enzymic Profiles of Fractions Obtained from Centrifugation of Kidney Homogenate in a Sucrose Step-Gradient

	Homogenate	Golgi	Plasma membranes
mg Protein/gm kidney	47.3	0.079	0.21
μg Phosphorus/mg protein	-	34.6	20.2
mg Sulfatide/mg protein	0.018	0.163	0.041
Sulfotransferase[a]	0.26	36.4	1.3
Galactosyltransferase[a]	12.5	568	14.3
Glucose-6-phosphatase	0.056	0.057	0.080
Succinate cyt. c red.	0.256	0.047	0.200
ATPase	0.048	0.41	1.51

[a]nmoles per hour per mg protein at 37°; all other activities expressed as μmoles per min per mg protein at 32° except glucose-6-phosphatase which was carried out at 37°. The values shown are averages of 3 preparations except the sulfatide values for Golgi and plasma membranes which are an average of 14 preparations (31).

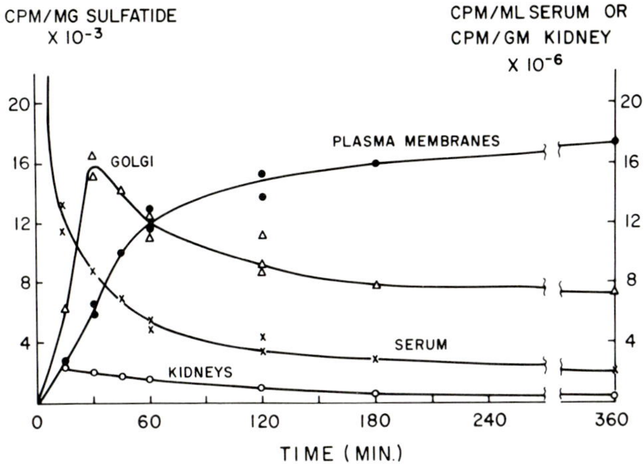

FIGURE 8 Rate of appearance of radioactive sulfatide in isolated Golgi apparatus and plasma membrane fractions from kidney after iv injection of ^{35}[S]-sulfate into rats. The values for serum and kidney are total counts and mainly reflect the presence of radioactive sulfate. The Golgi and plasma membrane values are counts found in the glycolipid fraction from these membranes and represent radioactive sulfatide.

dated. It is also not yet known whether glycosylation of glycoproteins located in the plasma membrane is also a function of the Golgi apparatus.

NOTATION

UDP	uridine diphosphate
UDP-Gal	uridine diphosphate galactose
GlcNAc	N-acetylglucosamine
PAPS	3'-phosphoadenosine-5'-phosphosulfate
Gal-Cer	galactosylceramide
Glu-cer	glucosylceramide
CMP-NAN	cytidine-5'-monophospho-N-acetyl neuraminic acid
RI	rotenone insensitive

ACKNOWLEDGMENTS

The author would like to thank Mr. Akitsugu Saito for his excellent assistance in all aspects of the electron microscopy presented in this article. The able technical assistance of Mr. Robert Katt and Mrs. Kathryn Dewey is also gratefully acknowledged. This study was aided by NIH Grants AM14632 and AM17223.

REFERENCES

1. Waechter, C. J., and Lennarz, W. J., *Ann. Rev. Biochem.* (1976) *45*, 95.
2. Roseman, S., *Chem. Phys. Lipids* (1970) *5*, 270.
3. Spiro, R. G., Spiro, M. J., and Adamany, A. M., *Biochem. Soc. Symp.* (1974) *40*, 37.
4. Redman, C. M., *J. Biol. Chem.* (1969) *244*, 4308.
5. Blobel, G., and Sabatini, D. D., *J. Cell Biol.* (1970) *45*, 130.
6. LeBlond, C. P., and Bennett, G., *in* "The Cell Surface in Development" (1974), Moscona, A. A., ed., Wiley, New York, pp. 29-49.
7. Palade, G., *Science* (1975) *189*, 347.
8. Fleischer, B., and Fleischer, S., *Biochim. Biophys. Acta* (1970) *219*, 301.
9. Fleischer, B., *Methods in Enzymology* (1974) *31*, 180.
10. Neutra, M., and LeBlond, C. P., *J. Cell Biol.* (1966) *30*, 137.
11. McGuire, E. J., Jourdian, G. W., Carlson, D. M., and Roseman, S., *J. Biol. Chem.* (1965) *240*, PC4113.
12. Schanbacher, F. L., and Ebner, K. E., *J. Biol. Chem.* (1970) *245*, 5057.
13. Fleischer, S. and Kervina, M., *Methods in Enzymology* (1974) *31*, 6.
14. Hamilton, R. L., Regen, D. M., Gray, M. E., and LeQuire, V. S., *Lab. Invest.* (1967) *16*, 305.
15. Zambrano, F., Fleischer, S., and Fleischer, B., *Biochim. Biophys. Acta* (1975) *380*, 357.
16. Van Golde, L. M. G., Fleischer, B., and Fleischer, S., *Biochim. Biophys. Acta* (1971) *249*, 318.
17. Van Golde, L. M. G., Raben, J., Batenburg, J. J., Fleischer, B., Zambrano, F., and Fleischer, S., *Biochim. Biophys. Acta* (1974) *360*, 179.
18. Peters, Jr., T., Fleischer, B., and Fleischer, S., *J. Biol. Chem.* (1971) *246*, 240.

19. Edwards, K., Fleischer, B., Dryburgh, H., Fleischer, S., and Schreiber, G., *Biochem. Biophys. Res. Commun.* (1976) *72*, 310.
20. Fleischer, B., and Zambrano, F., *J. Biol. Chem.* (1974) *249*, 5995.
21. Kornblatt, M. J., Knapp, A., Levine, M., Schachter, H., and Murray, R. K. (1974) *Can. J. Biochem.* *52*, 689.
22. Schacter, H., Jabbal, I., Hudgin, R. L., Pinteric, L., McGuire, E. J., and Roseman, S., *J. Biol. Chem.* (1970) *245*, 1090.
23. Smigel, M., PhD. Thesis (1976), Vanderbilt University, Nashville, Tennessee.
24. Fleischer, B., and Smigel, M., *Tenth Int. Cong. Biochem.* (1976) 05-5-260 Abstr.
25. Tanford, C., Nozaki, Y., Reynolds, J. A., and Makino, S., *Biochemistry* (1974) *13*, 2369.
26. Barker, R., Olsen, K. W., Shaper, J. H., and Hill, R. L., *J. Biol. Chem.* (1972) *247*, 7135.
27. Constantino-Ceccarine, E., and Morell, P., *J. Biol. Chem.* (1973) *248*, 8240.
28. Keenan, T. W., Morre, D. J., and Basu, S., (1974) *J. Biol. Chem.* *249*, 310.
29. Rouser, G., Kritchevsky, G., and Yamamoto, A., *in* "Lipid Chromatographic Analysis" (1967), Marinetti, G. V., ed., Marcel Dekker Inc., New York, Vol. 1, pp. 99-162.
30. Zambrano, F., Fleischer, S., and Fleischer, B., *Biochim. Biophys. Acta* (1975) *380*, 357.
31. Fleischer, B., Zambrano, F., and Fleischer, S., *J. Supramol. Struct.* (1974) *2*, 737, and unpublished observations.

STRUCTURAL STUDIES ON THE MAJOR GLYCOPROTEIN OF THE TA3-HA ASCITES TUMOR CELL

JOHN F. CODINGTON
*DIRK H. VAN DEN EIJNDEN**
ROGER W. JEANLOZ

Laboratory for Carbohydrate Research
Departments of Biological Chemistry and Medicine
Harvard Medical School and Massachusetts General Hospital
Boston, Massachusetts

The carbohydrate chains of epiglycanin, a glycoprotein implicated in masking antigens at the surface of the allo-transplantable TA3-Ha murine mammary carcinoma ascites cell, numbered more than 500 per molecule of 500,000 molecular weight. These were cleaved by alkali in the presence of sodium borohydride and fractionated by gel filtration. Based upon chemical and immunochemical investigations of the glycoprotein, such as periodate oxidation, alkaline borohydride reduction, and inhibition of hemagglutination induced by lectins, and upon studies with the purified reduced oligosaccharide chains, the structures of 5 distinct carbohydrate chains in epiglycanin were proposed. All possessed N-acetylgalactosamine. The linkages, α-NeuNAc-(2→3)-Gal and β-Gal-(1→3)-GalNAc, were proposed.

INTRODUCTION

Glycoproteins at the surfaces of mammalian cells participate in a variety of important functions (1, 2), and appear to play a unique role in transformation or malignancy (3, 4). On

*Present address: Department for Medical Chemistry, Vrije Universiteit, Van der Boechlorststraat 7, Amsterdam, The Netherlands.

the basis of chemical (5, 6), physical (7), and immunological data (8) it has been suggested (9, 10) that a large glycoprotein, epiglycanin, at the surface of the TA3-Ha mammary carcinoma ascites cell of the strain A mouse serves to protect this cell against attack by specific antibodies and immune lymphocytes by masking cell-surface histocompatibility antigens. From measurements of the surface area of the TA3-Ha cell from scanning electron micrographs (11), it has been calculated that epiglycanin molecules are probably situated at the cell surface on the average no more than 19 nm apart (10), a spacing of molecules consistent with a masking function for the material. Masking of surface transplantation antigens by such an abundance of long rodlike macromolecular structures is consistent with the capacity of this cell to grow in and kill both foreign mouse strains (8, 10, 12) and other species, such as the rat and the hamster (13).

The masking hypothesis is supported by the observation that an ascites subline, TA3-St, of the same solid tumor, which was only able to grow in syngeneic strain A mice (8, 12), possessed no epiglycanin at its surface (9, 10). The evidence for a masking role for epiglycanin was described in a recent review (10).

Although a general role in masking antigens at cell surfaces has been ascribed to glycoproteins (14), to our knowledge epiglycanin is the only characterized glycoprotein that has been implicated in masking surface antigens. Because of this unique biological function with its possible relevance to the behavior of certain types of cancer in humans (10) and the relatively large amount of epiglycanin available in comparison to other cell-surface macromolecules, investigations in our laboratory have been concentrated on the elucidation of the physicochemical properties of epiglycanin. An earlier review of chemical studies on epiglycanin has appeared (15).

OCCURRENCE AND ISOLATION OF EPIGLYCANIN

Epiglycanin was first isolated from the supernatant solution after incubation of viable TA3-Ha ascites cells with TPCK-trypsin (5). By this method it has been possible to remove 2-3 mg of this glycoprotein from 10^9 cells. Purification was achieved by gel filtration on columns of Biogels P-4, P-100 (5), and Sepharose 4B (7). Almost identical results were obtained by the use of papain (16). The products of proteolysis in these experiments produced fragments of molecular weight 80,000-460,000 (7). Epiglycanin may be detected at the surface of TA3-Ha cells by the quantitative adsorption of a lectin from the seeds of *Vicia graminea*, a lectin specific for a carbohydrate chain present in epiglycanin (17-19). On the basis of isolation and lectin adsorption studies it has been estimated that there are, on the average, approximately $3\text{-}5 \times 10^6$ molecules of epiglycanin per cell (19). Epiglycanin is shed from the cells during growth in the peritoneal cavity of the mouse and may be isolated from the ascites fluid and purified by gel filtration (10, 19). This material is larger than that cleaved by proteolysis. The bulk of this material possesses a molecular weight of approximately 500,000 although a small proportion is of yet larger size. Cells grown in suspension culture also possess epiglycanin at their surfaces, and this material may be cleaved by proteolytic enzymes and isolated by those procedures employed for the ascites cells (20). After a prolonged culture period, however, epiglycanin can no longer be detected at the surfaces of these cells (20). Epiglycanin may also be removed from TA3-Ha cells by the use of lithium diiodosalicylate or detergents, but the yields are low.

In recent studies it has been possible to demonstrate by electron microscopy, after glutaraldehyde fixation, at the surface of the TA3-Ha ascites cell, but not at the surface of the strain-specific TA3-St cell, an abundance of long extended macromolecular structures, some of these extending outward about 400 nm from the outer cell membrane (11). Although it cannot be categorically stated that such observed structures represent clusters of epiglycanin molecules, their lengths appear consistent with those of isolated epiglycanin molecules, as viewed by electron microscopy with a shadow casting technique (7). By this technique epiglycanin molecules isolated from the ascites fluid appeared to be in an extended conformation with dimensions of approximately 400 to 500 nm in length and 2.4 nm in width.

COMPOSITION OF THE CARBOHYDRATE MOIETIES IN EPIGLYCANIN

The carbohydrate compositions of various epiglycanin samples are presented in Table I. The amino acid compositions, characterized by exceptionally large proportions of serine and threonine, have been reported (5, 10). Significant variations have occurred in the relative proportions of the four major carbohydrate components isolated from cells of different transplant generations. These changes reflect differences in the proportion of the two major chain types that occur in epiglycanin, as will be described later. Sample I (Table I) was isolated from the trypsinate of a batch of cells

TABLE I Compositions of Samples of Epiglycanin Isolated from TA3-Ha Ascites Cells

	Method of isolation	Molecular wt.[a]	Gal	Man	GalNAc	GlcNAc	NeuNAc
I	Proteolysis, fractionation on Biogel P-100[c]	200,000 (av)	29% (10.7)[b]	0.4% (0.1)	31% (9.1)	3.4% (1.0)	10% (2.0)
IIA	Proteolysis, fractionation on Biogel P-100 and sepharose 4B[d]	460,000	24% (4.1)	0.4% (0.07)	18% (2.4)	7% (1.0)	14% (1.4)
IIB		-	27% (5.0)	0.4% (0.07)	24% (3.5)	7% (1.0)	13% (1.3)
IIC		100,000	31% (5.2)	0.4% (0.07)	28% (3.7)	7% (1.0)	7.3% (0.7)
III	$HClO_4$-soluble ascites fluid fractionation on sepharose 4B[e]	500,000	32% (8.1)	0.3% (0.08)	29% (5.8)	5% (1.0)	14% (2.0)

[a]Determined by sedimentation equilibrium (34).
[b]Values in parentheses represent relative proportions of components in comparison to GlcNAc.
[c]Reference 5. [d]Reference 7. [e]Reference 10.

harvested recently and was utilized in alkaline borohydride reduction studies described in detail below. In Sample II, A, B, and C represented the products of the fractionation of Sample I on a column of Sepharose 4B (7). The compositions of these fractions are significantly different and reflect variations in the relative proportions of different carbohydrate chains, which are presented in Fig. 1. Sample III represents the native molecule as isolated from the ascites fluid of syngeneic strain A mice bearing TA3-Ha cells. Based upon immunochemical and chromatographic studies (18) similar molecules are believed to exist in the serum of tumor-bearing mice.

Epiglycanin (III) contains about 75% carbohydrate and 25% protein and is characterized by large proportions of galactose (28-33%), *N*-acetylgalactosamine (27-32%), and sialic acid (12-14%) and contains lesser amounts of *N*-acetylglucosamine (3-7%). The sialic acid residues of epiglycanin, which represent approximately 50-60% of the total surface sialic acid, appear to possess approximately the same proportion of *N*-acetyl (90-93%) and *N*-glycolyl (7-10%) substituents as the residues cleaved from viable cells by the action of neuraminidase from *Vibrio cholerae*.

GLYCOPEPTIDE LINKAGES AND THE TYPES OF CARBOHYDRATE CHAINS

The large proportion of *N*-acetylgalactosamine in epiglycanin is consistent with the presence of *O*-glycosyl-linked carbohydrate chains, i.e., those containing an *O*-(2-acetamido-2-deoxy-α-D-galactopyranosyl)-L-serine (or L-threonine) sequence. Indeed, it has been shown by alkaline borohydride reduction that all *N*-acetylgalactosamine units in this macromolecule are probably involved in glycopeptide bonds to serine or threonine residues in the polypeptide backbone (6). No nonreduced *N*-acetylgalactosamine could be detected among the products of the reaction performed under "nonpeeling" conditions.

Partial methanolysis of Sample I with 0.5 *M* anhydrous hydrogen chloride in methanol at 65° for 60 min indicated that *N*-acetylgalactosamine was the terminal residue at the "reducing end" of the carbohydrate chains. Although more than 70% of the sialic acid, galactose, and *N*-acetylglucosamine were cleaved under these conditions, no free *N*-acetylgalactosamine could be detected (6). Alkaline borohydride reduction resulted in the loss of both serine and threonine residues and

Chain	
I	GalNAc-
II	β-D-Gal-(1→3)-GalNAc-
III	α-NeuNAc(2→3)-β-D-Gal-(1→3)-GalNAc-
IV	β-D-Gal-(1→3 or 4)-D-GlcNAc-(1→2,4, or 6)-β-D-Gal-(1→3)-GalNAc-
V	α-NeuNAc-(2→3)-β-D-Gal-(1→3 or 4)-D-GlcNAc-(1→2,4, or 6)-β-D-Gal-(1→3)-GalNAc-

FIGURE 1 Proposed structures for carbohydrate chains in epiglycanin.

the appearance of α-aminobutyric acid and a greater proportion of alanine. The proportion of all other amino acids remained constant (6).

As a result of these studies it has been estimated that more than 500 carbohydrate chains were attached to approximately 60% of the serine and threonine residues present in a single extended polypeptide chain of about 1300 amino acid residues. The width of epiglycanin, approximately 2.4 nm, as determined by shadow casting electron microscopy, suggests the presence of relatively short chains of about 1 to 5 residues (7), a conclusion consistent with chemical data, as described below.

Although no definitive evidence is available to support the presence of 2-acetamido-1-*N*-(L-aspart-4-oyl)-2-deoxy-β-D-glucopyranosylamine-linked carbohydrate chains, the small, yet consistent, amounts of mannose detected in all samples of epiglycanin, suggest that the existence of one or two such chains per molecule of molecular weight 500,000 is possible, although mannose residues (generally 0.3-0.4%) may also be attached to a contaminating glycoprotein.

IMMUNOCHEMICAL STUDIES

The results of inhibition studies with epiglycanin against hemagglutinations by a number of different lectins (see Table II) are consistent with chemical data (6). In addition, however, they have been useful in suggesting either the presence or absence of carbohydrate sequences not readily elucidated, due to the small quantities of material available. Results are presented in Table II. Epiglycanin in fragments A, B, and C (Sample II, Table I) demonstrated high, but different, activities against agglutination by five lectins known to be specific for sequences found in *O*-glycosyl-linked carbohydrate chains, *Vicia graminea*, *Bauhinea purpurea*, *Iberis amara*, *Arachis Hypogaea* (Japan), and *Wisterea floribunda*. Although the specificities for carbohydrate linkages in glycoproteins of high molecular weight, such as epiglycanin, have not been determined for each of these lectins, all appear to require a terminal galactose or *N*-acetylgalactosamine residue in the *O*-chain (21, 22). Of particular interest is the high inhibitory activity against the agglutination of human erythrocytes of blood group N specificity by the lectin of *V. graminea* seeds (17-19). It has been suggested that this lectin required a β-D-Gal(1→3)GalNAc linkage (22), a structure suggested for one type of carbohydrate chain in epiglycanin on the basis of chemical data, as described below (see Fig. 1). The enhanced

TABLE II Hemagglutination Inhibition Activities (before and after Treatment with Neuraminidase) of Purified Fractions Isolated after Proteolysis of TA3-Ha Mammary Carcinoma Ascites Cells

Lectin	Hemagglutination inhibition activity[a]					
	Fraction A		Fraction B		Fraction C	
	Nontreated	Treated	Nontreated	Treated	Nontreated	Treated
Arachis hypogaea (Japan)[b]	2	2	3	1	3	5
Bauhinia purpurea[b]	28	28	18	13	73	73
Iberis amara[b]			4	4		
Vicia graminea[c]	15	1	4	2	4	1
Wistaria floribunda	55	28	56	18	76	36
Ricinus communis	>440	220	320	80	>580	>580
Glycine max[b]	>440	>440	320	160	290	290
Phaseolus limensis (Anti A)			>125	>125		
Phaseolus vulgaris	>440	>440	>640	>640	>580	>580

[a]The values are expressed as μg of inhibitor per ml required to give complete inhibition of agglutination in 30 min.

[b]Tested against neuraminidase-treated, human erythrocytes.

[c]Inhibitory activities estimated from values obtained in an automated system and previously reported (19).

activity demonstrated in this system after removal of sialic acid by neuraminidase suggested the presence of terminal sialic acid residues in a significant proportion of these chains. Indeed, a chain of this probable structure (chain III) has been isolated after alkaline borohydride reduction, as illustrated in Fig. 1.

Significant inhibitory activity was observed against hemagglutination by lectins of *Ricinus communis* (castor bean) and glycine max (soybean). It has been reported (21) that the *R. communis* agglutinin is active against a 2-acetamido-2-deoxy-*O*-β-D-galactopyranosyl-D-glucopyranosyl linkage. The presence of this sequence in a terminal position (chain IV, Fig. 1), was indicated by the chemical studies described below, as was a similar sequence bearing a terminal sialic acid residue (chain V).

It is probably significant that no inhibitory activity was demonstrated against hemagglutination by two lectins that are specific for carbohydrate chains attached by a 2-acetamido-1-*N*-(L-aspart-4-oyl)-2-deoxy-β-D-glucopyranosylamine linkage. Chemical studies have not yet established the presence of such chains in epiglycanin.

PARTIAL CHARACTERIZATION OF THE CARBOHYDRATE CHAINS IN EPIGLYCANIN BY CHEMICAL METHODS

As indicated by the compositional data presented in Table I, one out of every 4-6 chains in epiglycanin possesses a sialic acid residue. Since similar amounts of sialic acid were released by the action of neuraminidase from *V. cholerae* as by dilute acid hydrolysis (0.05 *M* H_2SO_4, 80°, 60 min) or by methanolysis (1.0 *M*, HCl-methanol, 85°, 20 hr) (24, 25), it was concluded that all sialic acid residues occupied terminal positions. Periodate oxidation of epiglycanin fragments (sample I), after these had been treated with neuraminidase, destroyed 32% more galactose than was destroyed in untreated material. No oxidation of *N*-acetylglucosamine occurred either before or after removal of sialic acid. However, a very small, and perhaps insignificant decrease in the proportion of *N*-acetylgalactosamine occurred after the removal of sialic acid. Approximately 7% of the galactosamine in untreated epiglycanin was destroyed by periodate. From these results, it was concluded that sialic acid, galactose, and *N*-acetylgalactosamine residues occupy terminal positions in epiglycanin, and that most, if not all, of the sialic acid was attached to a penultimate galactose by an α-(2→3)-linkage. It is apparent that nonterminal galactose residues (i.e., those neither terminal

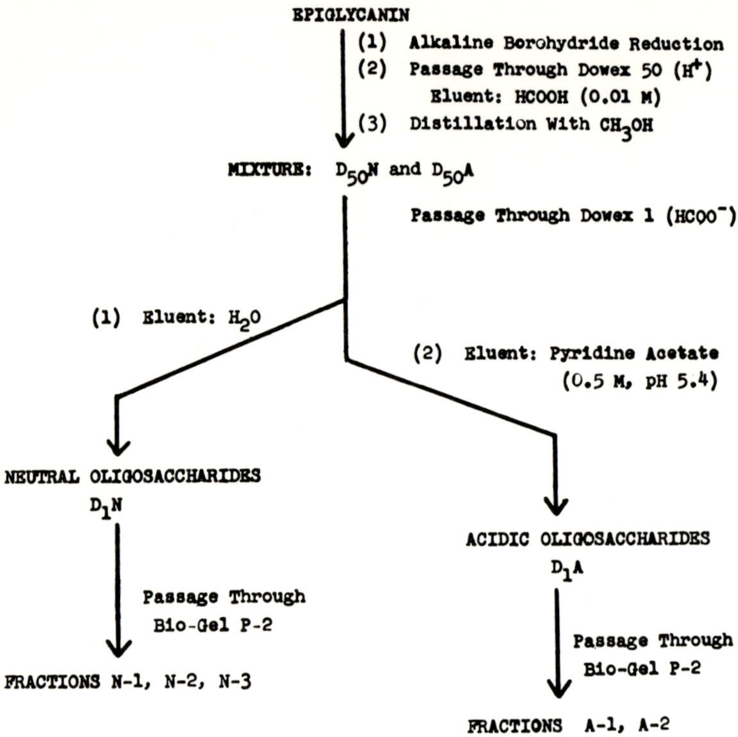

FIGURE 2 Procedure for the release and reduction of the carbohydrate chains of epiglycanin and their isolation as the reduced oligosaccharides.

nor penultimate to terminal sialic acid residues) possessed unsubstituted hydroxyl groups in either the 2, 3- or 3, 4-positions and were thus readily oxidized by periodate.

ISOLATION AND PARTIAL CHARACTERIZATION OF CARBOHYDRATE CHAINS

In order to characterize the individual chain types in proteolytically-derived fragments of epiglycanin, sample I (Table I) was subjected by van den Eijnden *et al.* (23), to alkaline borohydride reduction (0.9 *M* sodium borohydride, 0.09 *M* sodium hydroxide, 37°, 70 hr) (23). Reduced oligosaccharides were fractionated according to the scheme presented in Fig. 2. After removal of inorganic salts on a cation exchange column (Dowex 50, H^+) followed by the distillation of methyl borate, the residue ($D_{50}N$ and $D_{50}A$) was found to contain 87% of the original sialic acid and 76% of the original neutral carbohydrate, as determined by colorimetric procedures (26, 27) and glc (24, 25). Elution from Dowex I with water separated those reduced neutral oligosaccharides (D_1N) from reduced oligosaccharides containing sialic acid (D_1A), which were then eluted with 0.5 *M* pyridine acetate. The reduced neutral oligosaccharides (D_1N) and acidic oligosaccharides (D_1A) were fractionated separately on a calibrated column of Biogel P-2 (1.5 × 142 cm) with 0.05 *M* pyridine acetate, pH 5.4, as an eluent. The elution profiles of the two fractionations, based upon the phenol-sulfuric acid determination for neutral carbohydrates (27), are shown for D_1N and D_1A in Figs. 3 and 4, respectively. Fractions in regions N-1, N-2, and N-3 (the neutral reduced oligosaccharide fractions) and A-1 and A-2 (the acidic reduced oligosaccharide fractions) were pooled, as illustrated in the two figures, and lyophilized. Hexoses and hexosamines were determined by glc as their per-*O*-(trimethylsilyl)ated methyl glycosides, and sialic acid as the methyl ester of the per-*O*-(trimethylsilyl)ated methyl glycoside of *N*-acetylneuraminic acid (24, 25). The relative proportions of the carbohydrate components and the recoveries of material from epiglycanin are presented in Table III. Each fraction contained *N*-acetylgalactosaminitol and, based upon the effluent volumes from this calibrated column, each reduced oligosaccharide contains only one residue of this reduced hexosamine. Fraction N-3, which contains only *N*-acetylgalactosaminitol, was eluted in the identical effluent volume as an authentic sample of *N*-acetylgalactosaminitol (23). The presence of this monosaccharide chain (chain I, Fig. 1) had been suggested by periodate oxidation studies (6), as described above.

A reduced disaccharide, *O*-β-D-galactopyranosyl-2-acetamido-2-deoxygalactitol, was eluted as the major neutral fraction (23), as shown in Fig. 2, and represented a recovery of about 46% of the *N*-acetylgalactosamine from epiglycanin (Table III). The presence of this chain (chain II) had been suggested on the basis of composition (Table I) and hemagglutination inhibitory studies with *V. graminea* lectin (17, 19, 22). Fraction N-1 was a reduced tetrasaccharide that contained *N*-acetylglucosamine. This oligosaccharide (chain IV) represented 3.7% of the *N*-acetylgalactosamine present in the starting sample of epiglycanin or 5% of the *N*-acetylgalactosaminitol recovered.

The acidic fractions contained chains II and IV but with bound sialic acid, as represented by chains III and V. The presence of sialic acid as a terminal carbohydrate component attached to a penultimate galactose residue had been suggested

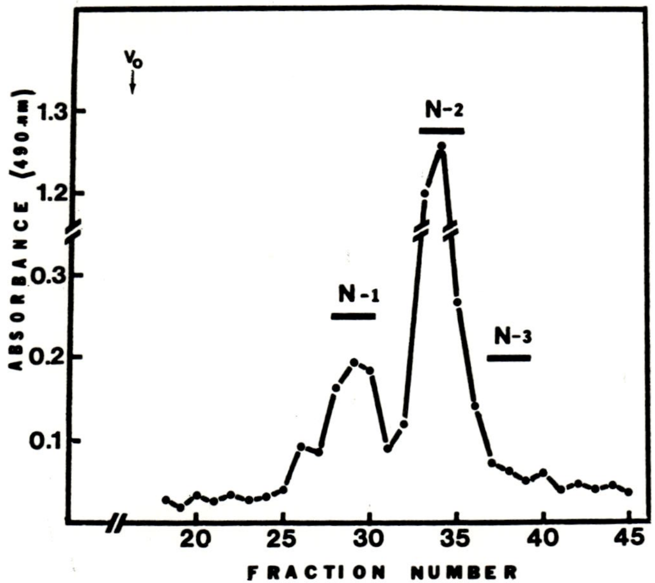

FIGURE 3 Fractionation of the neutral reduced oligosaccharides on a column of Biogel P-2, 1.2 × 142 cm.

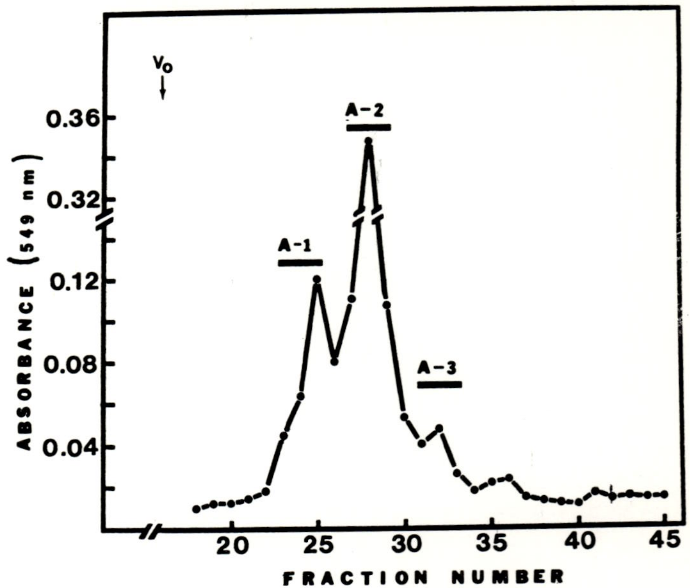

FIGURE 4 Fractionation of the acidic reduced oligosaccharides from epiglycanin on a column of Biogel P-2, 1.2 × 142 cm.

TABLE III Recoveries and Relative Proportions of Carbohydrate Components in Neutral (N) and Acidic (A) Reduced Oligosaccharides Released by Alkaline Borohydride Reduction of Epiglycanin

Reduced oligosaccharide	Molar proportions				Recoveries[a]	
	Sialic acid	Gal	GalNAc.ol	GlcNAc	Sialic acid	GalNAc.ol
N-1	-	1.75	1.0	0.93	-	3.7%
N-2	-	0.92	1.0	-	-	45.6%
N-3	-	-	1.0	-	-	14.2%
A-1	1.72	1.84	1.0	0.53	15.0%	2.2%
A-2	0.93	0.90	1.0	-	34.5%	9.7%

[a]Recoveries in % of epiglycanin used in experiment.

by the enhancement of inhibitory activity against hemagglutination both by the *V. graminea* lectin, possibly specific for chain II, and the *R. communis* lectin, possibly specific for chain IV (6), and by periodate oxidation studies, as described above. Only about 8% as much of fraction A-2 (chain III), the sialic acid-containing counterpart of the disaccharide N-2 (chain II), was recovered as fraction N-2. The composition of fraction A-1 (Table III) suggested the possible presence of two residues of sialic acid. However, the recovery of this oligosaccharide was low, and no firm evidence is yet available to suggest that sialic acid in epiglycanin is attached to any component other than a penultimate galactose moiety. Chain V (Fig. 1) is thus represented with only one sialic acid residue, as previously suggested (6).

FINE STRUCTURES OF THE CARBOHYDRATE CHAINS

Chemical, enzymatic, and immunochemical evidence just presented has suggested the presence of certain specific anomeric configurations in epiglycanin. Periodate oxidation before and after neuraminidase treatment indicated that the sialic acid → galactose linkage in epiglycanin, specifically in Chains III and V (Fig. 1), was α-(2→3). According to the suggestion (22) that the receptor for the *V. graminea* lectin is a β-(1→3)-linked galactose residue to an *N*-acetylgalactosamine residue, the high inhibitory activity found in epiglycanin probably reflects the presence of this linkage in chain II. In recent investigations van den Eijnden *et al.* (28) established a (1→3) linkage to *N*-acetylgalactosamine in fraction N-2 (chain II) by a microautoradiographic method. After periodate oxidation, reduction, and acid hydrolysis (Smith degradation) of fraction N-2 only 2-acetamido-2-deoxythreitol was detected (28). Efforts to establish the anomeric configuration in this material by enzyme cleavage have been disappointing. No cleavage of galactose in N-2 was observed after prolonged incubation with a purified β-galactosidase from *E. coli* at 37° and pH 7.0. Although complete cleavage, after 96 hr, of galactose from a reduced disaccharide, presumably of the β-Gal-(1→3)-GalNAc-Ol structure, prepared from fetuin had been reported (29), we were unable to observe cleavage in material prepared from fetuin in a similar manner. Of interest was the report (30) that no galactose cleavage occurred during prolonged incubation of the same enzyme with a reduced disaccharide of similar composition but unknown anomeric configuration prepared from brain tissue.

Attempts were made to establish the anomeric configuration in fraction N-2 by use of a method based upon the observation (31) that in reduced and peracetylated oligosaccharides, residues of the β-D-configuration were more readily destroyed by chromic oxide than similar residues in the α-D-configuration. Comparison of the proportion of D-galactose destruction by CrO_3 oxidation in glacial acetic acid at 50° for 30 min in peracetylated fraction N-2 with reduced and peracetylated α- and β-linked galactose in peracetylated melibiitol and lactitol, respectively, suggested, but did not prove, the presence of a β-linkage in fraction N-2. Approximately 75% of the D-galactose was destroyed under conditions that oxidized 85% of the galactose in peracetylated lactitol (β-D-configuration) and 34% of the galactose in peracetylated melibiitol (α-D-configuration).

Since α-galactosidase from ficin, an enzyme known to cleave α-linked D-galactose residues in molecules of high molecular weight (32), exhibited no activity against epiglycanin, either before or after neuraminidase treatment, or against fraction N-2, the probability of α-linked galactose residues seemed small. On the other hand, a crude preparation of β-galactosidase from *C. perfringens* (33), which exhibited no activity against the model substrate, p-nitrophenyl-α-D-galactopyranoside, slowly cleaved all carbohydrate residues, galactose, sialic acid, and *N*-acetylglucosamine, but not *N*-acetylgalactosamine, from epiglycanin. Although evidence in support of β-linked-D-galactose residues in epiglycanin is not conclusive, immunochemical, enzymic, and chemical data are in accord with a β-D-configuration.

CONCLUSIONS

Although individual carbohydrate-containing macromolecules may play important and specific roles in the complex environment of the mammalian cell surface, little is generally known regarding structure-function relationships in individual macromolecules. A biological role for epiglycanin, however, that of masking surface antigens, has been proposed, and evidence to support this hypothesis has been reported (8-10). Since the physicochemical requirements for antigen masking by endogenous cell-surface components have not been determined in other cell systems, epiglycanin may be considered a prototype for such a function.

Certain parameters of the epiglycanin-TA3 cell system appear consistent with a masking role. For example, single extended polypeptide chains of approximately 1300 amino acid residues extend 400-500 nm outward from membrane-bound surface antigens. The abundance of such structures (3-5 × 10^6 per cell) suggests a calculated average spacing of about 19 nm, a spacing considered small enough to block or retard antibody approach to the cell surface (10). Whether or not a relationship exists between the chemical structures of the carbohydrate chains in epiglycanin and a masking role has not been determined. It would be expected that the frequency of these chains, more than 500 individual chains attached to the peptide backbone in a molecule of 500,000 molecular weight, would aid the rigidity of this molecule. The presence of negatively charged sialic acid residues in a terminal position in about 20% of these chains would probably help to prevent matting or aggregation of epiglycanin molecules and possibly would serve to keep them in a direction vertical to the cell surface. Furthermore the sialic acid residues might serve to repel negatively charged immune lymphocytes from TA3-Ha cells.

The majority of the chains terminate in β-D-galactopyranosyl residues. Although these chains may serve as receptors for plant agglutinins, they do not appear to be antigenic. Efforts to produce antibodies to epiglycanin in either the rabbit or the mouse have not been successful (Cooper, A. G., and Codington, J. F., unpublished data). This may be due to tolerance to these carbohydrate structures in at least some mammals. Indeed, a lack of immunogenicity of carbohydrate sequences in epiglycanin would appear consistent with a masking role for epiglycanin. Whether or not specific carbohydrate structures in epiglycanin play a role in masking or in some other as yet unknown function of this glycoprotein must await further investigation.

ACKNOWLEDGMENTS

The authors wish to thank Mrs. Cyla Silber for technical assistance and Mr. Keyes B. Linsley for gas-chromatographic analyses. These investigations were supported by Research Grant CA-08418 from the National Cancer Institute, National Institues of Health, U.S. Public Health Service. One of us (Dirk H. van den Eijnden) was supported by a fellowship from the Netherlands Organization for the Advancement of Pure Research (Z.W.O.). This is publication No. 725 of the Robert W. Lovett Memorial Group for the Study of Diseases Causing Deformities, Harvard Medical School and Massachusetts General Hospital.

REFERENCES

1. Hughes, R. C., "Membrane Glycoproteins," Butterworth and Co., Reading, Massachusetts, 1976.
2. Cook, G. M. W., and Stoddart, R. W., "Surface Carbohydrates of the Eukaryotic Cell," Academic Press, New York, 1973.
3. Glick, M. C., *in* "Biology and Chemistry of Eucaryotic Cell Surfaces," Lee, E. Y. C., and Smith, E. E., eds., Academic Press, New York, 1974, pp. 213-240.
4. Fox, T. O., Sheppard, J. R., and Burger, M. M., *Proc. Nat. Acad. Sci. U.S.A.* (1971) *68*, 244-247.
5. Codington, J. F., Sanford, B. H., and Jeanloz, R. W., *Biochemistry* (1972) *11*, 2559-2564.
6. Codington, J. F., Linsley, K. B., Jeanloz, R. W., Irimura, T., and Osawa, T., *Carbohydr. Res.* (1975) *40*, 171-182.
7. Slayter, H. S., and Codington, J. F., *J. Biol. Chem.* (1973) *248*, 3405-3410.
8. Sanford, B. H., Codington, J. F., Jeanloz, R. W., and Palmer, P. D., *J. Immunol.* (1973) *110*, 1233-1237.
9. Codington, J. F., Sanford, B. H., and Jeanloz, R. W., *J. Nat. Cancer Inst.* (1973) *51*, 585-591.
10. Codington, J. F. *in* "Cellular Membranes and Tumor Cell Behavior," Williams and Wilkins, Baltimore, 1975, pp. 399-419.
11. Miller, S. C., Hay, E. D., and Codington, J. F., *J. Cell Biol.* (1972) *72*, 511-529.
12. Friberg, S., Jr., *J. Nat. Cancer Inst.* (1972) *48*, 1463-1476.
13. Hauschka, T. S., Weiss, L., Holdridge, B. A., Cudney, T. L., Zumpft, M., and Planinsek, J. S., *J. Nat. Cancer Inst.* (1971) *47*, 343-359.
14. Apffel, C. A., and Peters, J. H., *J. Theor. Biol.* (1970) *26*, 47-59.
15. Jeanloz, R. W., and Codington, J. F., *in* "Biology and Chemistry of Eucaryotic Cell Surfaces," Academic Press, New York, 1974, pp. 241-257.
16. Codington, J. F., Tuttle, B., and Jeanloz, R. W., *in* "Colloquiem on the Glycoconjugates," Montreuil, J., ed., Colloques Internationaux du C.N.R.S., Paris, 1974, pp. 793-801.
17. Springer, G. F., Codington, J. F., and Jeanloz, R. W., *J. Nat. Cancer Inst.* (1972) *49*, 1469-1470.
18. Cooper, A. G., Codington, J. F., and Brown, M. C., *Proc. Nat. Acad. Sci. U.S.A.* (1974) *71*, 1224-1228.
19. Codington, J. F., Cooper, A. G., Brown, M. C., and Jeanloz, R. W., *Biochemistry* (1975) *14*, 855-859.

20. Miller, D. K., Cooper, A. G., Brown, M. C., and Jeanloz, R. W., *J. Nat. Cancer Inst.,* (1975) *55*, 1249-1252.
21. Osawa, T., Terao, T., Kawaguchi, T., Fujuda, M., Toyoshima, S., and Irimura, T., *in* "Colloquiem on the Glycoconjugates," Colloques Internationaux du C.N.R.S., Paris, 1974, pp. 765-776.
22. Uhlenbruck, G., and Dahr, W., *Vox Sang* (1971) *21*, 338-351.
23. Van den Eijnden, D. H., Codington, J. F., and Jeanloz, R. W., in preparation.
24. Reinhold, V., *Methods Enzymol.* (1972) *25*, 244-249.
25. Chambers, R. E., and Clamp, J. R., *Biochem. J.* (1971) *125,* 1009-1018.
26. Warren, L., *J. Biol. Chem.* (1959) *234*, 1971-1975.
27. Dubois, M., Gilles, K. A., Hamilton, J. K., Rebers, P.A., and Smith, F., *Anal. Chem.* (1966) *28*, 350-356.
28. Van den Eijnden, D. H., Codington, J. F., and Jeanloz, R. W., *Carbohydr. Res.* (1977) *52*, 209-213.
29. Spiro, R. G., and Bhoyroo, V. D., *J. Biol. Chem.* (1974) *249*, 5704-5717.
30. Margolis, R. K., and Margolis, R. U., *Biochem. Biophys. Acta.* (1973) *304*, 421-429.
31. Hoffman, J., Lindberg, B., and Svensson, S., *Acta Chem. Scan.* (1972) *26*, 661-666.
32. Hakomori, S-I., Siddiqui, B., Li, Y-T, Li, S-C., and Hellerqvist, C. G., *J. Biol. Chem.* (1971) *246*, 2271-2277.
33. Chipowsky, S., and McGuire, *Fed. Proc.* (1969) *28*, 606 abs.

CELL SURFACE GALACTOSYLTRANSFERASE IN MITOSIS, DIFFERENTIATION, NEOPLASTIC TRANSFORMATION AND METASTASES*

MILTON M. WEISER
DANIEL K. PODOLSKY

Department of Medicine
Harvard Medical School
and Massachusetts General Hospital
Gastrointestinal Unit
Boston, Massachusetts

Several glycosyltransferase activities, usually associated with intracellular membranes, were detected as cell surface membrane activities of the undifferentiated intestinal crypt cell, fetal intestinal cell, and the chemically induced intestinal tumor cell. In contrast, the differentiated villus cell demonstrated only cell surface sialyltransferase activity. Subcellular fractionation of villus and crypt cells demonstrated a lateral membrane fraction of villus cells separable from Golgi membrane whereas crypt cell surface membrane had the enzyme features of Golgi membranes. These data suggest a similarity between crypt cell surface membrane and crypt Golgi membrane. In addition, crypt cells, fetal intestinal cells, and tumor cells were found to be preferentially agglutinated by Concanavalin A (Con A). Exposure of these cells to Con A appeared to specifically reduce galactosyltransferase. Virally transformed tissue culture cells and animal tumor cells were shown to have a number of cell surface glycosyltransferase activities but to release only galactosyltransferase into the medium or serum. A specific role is proposed for this enzyme in cell adhesion, cell recognition, and metastatic processes.

*This work was supported by grants from the National Institute of Health CA-16703, American Cancer Society BC-93.

INTRODUCTION

Glycosyltransferase activities were once considered to be characteristic of the microsomal pellet and the Golgi apparatus (1). These subcellular locations were also compatible with prevailing concepts on the sequence of glycosylation of secretory proteins, which envisioned glycosylation of the core carbohydrates, particularly the peptidoglycan linkage, to occur on the rough endoplasmic reticulum and the branched carbohydrate chains to be added within the smooth endoplasmic reticulum and Golgi apparatus (2). Papers concerned with Golgi purification emphasized the enrichment of the glycosyltransferase activities but often ignored a parallel enrichment for surface membrane enzyme markers (3). Other workers intent on purification of surface membrane totally ignored glycosyltransferase activities or other Golgi markers (4). In 1970, Roseman published his theory on the role of glycosyltransferases in cell adhesion, predicting that they should be found on the cell membrane (5). Subsequent work from Roseman's and Roth's laboratories appeared to present evidence for the existence of cell surface glycosyltransferases on tissue culture cells (6, 7). Despite some continuing controversy (8) most other papers, including work from our laboratory, have presented supportive evidence for a cell surface location of glycosyltransferases (9).

The suggested functions, if any, for these cell surface glycosyltransferases have varied with the interest of the investigator. Roseman's proposal was that cellular adhesion was the result of the preferential binding of a particular glycosyltransferase to an oligosaccharide acceptor substrate. He likened the interaction to that originally proposed by Tyler (10) and by Weiss (11), i.e., similar to antigen-antibody complexing. Mechanism of specificity would be analogous to enzyme-substrate interactions; addition of the second substrate, i.e., the nucleotide sugar,[1] would result in completion of the reaction and subsequent release of the acceptor:enzyme complex. This concept was extended by Roseman to suggest that various cellular recognition processes might be mediated by the same system (12). Roth has suggested that the adhesive specificity between embryonic cells, the migration of cells during development, and neuronal specificity are mediated by these cell surface glycosyltransferases (9).

When we made the observation that glycosyltransferase activities appeared to be present and characteristic of the mitotically active intestinal crypt cell (3), we reasoned that

[1]Isoprenoid intermediates have also been suggested.

their appearance on the cell surface membrane could be the result of the cell's sudden need, during mitosis, for new cell membrane. Since other investigators had previously suggested that the cell membrane was derived from the Golgi apparatus we concluded that the "early cell surface membrane is, in fact, externalized Golgi membrane" (13). Patt and Grimes have recently suggested a similar mechanism for the appearance of glycosyltransferases on cell surface membrane (14). They had been investigating the level of these cell surface glycosyltransferases in tissue culture systems as related to viral transformation and the cell cycle. There is even greater controversy whether there is any correlation between the appearance of cell surface glycosyltransferases and malignant transformation or the phase of the cell cycle (9).

In this paper we will review some of our data regarding cell surface glycosyltransferases in mitosis, differentiation, and malignant transformation. The evidence we present shows that not only are these enzymes present on the cell surface but they are secreted or shed into tissue culture medium and into the serum of tumor bearing animals. If these enzymes do play a role in modifying cell behavior or as recognition factors, their appearance in serum also suggests that they may prove to be important in the metastatic process.

GLYCOSYLTRANSFERASES OF THE UNDIFFERENTIATED INTESTINAL CRYPT CELL AND FETAL INTESTINE

With respect to differentiation, epithelial cells of the small intestine may be divided into three major types: (1) *lower crypt cells*, which are undifferentiated and the site of active mitosis, (2) the upper villus cells, which are differentiated, "mature," and actively engaged in nutrient uptake and transport, and (3) the upper crypt-lower villus cells, which are in varying stages of differentiation (Fig. 1). The turnover rate of the intestinal epithelium is rapid (36-72 hrs in the rat) and there is a constant loss of cells from the villus tip. This suggests that there must be an alteration in cell adhesion factors for the more mature villus cell so that it may be easily extruded into the lumen. Exposure to citrate was shown to release the intestinal villus tip cells first and subsequently to develop a gradient of cells from villus tip to crypt base (15), in a sense reflecting the gradient of increasing cell adhesion. The cells were obtained as isolated cells, which also permitted studies on lectin agglutinability of the cells and of their surface membrane glycosyltransferases.

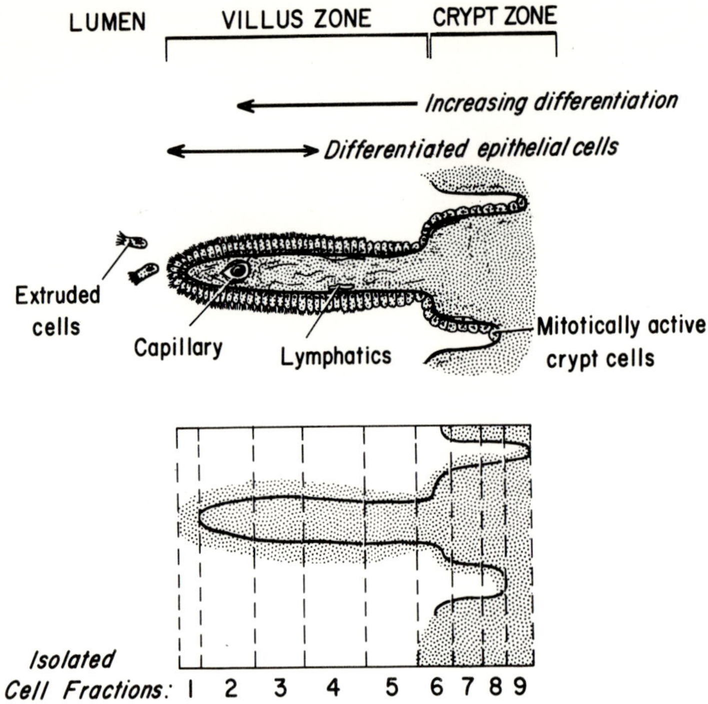

FIGURE 1 Schematic representation of an intestinal villus and the cell fractions obtained by the method of Weiser (15).

The mitotically active crypt cell surface membrane demonstrated the highest glycosyltransferase:endogenous acceptor activities being 10 times that of the activity detected on the villus cell (Fig. 2) (13). The one exception was sialyltransferase:endogenous acceptor activity, which appeared to be higher on the villus cell. Localization of these enzyme activities to the cell surface was supported by various experiments:

(1) The crypt cell in contrast to the villus cell has little hydrolytic activities associated with its surface membrane, and in fact did not demonstrate significant hydrolysis of the nucleotide sugars.

(2) The crypt cell, in contrast to the villus cell, had a very low rate of glycoprotein synthesis when measured by incorporation of simple sugars (^{3}H-fucose) and, therefore, if there had been breakdown of the nucleotide sugar it still

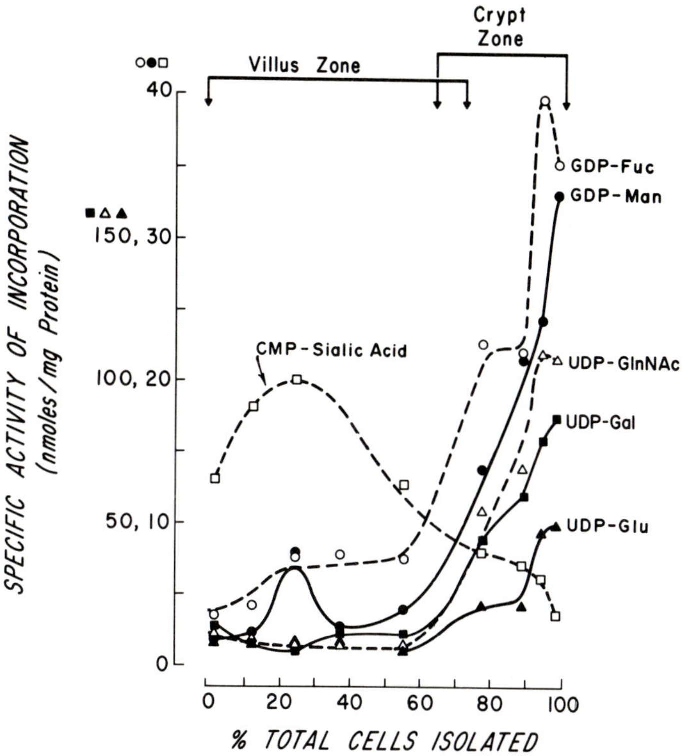

FIGURE 2 Glycosyltransferase:endogenous acceptor activities of isolated villus and crypt cells obtained as the gradient of differentiation represented in Fig. 1.

would not have shown significant incorporation into glycoproteins.

(3) The enzyme activities were capable of glycosylating exogenous acceptors, which remained outside the cell.

(4) No enzyme activity was detected in the medium at the end of incubation.

(5) The increases in glycosyltransferase activities consequent to homogenization were equivalent for villus and crypt cells.

At least 85-90% of the product was a glycosylated glycoprotein, presumably of the cell surface. Similar data were obtained with human fetal intestinal cells, which showed that the fetal cells demonstrated galactosyltransferase:endogenous and exogenous acceptor activities (15). Therefore, it ap-

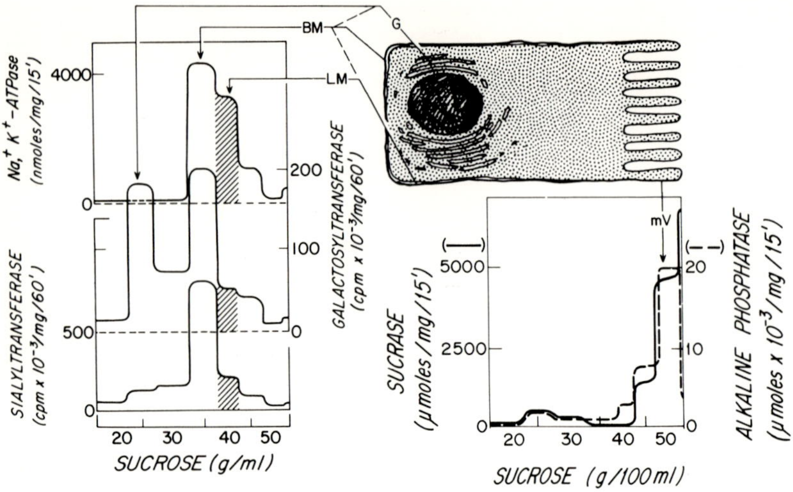

FIGURE 3 Isolation of intestinal epithelial cell surface membrane fragments and their separation from a purified Golgi membrane fragment.

peared that such surface membrane enzyme activities were characteristic of the undifferentiated intestinal epithelial cell.

More recently, the villus and crypt cell surface membranes have been isolated and studied (16). The villus cell surface membrane can be obtained as three separate fractions: (1) purified microvillus membranes, (2) a Na^+, K^+-ATPase rich, glycosyltransferase poor membrane presumed to be a part of or mostly lateral membrane (Fig. 3, LM), and (3) a Na^+, K^+-ATPase rich, sialyl- and galactosyltransferase rich membrane presumed to be part basal, part lateral, and possibly Golgi membrane (Fig. 3, BM). Another fraction lacking Na^+, K^+-ATPase but rich in galactosyltransferase activity was considered to be Golgi membranes (Fig. 3, G). This suggested that there may be little difference between Golgi membranes near the surface membrane and the surface membrane itself and that the surface membrane may partly consist of patches of Golgi membrane (17, 18). The question can then be raised as to whether any of these Golgi patches are there to serve as cell adhesive areas. When the crypt cell was analyzed, no significant Na^+, K^+-ATPase rich, glycosyltransferase poor fraction was observed, and most of the Na^+, K^+-ATPase activity co-purified with a glycosyltransferase rich fraction; this supported the previous isolated cell data that the crypt cell surface membrane had high glycosyltransferase activities (13).

There may be two explanations for the detection of glycosyltransferase activities on these cell surface membranes. Our original suggestion was that those cells requiring new cell membrane rapidly would exhibit a cell membrane with features characteristic of the internal cell membrane from which they were derived — namely the Golgi membrane (13). This explanation views the emergence of cell membrane glycosyltransferases as an epiphenomenon and of no functional significance. The subcellular fractionation data may also be explained by assuming that Roseman's hypothesis (5) is valid and that those parts of the membrane involved in cell adhesion should be rich in glycosyltransferase activities. The lateral membrane on the differentiated villus cell is apparently not adherent (except at the desmosomal and junction areas) to the adjacent cell and would be low in glycosyltransferase activities (Fig. 3), while the adherent basal membrane would have high glycosyltransferase activities. The crypt cell surface membrane, not yet differentiated into the columnar villus cell, has no lateral membrane. The presence of glycosyltransferase activities on the crypt cell could be considered a consequence of mitosis or that more of its cell surface was involved in cell adhesion.

LECTIN BINDING SITES, AGGLUTINATION, AND CELL MEMBRANE GLYCOSYLTRANSFERASE ACTIVITIES

The differences of the cell surface membranes between the differentiated and undifferentiated epithelial cell were also examined by using lectins (19, 20). It was demonstrated that Con A caused the agglutination of crypt cells while wheat germ agglutinin (WGA) agglutinated only villus cells. This was partly confirmed histochemically by Etzler (21). In addition, it was possible to specifically select out crypt cells with Con A derivatized nylon fibers (20). These cells were high in galactosyltransferase and mannosyltransferase activities but did not exhibit sialyl- or N-acetylglucosaminyltransferase activities. It was also observed that if the cells were assayed for glycosyltransferase activities, selective diminution of galactosyltransferase activity was observed after preincubation with nonagglutinating concentrations of Con A (22). This suggested a possible relationship between Con A binding to the cell surface and cell surface galactosyltransferase. In at least one system, the rabbit erythrocyte, this was indeed shown to be the case (23). It was demonstrated that this was due to the glycoprotein character of the enzyme itself, which was shown to have terminal α-linked D-mannose

TABLE I Surface Membrane Properties of Differentiated and Undifferentiated Intestinal Cells

Property	Cell type: Villus	Crypt	Fetal	Tumor (sm. intestine)
1. Surface membrane glycoproteins	complete	incomplete	incomplete	incomplete
2. Surface membrane glycosyltransferases				
a. Sialyltransferase	+++	0	0	0
b. Galactosyltransferase (and most other transferases)	+/-	+++	+++	+++
3. Con A agglutination	+/-	+++	+++	+++

residues. It was this terminal mannose that appeared to be necessary for the binding of Con A. These studies led us to survey a number of different cell types to see if the association of Con A agglutination and cell membrane glycosyltransferase activity was consistent. It was found that the presence of galactosyltransferase activity on the cell membrane did not necessarily confer Con A agglutination but cells agglutinated by Con A were found to have significant galactosyltransferase activities (22). In particular, chemically induced intestinal tumor cells were found to be agglutinated by Con A and to have high galactosyltransferase activities (24). At that point there appeared to be strong correlations between Con A agglutination and the presence of glycosyltransferase activities on the cell surface of undifferentiated, fetal and neoplastic intestinal cells as contrasted with the differentiated villus cell (Table I). In an attempt to more rigidly test this correlation we turned to tissue culture systems.

TISSUE CULTURE CELLS, CON A AGGLUTINATION, CELL SURFACE GALACTOSYLTRANSFERASES, AND THE RELEASE OF GALACTOSYLTRANSFERASE INTO THE TISSUE CULTURE MEDIUM

BHK cell lines were tested for the presence of cell surface galactosyltransferase activities and Con A agglutination (Table II). Both NIL and NILpy cells had been shown to be agglutinated by Con A and both demonstrated cell surface galactosyltransferase activity (25). However, BHKpy cells that were agglutinable by Con A demonstrated more cell surface galactosyltransferase activity than the nonagglutinable BHK cells. LaMont, while working in our laboratory, demonstrated that there was a marked increase in cell surface galactosyltransferase activity in all of the above cell lines when grown in serum (heat treated) stimulated conditions (25). There was a suggestion that this increase occurred with the entry of the cells into the S phase of the cell cycle. Another intriguing finding was the detection of soluble galactosyltransferase activity in the tissue culture medium, which increased with growth. No other glycosyltransferases were detected in this medium and the level of galactosyltransferase activity released into the medium appeared to be higher for the NILpy cells (25).

Gammon and Isselbacher, in our laboratory, studied a variant of BHKpy cells, the glutamine independent variant (giv) and they were able to show that the BHKpygiv variant was significantly more agglutinable by Con A. Interestingly, the giv variant was also more tumorigenic than the parent BHKpy

TABLE II Tumorigenicity, Con A Agglutination and Galactosyltransferase Activities of BHK Cells

			Galactosyltransferase activity (pmol/mg/hr)	
	Tumorigenicity	Con A agglutination	Isolated cells	Medium
BHK	0	+	140	3,570
BHKpy6	+++	++	208	12,400
BHKpygiv	++++	+++	220	6,460

when injected into hamsters even though their behavior in tissue culture was more characteristic of the nontransformed BHK cells (26). Both tumorigenic lines demonstrated higher cell surface galactosyltransferase activities than the nontumorigenic BHK cells. Equally impressive were the high levels of galactosyltransferase activity released into the medium (Table II). The findings of Gammon and Isselbacher stress an important parameter in the evaluation of neoplastic transformation: the final test for tumorigenicity is the ability to grow in a host animal. The increase in Con A agglutinability, the release of galactosyltransferase activity, and tumorigenicity suggested that these three factors might be related. Our previous demonstration that Con A agglutinability and cell surface galactosyltransferase activity were related to the oligosaccharide composition of the galactosyltransferase suggests that the galactosyltransferase associated with Con A agglutination and tumorigenicity might be structurally different than other galactosyltransferases.

ANIMAL TUMOR MODELS AND THE RELEASE OF AN ELECTROPHORETICALLY DISTINCT GALACTOSYLTRANSFERASE

When the tissue culture medium of BHKpy and the giv variant cells were subjected to polyacrylamide electrophoresis, gels sliced, and each slice evaluated for galactosyltransferase activity, a single slow moving peak near the cathode was observed (27). Galactosyltransferase activity solubilized (by sonication) from the tissue culture cells themselves dem-

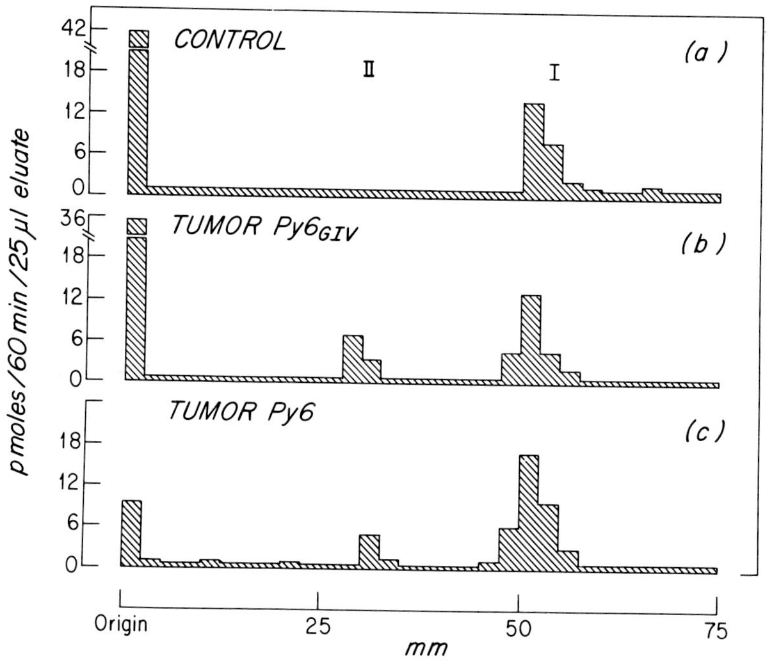

FIGURE 4 Galactosyltransferase activities eluted from polyacrylamide electrophoresis of hamster serum (25 μl). (a) Normal, non-tumor-bearing hamster, (b) hamster with tumor derived from BHKpygiv cells, (c) hamster with tumor derived from BHKpy cells.

onstrated a broad area of activity extending more anodally. These transformed BHK cells were then injected into hamsters and tumors permitted to develop. When the sera of these hamsters were assayed for total galactosyltransferase activity, no significant difference was found between tumor bearing animals and controls. However, when the hamster serum was analyzed by polyacrylamide electrophoresis, a new peak of activity was observed in tumor bearing animals (Fig. 4). This new peak was detectable before the detection of an obvious tumor and the levels of this new peak appeared to increase as the tumor grew. This new peak was similar to the peak found in the medium of the original tissue culture. When the tumors themselves were examined it was found that the majority of the solubilized galactosyltransferase activity appeared to co-electrophorese with the new peak found in the serum as well as that found in the tissue culture medium of the original transformed cells (27).

Another animal tumor model in which human mammary adenocarcinoma cells were grown in nude mice also demonstrated the emergence of a tumor-associated electrophoretically distinct peak of galactosyltransferase in the serum of the tumor-bearing mice. This galactosyltransferase isoenzyme was detected in the tumor itself, suggesting again that tumor was the source for the serum isoenzyme (28).

Finally, we have reported on the emergence of an electrophoretically distinct peak of galactosyltransferase activity in the serum of cancer patients (29). Approximately 70% of patients with various cancers showed this new more cathodal peak of activity.

CONCLUSIONS

Experimentally there does appear to be an association of cell surface galactosyltransferase activity and Con A agglutination with the undifferentiated, the fetal, and the cancerous cell. It is not clear to what extent Con A agglutination and cell surface galactosyltransferase activities are molecularly related except in the case of the rabbit erythrocyte (23). Evidence has also been presented which strongly suggests that a different form of galactosyltransferase is released from tumorigenic cells into the medium in which they are grown, including into the serum of tumor-bearing animals. Whether this galactosyltransferase is tumor specific remains to be determined.

At present we interpret our data on cell surface glycosyltransferases in terms of cell adhesion forces and assume that these glycosyltransferases are introduced into the surface membrane as patches of Golgi membrane. This interpretation stems from the works of Wallach (30) and Varga *et al.* (18), who have suggested a patchwork organization to the surface membrane. The interpretation also superimposes upon this concept Roseman's hypothesis regarding the role of cell surface glycosyltransferases in cell adhesion. We would tend to give galactosyltransferase, in contrast to other glycosyltransferases, an especially important role in cell adhesion and recognition. This partly stems from the work of Morell *et al.* (31), who demonstrated the significance of the penultimate galactose of glycoproteins as a recognition determinant in the clearance of serum glycoproteins. A special role for galactosyltransferase is also supported by our work demonstrating that it was the only glycosyltransferase released into tissue culture medium.

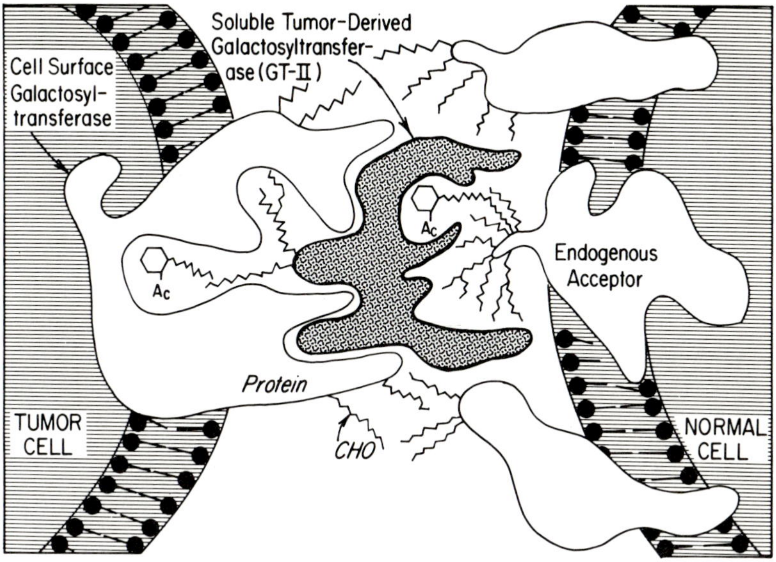

FIGURE 5 Possible molecular mechanism for galactosyltransferase as a mediator for cell adhesion and the metastatic process.

The interpretation of our findings of a distinct galactosyltransferase activity being released by tumor cells must be extremely speculative at this time. The speculation, however, is based on concepts suggested by Abercrombie and Ambrose (32) and later by Winterburn and Phelps (33). Our working hypothesis is that this glycoprotein enzyme when released by the cell serves as a recognition factor in the growth and metastases of tumors. We postulate (Fig. 5) that at the initial site of growth the tumor cells release galactosyltransferase, which binds to the surface of normal cells. In this way a more compatible surface is developed at which the tumor cell can bind, multiply, and grow. A similar process would then be operative during metastases. This glycoprotein enzyme, being continually released, would bind to cells in distant organs. Tumor cells that break away from the original tumor would then find a compatible area on which to adhere to those normal cells (or endothelial surfaces) that have tumor-released galactosyltransferase bound to their surface. We would invoke Roseman's and Roth's concepts of cell adhesion to explain the mechanism on a molecular level. That is, the tumor associated galactosyltransferase binds to any cell surface glycoprotein that can be galactosylated, i.e., serve as an acceptor. Thus, the active site of the enzyme must be available; the enzyme must be active at this stage. Once bound, tumor recognition may also occur through the glycoprotein character of the enzyme itself. In this regard, the *cell surface* galactosyltransferases may be the means by which the cell binds to the new cell surface. This would imply that the circulating tumor-associated galactosyltransferase should have an oligosaccharide fraction lacking a penultimate galactose and thus can itself serve as a galactosyltransferase acceptor. The specificity would result from the acceptor specificity of the released tumor cell surface galactosyltransferase. This might reside in the nature of the oligosaccharide or in the protein structure as a steric property of the active site of the enzyme.

One might argue that such an intermediary binding factor is superfluous in that cell surface bound galactosyltransferase could serve as the binding factor to normal cells or endothelial surfaces. This indeed could happen but the binding would not be as firm due to steric hindrance factors. The soluble galactosyltransferase allows for an intermediary "divalent" binding in exactly the way lectin agglutination is supposed to occur (33). Instead of "trypsinizing" the normal cell surface and enhancing agglutination as has been demonstrated for lectin agglutination (34), the soluble galactosyltransferase builds up on the cell surface and gets away from factors that might inhibit cell-to-cell interaction.

Extending this concept we postulate a mechanism by which tumors overcome host defense according to present concepts in tumor immunology (35). A circulating tumor-released glycoprotein antigen such as this galactosyltransferase, with an ability to bind to normal cell surfaces, would confuse and overwhelm the host's immune response against the tumor. As has been suggested by Currie (35), "Although undoubtedly an oversimplification, the immunological escape of tumors could be explained by the shedding of antigens . . . the local tissue fluid could become flooded by soluble TSTA [tumor-specific transplantation antigens] . . . neutralize . . . effector cells and humoral antibody . . . until the TSTA are present in a free form in the circulation . . . manifested as the failure of concomitant immunity and the development of metastases."

REFERENCES

1. Bergeron, J. J. M., Ehrenreich, J. H., Siekevitz, P., and Palade, G. E., *J. Cell Biol.* (1973) *59*: 73.
2. Zagury, D., Uhr, J. W., Jamieson, J. D., and Palade, G. E., *J. Cell Biol.* (1970) *46*: 52.

3. Cheetham, R. D., Morré, D. J., and Yunghaus, W. N., *J. Cell Biol.* (1970) *44*: 492.
4. Ray, T. K., *Biochim. Biophys. Acta* (1970) *196*: 1.
5. Roseman, S., *Chem. Phys. Lipids* (1970) *5*: 270.
6. Roth, S., McGuire, E. J., and Roseman, S., *J. Cell Biol.* (1971) *51*: 536.
7. Roth, S., and White, D., *Proc. Nat. Acad. Sci.* (USA) (1972) *69*: 485.
8. Keenan, T. W., and Morré, D. J., *FEBS Letters* (1975) *55*: 8.
9. Shur, B., and Roth, S., *Biochim. Biophys. Acta* (1975) *415*: 473.
10. Tyler, A., *Growth (Suppl.)* (1947) *10*: 7.
11. Weiss, P., *Yale J. Biol. Med.* (1947) *19*: 235.
12. Ohman, R., Barker, R., Hill, R., and Roseman, S., *Birth Defects: Original Article Series* (1973) *9*: 198.
13. Weiser, M. M., *J. Biol. Chem.* (1973) *248*: 2542.
14. Patt, L. M., and Grimes, W. J., *J. Biol. Chem.* (1974) *249*: 4157.
15. Weiser, M. M., *J. Biol. Chem.* (1973) *248*: 2536.
16. Weiser, M. M., Neumeier, M., Quaroni, A., and Kirsch, K., submitted for publication.
17. Whaley, W. G., Dauwaldes, M., Kephart, J. E., *Science* (1972) *175*: 596.
18. Varga, J. M., Moellman, G., Fritsch, P., Godawska, E., and Lerner, A. B., *Proc. Nat. Acad. Sci.* (USA) (1976) *73*: 559.
19. Weiser, M. M., *Science* (1972) *177*: 525.
20. Podolsky, D. K., and Weiser, M. M., *J. Cell Biol.* (1973) *58*: 497.
21. Etzler, M. E., and Branstrator, M. L., *J. Cell Biol.* (1974) *62*: 329.
22. Podolsky, D. K., Weiser, M. M., LaMont, J. T., and Isselbacher, K. J., *Proc. Nat. Acad. Sci.* (USA) (1974) *71*: 904.
23. Podolsky, D. K., and Weiser, M. M., *Biochem. J.* (1975) *146*: 213.
24. LaMont, J. T., Weiser, M. M., and Isselbacher, K. J., *Cancer Res.* (1974) *34*: 3225.
25. LaMont, J. T., Gammon, M., and Isselbacher, K. J., *Proc. Nat. Acad. Sci. (USA)* (1977) *74*: 1086.
26. Gammon, M., and Isselbacher, K. J., *J. Cell Physiol.*, (1976) *89*: 759.
27. Podolsky, D. K., and Weiser, M. M., *Fed. Proc.* (1976) *35*: 1442.
28. Podolsky, D. K., and Weiser, M. M., unpublished observation.
29. Podolsky, D. K., and Weiser, M. M., *Biochem. Biophys. Res. Comm.* (1975) *65*: 545.

30. Wallach, D. F. H., *in* "The Specificity of Cell Surfaces" (eds, B. D. Davis & L. Warren) (1967), Prentice Hall, Englewood Cliffs, New Jersey.
31. Morell, A. G., Gregoriadis, G., Scheinberg, I. H., Hickman, J., and Ashwell, G., *J. Biol. Chem.* (1971) *246*: 1461.
32. Abercrombie, M., and Ambrose, E. J., *Cancer Res.* (1962) *22*: 525.
33. Winterburn, P. J., and Phelps, C. F., *Nature* (1972) *236*: 147.
34. Burger, M. M., *in* "Macromolcules Regulating Growth and Development" (eds., E. D. Hay, T. J. King, and J. Papaconstantinov) (1974), Academic Press, New York.
35. Currie, G., *Biochim. Biophys. Acta* (1976) *458*: 135.

PERTINENCE OF SURFACE MEMBRANE CHANGES IN SPONTANEOUSLY AND VIRALLY TRANSFORMED CELLS TO THE BALANCE BETWEEN TUMORIGENICITY AND IMMUNE REJECTION

P. T. MORA
S. J. PANCAKE
S. W. LUBORSKY
C. CHANG
J. COLL

National Cancer Institute
National Institutes of Health
Bethesda, Maryland

Spontaneously transformed highly tumorigenic AL/N strain mouse cells and their SV40 virus transformed clonal derivatives were compared. Glycoproteins were labeled by various metabolic precursors (amino acids, glucosamine, mannosamine, etc.). The flow of such labels into various subcellular components was followed. Cell surface membranes were obtained and the components of these membranes were separated by polyacrylamide gel electrophoresis. The biosynthesis of a prominent glycoprotein component with an apparent molecular weight of 180 K dalton was reduced in the SV40 transformed cells, when all the different kinds of cells were grown in their logarithmic phase. Three types of SV40 specific antigens were detected by immunologic assays: (1) T antigens, (2) tumor specific surface antigens (TSSA), and (3) tumor specific transplantation antigens (TSTA). A rapid and sensitive microassay for TSSA guided the development of a mild and efficient detergent (Triton X-100) solubilization technique, and the purification of the antigens that appear to co-purify in various procedures. The antigens expressed on the cell surface that are responsible for immunologic recognition and rejection of the SV40 transformed cells represented a very minor fraction of the cell surface membrane components. The microassay for TSSA allowed partial purification and characterization of the SV40 specific cell surface antigen(s). The(se) SV40 spe-

cific cell surface antigen(s) in the AL/N strain mouse functionally opposes cellular tumorigenicity.

What cell surface changes may correlate with the acquisition of tumorigenicity by a cell? The surface of the cell through which the cell is in contact and communicates with other cells and with factors that control their growth must be changed in a heritable way so that the progeny cells become the tumor mass. There should be some kind of a signal from the cell surface to the genes to upset heritably the normal regulatory functions of the genes, and the genes in turn must instruct the progeny cells to perpetuate the change in the cell surface. Thus, a misdirection of such interrelated controls must exist between cell surfaces and genes in malignancy, which is indeed a much studied field currently (1). From a biochemical point of view, any cell surface functional change eventually must be reducible to some constitutional change, to either a qualitative or quantitative difference in the chemical constitution of the malignant cell surface membrane. But how can one go about such a study, especially if one wishes to be pertinent to the appearance of malignancy in humans?

All human tumors are "spontaneous" even if the causative agent is known or suspected. When we began our studies on cells, we chose normal cells from an AL/N strain mouse embryo (the N AL/N cells) (2); we observed that after some time in tissue culture they became spontaneously and highly malignant. After about 40 tissue culture transfers, only 100 tumorigenic cells (designated then as T AL/N) produced a progressive and lethal tumor (fibrosarcoma) at the site of the injection (hind leg muscle) in the syngeneic AL/N mouse (3, 6, 8, 9).

In this communication we wish to report a change that we induced in the already tumorigenic T AL/N cells by infecting these cells in culture with SV40, which is a small DNA virus. The SV40 DNA (3×10^6 molecular weight) is known to be inserted covalently into the DNA of certain cells, such as mouse embryo cells. This is known as "transformation by SV40"; it is a rare event and a high ratio of virus to cell is needed to achieve transformation by SV40. The spontaneously transformed cells, further transformed by SV40, may have then two properties. They should continue to exhibit the properties of the spontaneous transformation to tumorigenicity and properties that are endowed by the SV40 genome, which will be continuously replicated as an integrated part of the cellular DNA.

The SV40 virus is tumorigenic only in the hamster (10, 11) and the Mastomys (12). The SV40 virus or cells transformed in culture by the virus is not tumorigenic in the rat, mouse, rabbit, etc. (13, 14). It is true that after transformation by SV40 the normal N AL/N cells, or any other type of mammalian cells, generally exhibit changes in their pattern of growth in tissue culture such as increased "saturation density," lower serum requirements, ability to overgrow a monolayer of normal cells, and ability to form clones in agar or methylcellulose viscous suspension media. But in addition to these changes SV40 transformation leads to the appearance of new SV40 specific antigens, including cell surface antigens in any mammalian cell. Due to such cell surface antigens the SV40 transformed cells may be recognized and rejected by the immunologically competent host such as the syngeneic AL/N mice (2, 6, 8, 9). Thus in vivo the SV40 transformed T AL/N cells will have two properties that functionally oppose each other: the tumorigenicity and the SV40 induced antigenicity. This communication thus deals with the balance between tumorigenicity and antigenicity in the SV40 transformed T AL/N cells, and it attempts to relate these to biochemical observations on the cell surface.

ORIGIN AND BIOLOGIC PROPERTIES OF THE VARIOUS CELL LINES

To study biochemical correlates of cell properties, one needs reproducible and uniform cell populations related to each other in a defined way. Figure 1 gives the outline of the geneologic relationship of the cell lines employed. All cells originated from a normal embryo cell line (N AL/N) established in tissue culture from the inbred AL/N strain mouse (2). The cells were cultivated as described before (3). After 68 tissue culture transfers, the cells were highly tumorigenic [T AL/N cells (9)], when a single T AL/N cell was cloned out (T AL/N clone 3). A culture of the T AL/N clone 3 cells was infected by SV40 and was recloned. The sister subclones thus obtained belonged to two subgroups: subclones 1 and 2, which were SV40 transformed [as detectable by a T antigen test that is specific to SV40 transformed cells and manifests itself by immunofluorescence of the cell nuclei by an appropriately prepared specific and fluorescein labeled antibody (9)], and subclones 5, 6, and 8, which were T antigen negative (not transformed by SV40, and therefore considered simply as reclones of the T AL/N clone 3 parent cell). From the N AL/N mass cell line, at an early tissue culture transfer and after SV40 transformation, the SV AL/N mass cell line was obtained (2) and was also used in these studies.

The tumorigenicity and the SV40 antigens of these cell lines are summarized in Table I. Note that the SV40 antigen containing SV AL/N cells were nontumorigenic; also the sub-

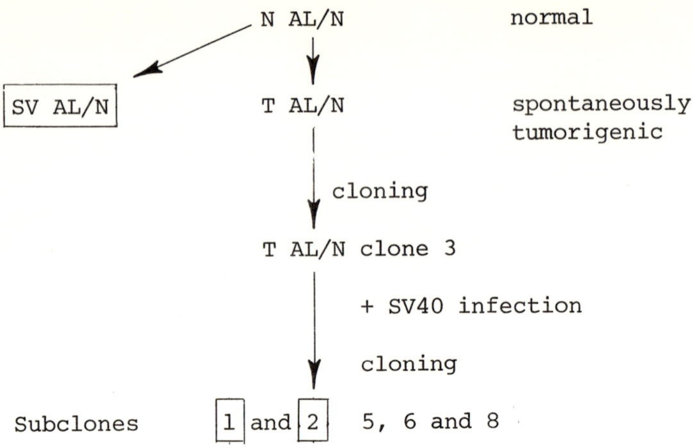

FIGURE 1 Geneologic relationships of the cell lines. Cloning was in microwells; the subclones came from single cells. The box indicates SV40 antigen positive cell lines. In the SV40 infection of T AL/N clone 3 cells the ratio of virus/cell was 360, to give 30-50% SV40 transformed cells.

clone 1 and 2 cells were less tumorigenic (about one hundred fold) than the SV40 antigen negative sister subclones or the parent T AL/N clone 3. The absence or decrease in tumorigenicity is due to recognition and rejection of the SV40 specific TSTA on the cell surface (4-9).

LABELING AND PREPARATION OF CELL SURFACE MEMBRANES

The external chemical labeling of cells was based on a reaction of pyridoxal phosphate, which forms a Schiff base preferentially with the lysine ε-amino groups of proteins and glycoproteins, followed by reduction with KB^3H_4. The method was a modification of procedures employed previously to label cell membrane enclosed RNA viruses (15) and HeLa cells grown in tissue culture (16). The following method allowed cells to remain in near physiological condition, exposed the cells to the reagents only for short periods, and did not result in detectable change in cell appearance or viability (the latter as determined by exclusion of vital dyes). Cells were plated out in modified Eagles medium with added 10% fetal calf serum as described (3), 20-24 hr before the labeling, plating $\sim 5 \times 10^6$ cells per 10 cm diam Falcon tissue culture plate. When the

TABLE I Tumorigenicity and Antigenicity of the Cell Lines

Cells	Tumorigenicity, TD_{50}[a]	SV40 Antigen		
		T[b]	TSSA[c]	TSTA[d]
N AL/N	$>10^7$	–	–	NT[e]
SV AL/N	$>10^7$	+	+	+
T AL/N	$<10^2$	–	–	NT
T AL/N clone 3	$10^{1.7}$	–	–	–
Subclone 1	$10^{3.8}$	+	+	+
Subclone 2	$10^{3.9}$	+	+	+
Subclone 5	$10^{1.5}$	–	–	–
Subclone 6	<10	–	–	–
Subclone 8	$10^{2.5}$	–	–	NT

[a]TD_{50} is the cell dose in which 50% of the injected mice (i.p.) develop large (>1 cm Ø) and lethal tumors; duplicate groups of 10 animals each were injected with 3-6 cell doses differing tenfold.

[b]The T antigen was detected by fluorescence of cell nuclei after reacting the cells with a 1:5 dilution of serum of a hamster bearing tumors induced by SV40 transformed cells (H-50) followed with fluorescein conjugated antihamster gamma globulin (Antibodies Inc.); +, 95-100% cells with strongly fluorescent nuclei; -, none.

[c]Tumor specific surface antigen that was detected by adsorption with preconfluent cells (2×10^6 cells, 3 times) of an SV40 specific antiserum prepared against the SV AL/N cell, and by titration of the residual lytic activity against ^{51}Cr labeled SV AL/N target cells (4-6); +, 0-15% residual cytolytic activity; -, 80-100% residual cytolytic activity at 1/200-1/400 dilution of the antiserum.

[d]Tumor specific transplantation antigen detected by immunization of Balb/c mice with extracts of the given cells, followed by challenge by a tumorigenic SV40 transformed Balb/c mouse cell line, as described before (7).

[e]NT, not tested.

cells were between 50 and 80% confluent, the plates were rinsed once with HSS (0.02 *M* Hepes buffer solution, pH 8.0, containing 0.0056 *M* KCl, 0.0023 *M* $CaCl_2$, 0.0008 *M* $MgSO_4$, and

0.134 *M* NaCl). Then 2 ml of HSS solution containing 5 m*M* pyridoxal phosphate, readjusted to pH 8, were added to each plate. The plates were incubated at 37°C for 20 min with occasional gentle swirling, then rinsed three times with additional HSS, and 1 ml of HSS was added to each plate. Then 1 mCi of KB^3H_4 (in 5 λ volume of 0.5 *N* NaOH) was added to each plate and the plates were kept at room temperature for 20 min with occasional gentle swirling. A few (∿5) milliliters of HSS solution containing 1 m*M* pyridoxal phosphate and 0.5 m*M* PMSF (phenylmethylsulfonylfluoride, Sigma Corp.) were added to each plate to stop the reaction. The cells were harvested by gentle scraping with a rubber policeman, centrifuged (300 × g), diluted to 10 ml in the same solution, and recentrifuged.

Metabolic or internal labeling of cellular glycoproteins was done with either 2 μCi/ml of ^{3}H-Glc-NH_2 or ^{3}H-Man-NH_2, or with 5 μCi of ^{14}C labeled mixed amino acids, as described before (17, 18).

The procedure for plasma (cell surface) membrane isolation was designed to preserve the intact "ghost" structure and to minimize the alteration of the plasma membrane. This procedure, outlined in Fig. 2, was based on a technique employed for tissue culture grown HeLa cells (19, 20), avoids the use of cross linking reagents or fixatives, and employs isopycnic banding in sucrose density gradients (18). The methods for cell disruption and for the preparation of the various subcellular fractions are described in detail separately (17).

Representative examples of distribution of radioactivity in the subcellular fractions from cells labeled by the precursor carbohydrates for glycoproteins or by the external chemical labeling with the KB^3H_4 method are given in Table II. There are generally similar distributions of radioactivity from both the T antigen positive subclones (as represented by subclone 1 cells) and from the T antigen negative subclones (as represented by subclone 5 cells). Distribution of ^{14}C amino acid labeled proteins in the subcellular fractions of the different cell lines were also found to be generally alike (17).

The membrane pellets were found to be (Table II, also reference 17) similar to those of pure plasma membranes from HeLa cells (19, 20). Their buoyant density was 1.16 g/ml (Fig. 3), which is characteristic of plasma membranes (22). From the chemically labeled intact cells, up to 80% of the ^{3}H can be removed by protease treatment without loss of cell viability (Table III).

(0.5-5) × 10^8 CELLS, in culture

| scrape, wash with TD

CELL PELLET

| "hypotonic buffer" for 45-60 min

CELL SUSPENSION

| Dounce, 2×-4×
| add Mg^{+2}

HOMOGENATE

| 230×g, 90 sec ——— PELLET

SUPERNATANT

| on 30%, 45% sucrose layers
| SW41, 7000×g, 40 min

LOWER INTERFACE

| dilute; on sucrose & repeat centrifugation

LOWER INTERFACE

| dilute; SW50.1, 10,000×g, 20 min

MEMBRANE PELLET

FIGURE 2 Scheme of cell membrane isolation. The conditions were described (19), except as specified above or as noted: 0.001 M phenylmethylsulfonylfluoride was included in all solutions. A tight fitting glass Dounce homogenizer was used. The membrane pellet was dissolved by boiling for 4 min in a buffer previously described for gel electrophoresis (21), and small aliquots containing ∿20-50 kcpm were stored at -60°.

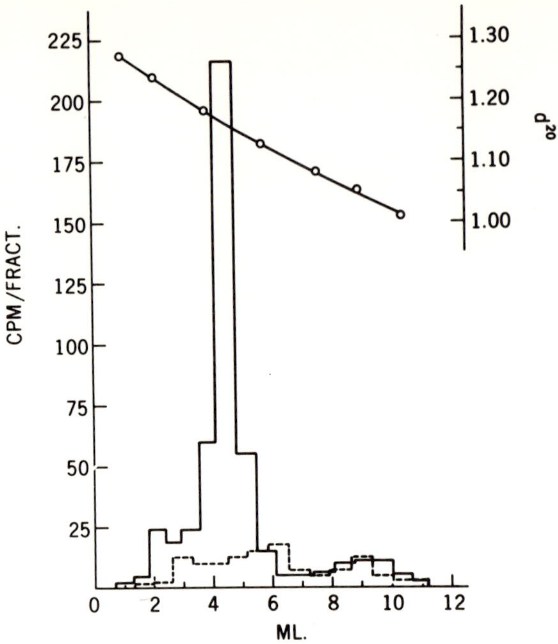

FIGURE 3 Distribution of labeled membrane components after equilibrium centrifugation into a linear sucrose gradient. Subclone 1 cells were labeled with ^{3}H-Man-NH$_2$ and fractionated on the first discontinuous sucrose gradient, as described in the text. The two interfaces obtained were diluted with "Mg hypotonic buffer" and pelleted in the SW41 rotor at 12,000 ×g for 10 min. The pellets were suspended in 1 ml "Mg hypotonic buffer," layered on a 10-60 wt% linear sucrose gradient, and centrifuged overnight in a SW41 rotor at 74,000 ×g. Fractions (0.6 ml) were collected after bottom puncture. Left ordinate: radioactivity; right ordinate: density, based upon refractive index measurements. Broken line: solution from upper interface; solid line: solution from lower interface. The sharp peak at 1.16 gm/ml is characteristic of plasma membranes (19, 22).

ANALYSIS OF THE CELL MEMBRANE PREPARATIONS FROM THE VARIOUS CELL LINES

The cell membrane pellets were analyzed by SDS acrylamide gel electrophoresis either after disc gel or slab gel electrophoresis.

TABLE II Distribution of Radioactivity in Subcellular Fractions from T-antigen Positive and Negative Subcloned Cells

	Internal Labeling: Metabolic Uptake						External Labeling: Chemical Reaction with KB^3H_4			
	of ^3H-Glc-NH_2				of ^3H-Man-NH_2					
Subclone no.	1		5		1		1		5	
Initial cell										
number × 10^{-7}	3.8		5.2		-	18	4.0		5.0	
packed vol (cc)	.3		.35		.25	.7	.2		.45	
protein (mg)[a]	6.3		5.8		-	12.6	5.2		6.5	
^3H-uptake (CPM × 10^{-6})	2.38		8.08		.027	.53	2.59		8.46	
	%CPM[b]	%mg[a]	%CPM[b]	%mg[a]	%CPM[b]		%CPM[b]	%mg[a]	%CPM[b]	%mg[a]
Samples from:										
cell suspension	100	100	100	100	100	100	100	100	100	100
homogenate	70	88	92	73	92	109	89	88	94	77
230×g supernatant	79	71	40	68	67	76	80	89	75	50
Sucrose interface:										
1st centrifugation	31	24	17	23	11	34	30	31	17	8
repeat	16	19	7	16	3[f,g]	10	17	24	10	2
Membrane pellet[c]	12	15	6	10	-	4	10[d]	19	8	5
(CPM/mg) × 10^{-3}[e]	272		356		-	12	119		322	

[a]Measured by Lowry method.
[b]Percentage of nondialyzable CPM at each step, after dissolving in gel sample buffer (contains 2% SDS, 5% mercaptoethanol), and dialyzing against 1/10 concentration of this buffer.
[c]Estimated, assuming 80% nondialyzable, as observed in all 3 preceding purification steps.
[d]Assume 70% nondialyzable, cf. note c. [e]Observed specific activity. [f]Run in linear sucrose gradient, Fig. 3; CPM of peak fractions were used to calculate % yield. [g]Not dialyzed.

TABLE III Removal of Radioactivity from Externally Labeled Cells by Proteolytic Digestion

Cells	Treatment, enzyme (conc.)	Percentage of Control			
		Viability[a]		SDS-dissolved cell pellet, reduction of CPM	
		Before	After	In TCA ppt.	Nondialyzable
Subclone 1 preconfluent	Trypsin (1.0 mg/ml) 10 min, pH 7.4, PBS[b]		90	49	
Subclone 1 preconfluent	Pronase (0.5 mg/ml) 10 min, pH 7.4, PBS	70[c]		58[c]	
Subclone 1 postconfluent	Pronase (0.5 mg/ml) 30 min, pH 8.0 PBS		70		80
N AL/N postconfluent	Pronase (0.5 mg/ml) 30 min, pH 8.0 PBS		70		78

[a]Trypan blue dye exclusion tested before or after incubation with enzyme.
[b]PBS is phosphate buffered saline.
[c]Cells treated in suspension; all other data for cells attached to substratum.

In the disc gel technique the cell membrane pellets were solubilized by heating (2 min, 100°C) in 2% SDS in 0.0625 *M* Tris buffer, pH 6.8, containing 0.001% bromophenol blue as tracking dye, 10% glycerol, 5% mercaptoethanol, and 0.001 *M* PMSF. The SDS 8% polyacrylamide gels were prepared and run overnight at 0.5 to 1 mA per gel, as described (21, 23). The gels were removed from the tubes, placed inside a length of a larger diameter glass tube, and were frozen by submerging in a dry ice-methanol mixture; 1 mm thick slices were then obtained. The slices were extracted overnight at 37°C in 0.6 ml of 9 parts of NCS (Amersham and Searle) to 1 part of H_2O, then 15 ml of Aquasol containing 0.15 *M* acetic acid was added. One day's storage in the dark preceded the counting. Figure 4 gives representative gel electrophoresis patterns, contrasting the T antigen positive subclone 1 with the T antigen negative subclone 5, when the cells were labeled either internally for glycoproteins with ^{3}H-Glc-NH_2 or externally for proteins with KB^3H_4. As can be seen, the disc gel electrophoresis resolution of the components was not sharp, and it was difficult to discern compositional differences between the two sister subclone lines. Upon close inspection and repeated studies the regular disappearance was noted of components from the high molecular weight region when the cells were labeled either internally with ^{3}H-Glc-NH_2 or externally by KB^3H_4 in the T antigen positive subclone 1 (and also in subclone 2, data not shown), as compared to the T antigen negative subclone 5 (and subclone 6, data not shown).

To carefully compare the composition of the membrane preparations in SDS acrylamide electrophoresis, slab gels were employed, and the optimal conditions for such comparative analysis were carefully worked out (17). The conditions of the SDS polyacrylamide slab gel electrophoresis and the detailed results on unlabeled and ^{14}C mixed amino acid labeled cell membrane preparations are reported separately (17). The main findings can be summarized as follows: All the cell lines were compared when in logarithmic phase of growth; they all had very similar unlabeled membrane protein distributions in which sharp and numerous (∿50) protein bands were detected by Commassie blue staining. The membranes from T antigen negative N AL/N and T AL/N cells, when compared to those from the SV AL/N cells, and also the membranes from the T antigen negative subclone 5 cells as compared to those of the membranes from the T antigen positive sister subclone 1 cells, had a decreased or missing (glyco)protein band [the only major component in the acrylamide gels that stained with periodic acid Schiff (PAS) stain] with a nominal molecular weight of 180,000. This was detectable both by careful comparison of the radioactive profiles, which gave sharp resolutions of at least a dozen peaks, and also by autoradiography. Since all

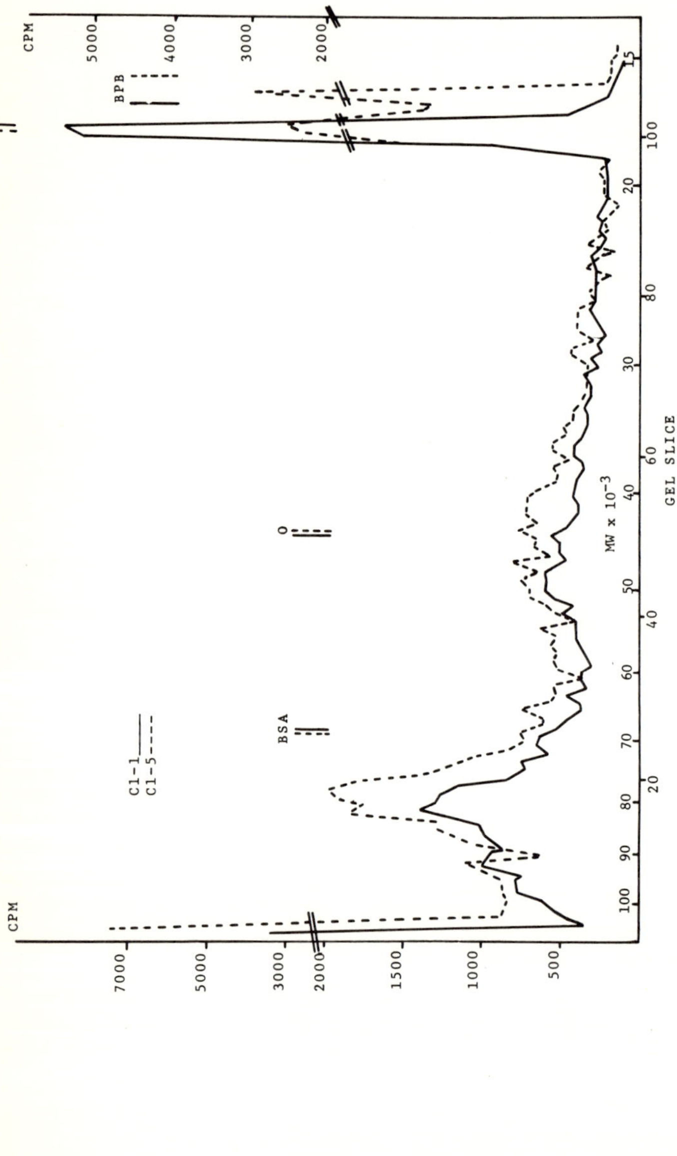
CPM
7000
5000
3000
2000
1500
1000
500
5000
4000
3000
2000
BPB
M
O
BSA
C1-1
C1-5
MW x 10^{-3}
GEL SLICE
(a)

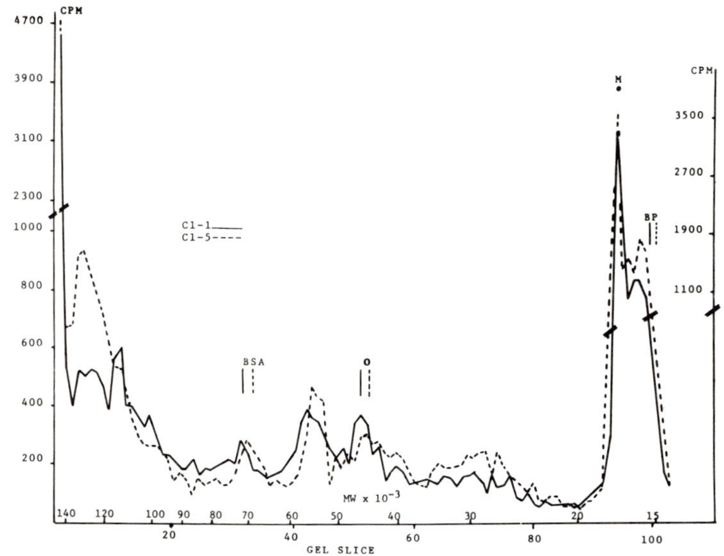

(b)

FIGURE 4 Representative dic gel electrophoresis patterns of ^3H labeled membrane pellet components obtained from subconfluent T antigen positive (subclone 1) and negative (subclone 5) cells labeled with either ^3H-Glc-NH_2 or KB^3H_4 as given in text. Molecular weight (M) markers of ^{14}C labeled proteins prepared as described (24) were added to all samples prior to electrophoresis: BSA, bovine serum albumin (68,500); O, ovalbumin (45,000); M, myoglobin (17,000); and BPB, bromophenol blue tracking dye. Their positions on each gel are shown by full or dashed vertical lines. An approximate molecular weight scale derived from these marker positions is given. The upper portion of the right ordinate scale refers only to the upper portion of the leading peaks on the gel patterns. Protein was determined by the method of Lowry. The input quantities used on these gels were as follows: (a): ^3H-Glc-NH_2 labeled cells: subclone 1, 34 × 10^3 CPM, 310 μg protein; subclone 5, 47 × 10^3 CPM, 130 μg protein; both in 0.2 ml aliquots. (b): KB^3H_4 labeled cells: subclone 1, 21 × 10^3CPM, 420 μg protein; subclone 5, 27 × 10^3 CPM, 160 μg protein; determined in 0125 and 0.18 ml aliquots respectively.

the different kinds of cells were compared in their logarithmic phase of growth, the decrease in the biosynthesis of this (glyco)protein component established this component as a different glycoprotein from the so called LETS (large external transformation sensitive) glycoproteins, from fibroblast surface (SF) antigens, and other fibroblast surface proteins that were recently reviewed (25). In these latter glycoproteins no differences were detectable when the nontransformed or the virally transformed cells were in logarithmic growth. Clearly these latter glycoproteins (25) pertain to some differences in growth in tissue culture between postconfluent cells only. They may pertain to secondary differences since the virally transformed cells require different serum and hormonal factors for proliferation; a proliferation that continues in the virally transformed cells but not in the normal cells when the former cells pass the confluent stage of growth on the substratum. In our study the decrease in the synthesis of the approximately 180,000 molecular weight (glyco)protein was specific to the SV40 transformation and it was observed when all the different types of cells were compared in their optimal, exponential phase of growth. Obviously, only changes found under such growth conditions could relate to the cause, rather than to the consequence of a difference in growth in tissue culture.

Note that all the above differences indicate a decrease (or loss) in a component upon viral transformation. However, very careful comparison of the slab gels both stained with Commassie blue for protein, or compared in their ^{14}C amino acid labeled profile, or in autoradiography, revealed a very minor membrane protein component in the 50-90,000 molecular weight region that appeared to be present (or increased) in the T antigen positive cells. The biochemical methods employed are apparently on the borderline of sensitivity to show unambiguously the newly synthesized SV40 specific cell surface antigens.

ANTIGENIC DIFFERENCES ON THE CELL SURFACE SOLUBILIZATION AND PURIFICATION OF THE SV40 SPECIFIC ANTIGENS

The virus specific antigenic changes on the cell surface that accompany transformation of mammalian cells by SV40 can be detected by various immunological methods. One of these changes can be detected by transplantation rejection tests *in vivo*, which indicate the presence of tumor specific transplantation antigens (TSTA). Several *in vitro* assays have also been introduced, among these a sensitive serum mediated complement dependent cytolytic microassay. Under appropriate conditions this assay is specific to SV40 induced tumor spe-

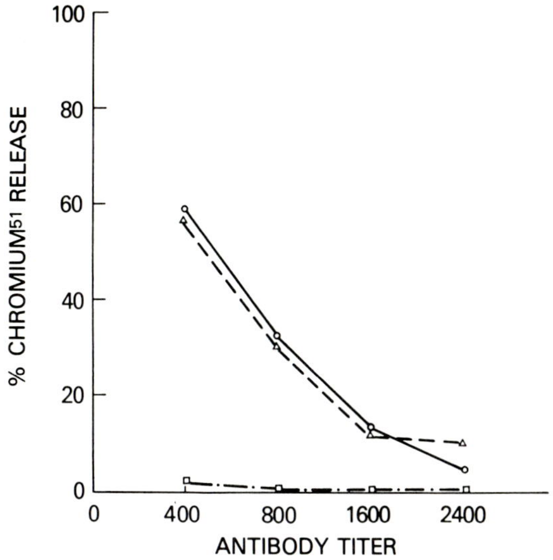

FIGURE 5 Inhibition of serum mediated cell lysis of 50 µg of 30-50% ammonium sulfate fractions of solubilized cells. Fraction from SV AL/N cells (□), T AL/N cells (Δ), TBS control (o). [See Pancake and Mora (5) for details of the assay and Chang et al. (7) for the extraction and $(NH_4)_2SO_4$ fractionation.] (% ^{51}Cr release = % cell lysis.)

cific surface antigens (TSSA) (4-8). This cytolytic assay has been adapted to an inhibition type microassay (5) suitable for rapid and relatively simple monitoring of TSSA.

By this assay, the cytolytic antiserum was inhibited by SV40 transformed T antigen positive cells such as the SV AL/N cells and any other SV40 transformed T antigen positive cells tested (such as hamster or human cells), but not by the N AL/N, T AL/N, T AL/N clone 3, subclone 5, 6, and 8 cells, or any other mouse, hamster, or human T antigen negative, not SV40 transformed, control cells tested (4, 5). This assay thus detected SV40 specific TSSA on the cell surface. By altering the antiserum concentration and the amounts of the SV40 transformed cells added, and interpolating to find the 50% inhibition values, it was also possible to estimate the relative amounts of the TSSA (5, 7).

Based on the inhibition type TSSA microassay, we recently developed a method of detergent solubilization and fractionation (partial purification) of the TSSA. Employing 0.5% Triton X-100, high yields (∿50%) of the TSSA were recovered in whole cell extracts and after ammonium sulfate fractionation

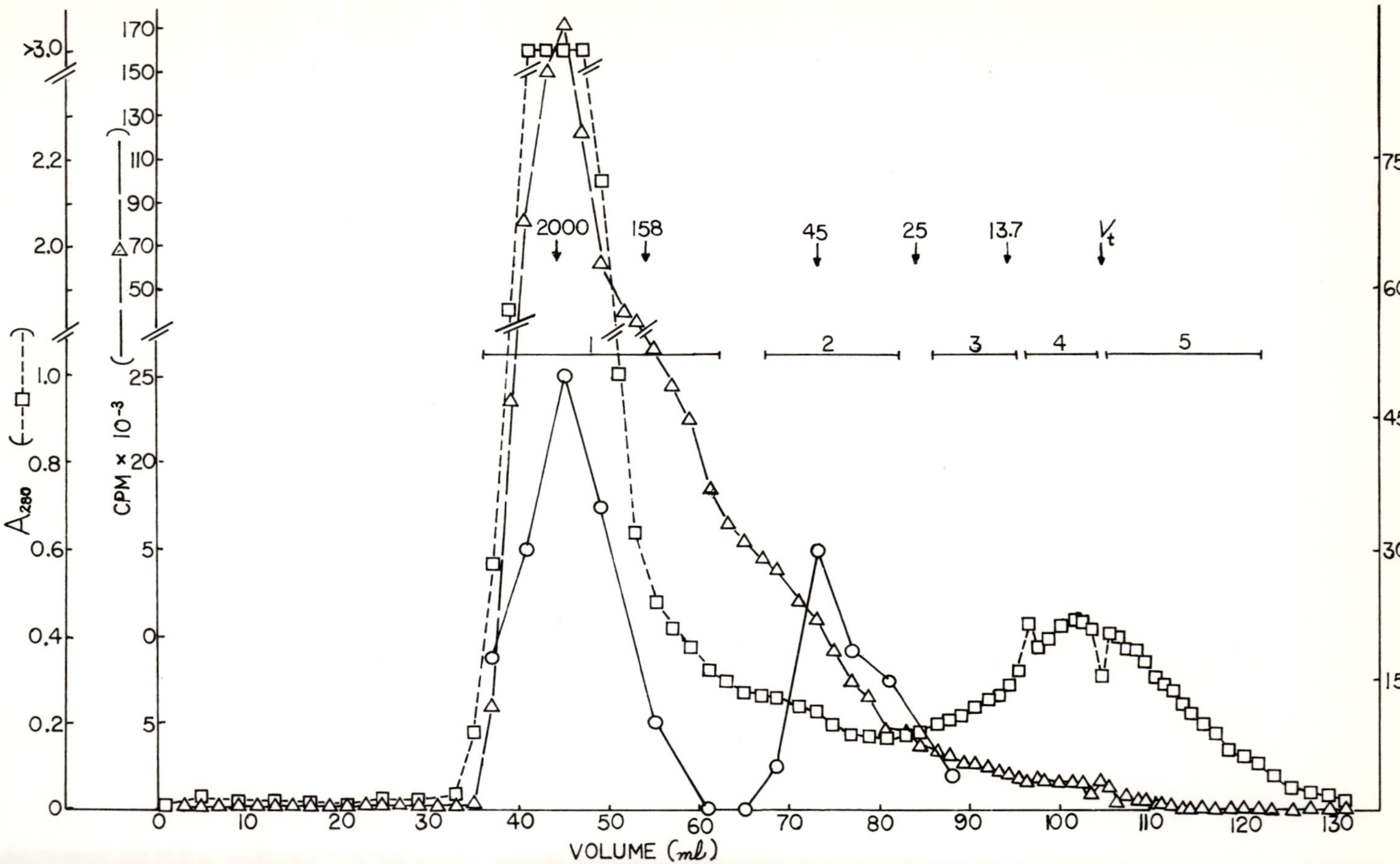
A280 (--□--)
>3.0
2.2
2.0
1.0
0.8
0.6
0.4
0.2
0
CPM x 10^-3 (—△—)
170
150
130
110
90
70
50
25
20
5
0
5
2000
158
45
25
13.7
Vt
1
2
3
4
5
%INHIBITION OF CELL LYSIS (—○—)
75
60
45
30
15
VOLUME (ml)
0
10
20
30
40
50
60
70
80
90
100
110
120
130

FIGURE 6 Gel exclusion chromatography on Sephadex G-150 of a 30-50% $(NH_4)_2SO_4$ fraction from SV AL/N cell extract. The cells were labeled with ^{14}C mixed amino acids. Arrows show elution positions of (Pharmacia) molecular weight standards ($\times 10^{-3}$), in decreasing order: Blue Dextran, aldolase, ovalbumin, chymotrypsinogen A, ribonuclease A; V_t = total bed volume. Bars show fractions pooled for further TSTA study with numbers assigned to each. Percent inhibition of cell lysis was measured using the ^{51}Cr release based cytolytic microassay, employing 25 λ aliquots of each fraction (●) (data from duplicate experiments). Effluent after 90 ml was cytolytic. Input: 2.5 ml of 40 A_{280}, 9.11×10^5 cpm/ml; recovery of input: >34% A_{280} (□), 82% cpm (Δ), cpm values from 25 λ aliquots).

(7, 18). For example, an aliquot of Triton X-100 solubilized SV AL/N cell filtrate that contained 30 μg protein, under the appropriate TSSA assay inhibited 43% of the serum mediated cytolytic activity, while the same amount of protein from a 0-30% $(NH_4)_2SO_4$ precipitated fraction inhibited 61%, and from the 30-50% $(NH_4)_2SO_4$ fraction inhibited 77% of the cytolytic activity. In contrast, in T AL/N cell filtrates, and all of its $(NH_4)_2SO_4$ precipitates, there were no detectable specific TSSA activities (7, 18). Figure 5 gives an example in which the percentage ^{51}Cr release, as the measure of the lysis of ^{51}Cr labeled SV AL/N target cells, is inhibited completely by 50 μg of 30-50% $(NH_4)_2SO_4$ fraction from the SV AL/N cells, but not by an equivalent fraction from the T AL/N cells. Figure 6 gives an example of further fractionation of TSSA by gel exclusion chromatography. Thus the inhibition type TSSA microassay provided a convenient way of monitoring further purification and characterization of the TSSA. By rate zonal centrifugation in a sucrose gradient, 50-60,000 daltons was obtained as the estimated molecular weight of the detergent solubilized TSSA (18). Current purification steps include fractionations on DEAE Sephadex, on phosphocellulose, and on DNA columns, following similar purification steps developed for SV40 T antigen (26).

The *in vivo* TSTA activity from SV AL/N cells has been found in the fractions that possessed TSSA activity, but not in the fractions that lacked it (7). The TSTA activity was not present in whole T AL/N cell extracts (7), or in extracts or whole cells of any non-SV40-transformed control cells. From SV40 transformed cells T, TSSA, and TSTA activities were found to co-purify (27). Current studies are in progress for further purification and biochemical characterization of these three antigens.

INTERPRETATION OF THE BIOLOGIC, IMMUNOCHEMICAL, AND BIOCHEMICAL RESULTS

The SV40 induced antigens that occur on the surface of the SV40 transformed AL/N cells and are detected by special immunologic assays apparently cause recognition and rejection of such cells in the immunologically competent syngeneic mouse (6-9). The absence of tumorigenicity of the SV AL/N cells, the reduced tumorigenicity of the T AL/N clone 3 cells after SV40 transformation (see the subclones 1 and 2 in Table I), thus is due to a dominant expression of the SV40 specific TSTA on the surface of SV40 transformed AL/N cells. A sensitive microassay developed to measure SV40 specific surface antigens (TSSA) allowed solubilization and provisional characterization of this cell surface antigen and provides a convenient tool to further define the nature of TSSA, its relation to SV40 TSTA, and also to T antigen. On the basis of current results, a partial constitutional identity can be postulated for these three antigens (27). Work is in progress to apply specific immunochemical (precipitation) techniques to further purify these antigens, which would allow chemical comparisons, and biologic assays such as TSTA assays on well purified preparations.

The direct biochemical analyses, as exemplified by cell membrane analysis by acrylamide gel chromatography, were not sensitive to show unambiguously the presence of the (new) SV40 specific cell surface antigens. Neither were biochemical differences observed between the N AL/N and the T AL/N cells, which may have then related to the tumorigenic potential of the T AL/N cells. In fact, our approach and the experimental model (cell-animal system) selected could only start to elucidate the factors that functionally oppose tumorigenicity, specifically the nature of the SV40 TSTA. It is important to realize, however, that such TSTA is specific only to SV40 transformed cells and in the AL/N mouse (and probably also in most mammals with the exception of hamster and the Mastomys) it may not be pertinent to the tumorigenic transformation; in fact under certain conditions it may result in the complete rejection of the spontaneously transformed tumorigenic cells. There are current findings that indeed show this and will be reported separately.

REFERENCES

1. Mora, P. T., "Cell Surfaces and Malignancy," Fogarty International Center Proceedings No. 24, Washington, D. C., U. S. Government Printing Office, 1976.
2. Takemoto, K. K., Ting, R. C. Y., Oser, H. L., *J. Nat. Cancer Inst. 41,* 1401-1409 (1968).
3. Mora, P. T., Cumar, I. A., Brady, R. O., *Virology 46,* 60-72 (1972).
4. Pancake, S. J., Mora, P. T., *Virology 59,* 323-327 (1974).
5. Pancake, S. J., Mora, P. T., *Cancer Res. 36,* 88-94 (1976).
6. Smith, R. W., Morganroth, J., Mora, P. T., *Nature 227,* 141-145 (1970).
7. Chang, C., Pancake, S. J., Luborsky, S., Mora, P.T., *Int. J. of Cancer,* in press.
8. Smith, R. W., Mora, P. T., *Virology 50,* 233-246 (1972).
9. McFarland, V. W., Mora, P. T., Schultz, A., Pancake, S., *J. Cell. Physiol. 85,* 101-112 (1975).
10. Eddy, B. E., Borman, G. S., Grubbs, G. E., Young, R. D., *Virology 17,* 65-75 (1962).
11. Girardi, A. J., Sweet, B. H., Slotnick, V. B., Hilleman, M. R., *Proc. Soc. Exp. Biol.* (New York) *109,* 649-660 (1962).
12. Rabson, A. S., O'Conor, G. T., Kirschstein, R. L., Branigan, W. J., *J. Nat. Cancer Inst. 29,* 765-787 (1962).
13. Altstein, A. D., Vassiljeva, N. N., Sarycheva, O. F., *Nature 213,* 931-932 (1967).
14. Black, P. H., Rowe, W. P., *Proc. Soc. Exp. Biol. Med. 114,* 721-727 (1963).
15. Rifkin, D. B. R., Compans, R. W., Reich, E., *J. Biol. Chem. 247,* 6493 (1972).
16. Hunt, R. C., Brown, J. C., *Biochem. 13,* 22 (1974).
17. Coll, J. C., Luborsky, S. W., Mora, P. T., Mss. submitted.
18. Luborsky, S. W., Chang, C., Pancake, S. J., Mora, P. T., *Biochem. Biophys. Res. Comm. 71,* 990-995 (1976).
19. Atkinson, P. H., Summers, P. F., *J. Biol. Chem. 246,* 5162-5175 (1971).
20. Atkinson, P. H., *J. Biol. Chem. 250,* 2123-2134 (1975).
21. Laemmli, U. K., *Nature 227,* 680-685 (1970).
22. Mora, P. T., Luborsky, S. W., *J. Macromol. Science, Chemistry, A4* (5), 1067-1068 (1970).
23. Maizel, J. V., *in* "Methods of Virology," *5,* p. 179 (1971), ed. by Maramarosch, K., and Koprowsky, H., Academic Press, New York.
24. Rice, R. H., Means, G. E., *J. Biol. Chem. 246,* 831 (1971).

25. Hynes, R. O., *Biochim. Biophys. Acta 458*, 73-107 (1976).
26. Henderson, I. C., Livingston, D. M., *Cell 3*, 65 (1974).
27. Anderson, J. L., Martin, R. G., Chang, C., Mora, P. T., Livingston, D. M., *Virology* (1977) in press.

MEMBRANE GLYCOLIPIDS AND THEIR RELATIONSHIP TO THE STRUCTURE AND FUNCTION OF CELL SURFACE RECEPTORS FOR GLYCOPROTEIN HORMONES, BACTERIAL TOXINS, AND INTERFERON

LEONARD D. KOHN
G. LEE
EVELYN F. GROLLMAN
FRED D. LEDLEY
BRIAN R. MULLIN
ROBERT M. FRIEDMAN
MARIA F. MELDOLESI
SALVATORE M. ALOJ

National Institute of Arthritis,
Metabolism, and Digestive Diseases
National Institutes of Health
Bethesda, Maryland

Gangliosides inhibit ^{125}I-thyrotropin (TSH) binding to the thyrotropin receptor. This inhibition, which appears to be hormonally specific, is critically altered by the number and location of the sialic acid residues within the ganglioside structure. The inhibition results from the interaction of gangliosides with TSH, rather than with the membrane receptor. The ganglioside-TSH interaction is associated with a distinct conformational change of the TSH molecule. The possibility that a ganglioside or ganglioside-like structure is a component of the thyrotropin receptor is suggested by the finding that gangliosides more complex than G_{M3} are present in bovine *thyroid membranes in much higher quantities than have been previously found in extraneural tissue and are absent in a thyroid tumor with no TSH receptor activity. The finding that the B protein of cholera toxin, which also interacts with gangliosides, has a peptide sequence in common with the β subunit of TSH, suggests that TSH and cholera toxin may be analogous in their mode of action. It is suggested, therefore, that a ganglioside or ganglioside-like structure is a basic component of the thyrotopin receptor and that cholera toxin and TSH have a common mechanism by which their message is transmitted to the cell machinery.*

The implications of these results are related to the structure and function of cell surface receptors for other glycoprotein hormones (luteinizing hormone, human chorionic gonadotropin, and follicle-stimulating hormone), to interferon, and to other bacterial toxins such as tetanus toxin.

INTRODUCTION

In a report presented earlier in this symposium (1), Fleischer showed that glycosphingolipids (gangliosides) are synthesized by the Golgi apparatus but are subsequently transferred to the plasma membrane. The present report reviews data that suggest that one role of the gangliosides in the plasma membrane is as a structural or functional component of receptors for the glycoprotein hormones: thyrotropin (TSH),[1] human chorionic gonadotropin (hCG), luteinizing hormone (LH), and follicle-stimulating hormone (FSH) (2-9). In addition, data are summarized that show that there is a primary sequence homology (3, 4, 10) as well as several functional similarities (3) between the glycoprotein hormones and cholera toxin sufficient to suggest that the glycoprotein hormones and cholera toxin have common mechanisms for interacting with the cell and transmitting their message to the cell machinery. Last, data will be presented to show that the common mechanism shared by the glycoprotein hormones and cholera toxin appear to extend to other bacterial toxins such as tetanus toxin (11) and to the antiviral agent interferon (7, 8).

Integral to this presentation, comment will be made concerning the changes in glycolipid orientation that occur within the membrane when these agents interact with their cell

[1]Abbreviations: TSH, thyrotropin; hCG, human chorionic gonadotropin; LH, luteinizing hormone; FSH, follicle-stimulating hormone.

surface receptors and the significance of these changes to the message transmission process. Speculation will center upon the role in the message transmission process of the ability of these agents to alter membrane permeability or affect the transmembrane electrochemical gradient, an ability that may be their mechanistic common denominator. To conclude, a structural relationship will be described between glycoprotein hormones, cholera toxin, and nonapeptide neurohypophyseal hormones such as vasopressin and oxytocin that supports these speculations and predicts analogies in their mode of modulating cell function and in their interactions with the cell membrane.

THE ROLE OF GANGLIOSIDE OR GANGLIOSIDE-LIKE COMPOUNDS IN THE STRUCTURE AND FUNCTION OF THE THYROTROPIN RECEPTOR

Gangliosides are a family of glycolipids that contain sialic acid. The lipid moiety of the gangliosides is ceramide, a nonpolar molecule presumably buried in the lipid bilayer of the cell membrane. The hydrophilic carbohydrate portion is attached to one end of the lipid moiety and sticks out of the cell surface, where it is potentially available for interaction with various effectors of cell function. Several gangliosides present in mammalian tissues are shown in Table I. The simplest is G_{M3}, consisting of ceramide, glucose, galactose, and sialic acid; the most complex is G_{T1}, which contains three sialic acid residues. G_{D1a} and G_{D1b} contain two sialic acid residues and are structural isomers differing only in the location of a single sialic acid residue. "G_{M1}," which has one sialic acid, has been postulated to be the surface receptor of cholera toxin on cells (12-16); it is a biosynthetic derivative of G_{M3} and G_{M2}.

When these individual pure gangliosides are added to a binding assay containing ^{125}I-labeled TSH and bovine thyroid membranes, the following is seen (Fig. 1):

(1) G_{D1b} is a potent inhibitor of the binding of TSH to thyroid membranes, whereas its isomer, G_{D1a}, is a minimal or weak inhibitor. The results with these two gangliosides suggest that the location of sialic acid residues on the ganglioside molecule is a critical factor in the inhibition, and that this inhibition is not a simple charge effect.

(2) G_{M1}, which interacts strongest with cholera toxin, shows an intermediate inhibition of TSH binding. In contrast, G_{D1b}, which is the best inhibitor of TSH binding, has no significant effect on cholera toxin binding (12-16).

(3) Sialic acid alone has no effect on TSH binding, nor does fetuin, a sialic-containing protein.

TABLE I *Structure of the Gangliosides*

```
                        β        β
G_M3    Cer → 1Glc4 → 1Gal3 → 2NANA

                        β        β
G_M2    Cer → 1Glc4 → 1Gal4 → 1GalNAc
                           3
                           ↑α
                           2NANA

                        β        β                β
G_M1    Cer → 1Glc4 → 1Gal4 → 1GalNAc3 → 1Gal
                           3
                           ↑α
                           2NANA

                        β        β                β
G_D1a   Cer → 1Glc4 → 1Gal4 → 1GalNAc3 → 1Gal3
                           3                         ↑α
                           ↑α                        2NANA
                           2NANA

                        β        β                β
G_D1b   Cer → 1Glc4 → 1Gal4 → 1GalNAc3 → 1Gal
                           3
                           ↑α      α
                           2NANA8 → 2NANA

                        β        β                β
G_T1    Cer → 1Glc4 → 1Gal4 → 1GalNAc3 → 1Gal3
                           3                         ↑α
                           ↑α      α                 2NANA
                           2NANA8 → 2NANA
```

When gangliosides were preincubated with membranes followed by centrifugation to remove any free ganglioside (experiment No. 5, Table II), no inhibition of ^{125}I-TSH binding was measured. This experiment indicated that the gangliosides were not interacting with the thyroid membranes to cause the inhibition of binding seen in Fig. 1, but rather were interacting with the TSH. Ultracentrifugation studies of TSH, run in the presence and absence of a ganglioside (2), confirmed this condlusion (Fig. 2). In this case, G_{M1} caused a shift in the sedimentation coefficient of TSH from 2.5 to 12.3 *S*; i.e.,

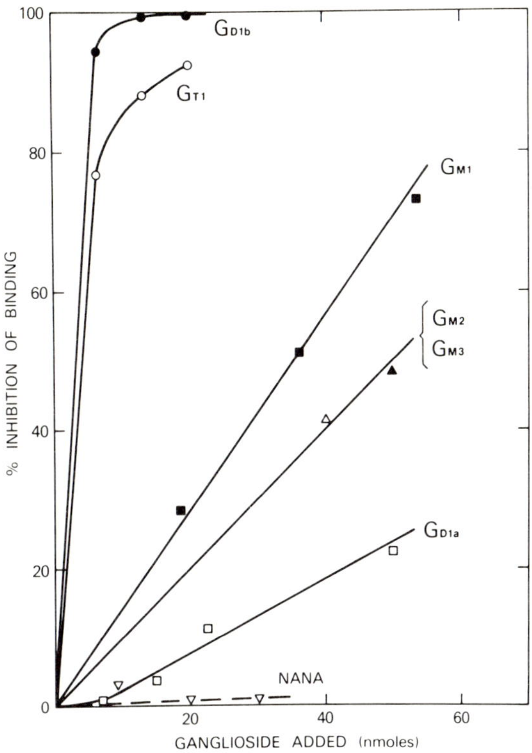

FIGURE 1 Inhibition of ^{125}I-labeled TSH binding to bovine thyroid plasma membranes by gangliosides or NANA (N-acetyl-neuraminic acid). Addition of 1 nmole of fetuin does not inhibit ^{125}I-labeled TSH binding (data not shown); this amount of fetuin is equivalent to 13 nmoles of protein-bound NANA. (Data are taken from reference 2.)

a significant portion of TSH was associated with a complex having a molecular weight of approximately 250,000.

Fluorescence studies of TSH in the presence of individual gangliosides (2) indicated a conformational change in the TSH molecule under the influence of gangliosides (Fig. 3). The change in conformation of TSH induced by the most potent inhibitor of TSH binding, G_{D1b}, was distinct from that of a minimal or weak inhibitor, its isomer, G_{D1a}. The progression from a minimal inhibitory conformation to a maximal inhibitory conformation exactly paralleled the binding inhibition data of Fig. 1.

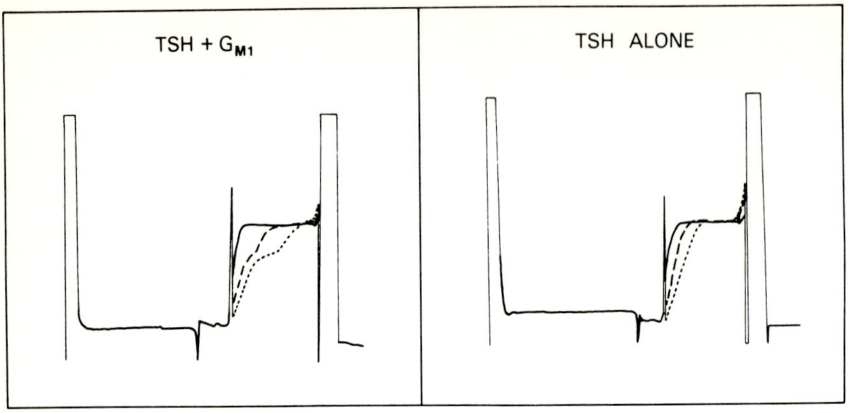

FIGURE 2 Ultracentrifugation pattern of TSH alone (right) compared with TSH plus G_{M1} (left). The patterns were obtained using a Model E ultracentrifuge equipped with a scanning device; absorbance at 280 nm is presented. The bottom of the cell is to the right and the top to the left of each frame. Centrifugation was at 40,000 rpm. The times of the scans (———), (-----), and (.....) were at 8, 24, and 40 min after speed was reached. TSH and G_{M1} concentrations were in the 1×10^{-5} M range.

In the light of current views concerning the structure and function of the cholera toxin receptor (12-16), these results were sufficient to consider the possibility that TSH and cholera toxin might have a similar mechanism of cell receptor interaction and a similar mechanism by which their "message" is transmitted to the cell machinery. Specifically, the following could be predicted:

(1) In the case of TSH as well as cholera toxin the initial interaction with the cell membrane would be through the β subunit, and the initial interaction would be with a ganglioside or ganglioside-like molecule that conferred specificity through its oligosaccharide structure.

(2) The interaction with the ganglioside or ganglioside-like receptor would cause a discrete and specific change in the conformation of the toxin or hormone that is necessary for these agents to transmit their message through the membrane.

(3) The change in toxin or hormone conformation would allow the A protein or α subunit, respectively, to translocate within or through the membrane.

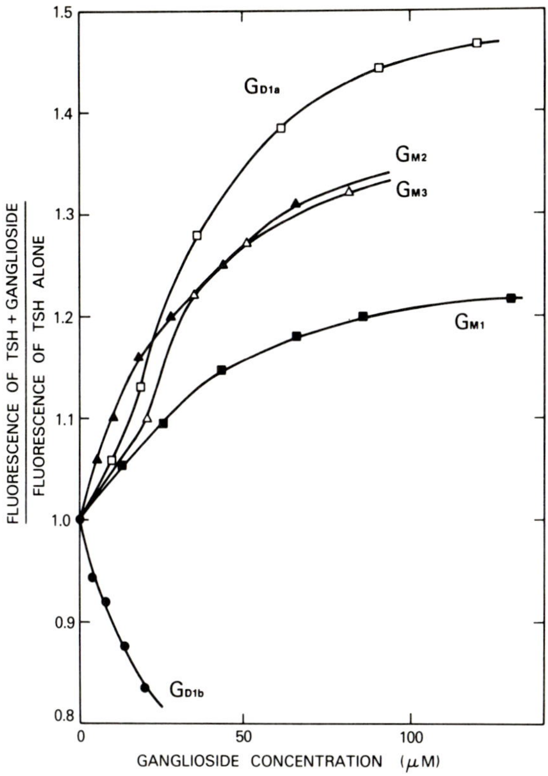

FIGURE 3 Fluorescence changes in the TSH molecule as measured after interacting with specific gangliosides. The fluorescence studies measure the increase or decrease in quantum yield of the tyrosine fluorescence of TSH when an interacting molecule alters the environment of the phenolic chromophore. The altered environment thus reflects a conformational change in the TSH molecule independent of whether this results from internalization of the tyrosine residues to a more "buried" location or from externalization to a more "exposed" location. The TSH concentration was 2 × 10^{-6} M. (Data are taken from reference 2.)

(4) As a consequence of the appropriate B protein or β subunit interaction with its specific receptor, as a consequence of the translocation of the A protein or α subunit within the membrane, or as a consequence of both events, a change in membrane state would ensue, and adenylate cyclase activity would be stimulated by direct or indirect means.

TABLE II Effect of Preincubation of Membrane and Ganglioside on Inhibition of ^{125}I-Labeled TSH Binding to Thyroid Plasma Membranes[a]

Experiment	Preincubation components	Inhibition (%)
1	None	95
2	Membranes + ^{125}I-labeled TSH	87
3	Ganglioside + ^{125}I-labeled TSH	97
4	Ganglioside + membranes without centrifugation before assay	93
5	Ganglioside + membranes followed by centrifugation before assay	0

[a]In the control experiment where no preincubation was performed (experiment 1), all components (membranes, gangliosides, and ^{125}I-labeled TSH) were added within 10 sec, mixed, and incubated for a total of 75 min prior to filtration. In experiments 2, 3, and 4, the noted components were preincubated in assay buffer for 15 min before the missing component, ganglioside, membranes, and ^{125}I-labeled TSH, respectively, were added; the binding assay then proceeded for 60 min before filtration or a total of 75 min from the onset of preincubation. In experiment 5, after the ganglioside and membranes were preincubated for 15 min, the mixture was centrifuged at 12,000 × g for 15 min to sediment the membranes. The membranes were then resuspended in buffer and the missing component, ^{125}I-labeled TSH, was added. The ganglioside preparation used in these experiments was a mixed preparation (from bovine brain) containing 47% G_{D1a}, 25% G_{T1}, 16% G_{D1b}, and 12% G_{M1}; 70 nmole were added. In addition to serving as a control for experiment 5, experiment 2 shows that gangliosides can "chase" bound TSH off the membrane. All procedures were performed at 2-4°. (Data are from reference 2.)

Experimental evidence that supports these predictions is summarized below.

Gangliosides are present on thyroid plasma membranes. If gangliosides themselves are part of the functional thyrotropin receptor on thyroid plasma membranes, they should at least be present in thyroid plasma membranes. When bovine thyroid membranes were subjected to ganglioside extraction procedures (2) and the extracted gangliosides subjected to thin-layer chromatography, these membranes were shown to be *rich* in gangliosides, including many with the chromatographic characteristics of the higher order gangliosides, i.e., G_{M1}, G_{D1a}, G_{D1b}, and G_{T1}. It is noteworthy that these more complex gangliosides are not usually present in extraneural tissues.

A TSH receptor defect correlates with the absence of the "higher order" gangliosides in plasma membranes of a thyroid tumor. A rat thyroid tumor has been described (17) that is unresponsive to TSH but is responsive to dibutyryl cyclic AMP. The plasma membranes of this tumor were unable to bind TSH, yet contained an adenylate cyclase activity normally responsive to fluoride and prostaglandins. When these tumor thyroid membranes were examined (6), the TSH receptor deficiency was correlated with an alteration in their plasma membrane content of gangliosides, i.e., the data supported the view that gangliosides as well as membrane glycoproteins with potentially analogous oligosaccharide structures (18, 19) were important to receptor function.

The lack of ganglioside G_{M1} in the rat tumor was dramatically demonstrated after labeling the membranes by sequential exposure to galactose oxidase and sodium borotritide and after chromatographing the gangliosides of normal and tumor membranes on thin-layer plates (6). Consistent with these findings is the observation that the tumor is deficient in enzymes (Table III) responsible for the synthesis of the "higher order" gangliosides that might be presumed to be related to the function of the TSH receptor, i.e., G_{D1b}, G_{T1}, and G_{M1}.

Sequence homologies exist between cholera toxin, thyrotropin, and other glycoprotein hormones. The B chain of cholera toxin is an 8000-10,000 molecular weight peptide composed of approximately 100 amino acids, of which the sequence of only 42 from the amino terminal end has been reported (20, 21). An initial screening was performed for sequences with any possible homology to this portion of the cholera toxin B chain using a computer program and the data bank of sequences in the "Atlas of Protein Sequence and Structure" (22). Of the hundreds of sequences examined, the search showed that a CAEY region in the cholera toxin sequence (residues 9-12) was significantly homologous to a highly preserved CAGY region of the glycoprotein superfamily (TSH residues 27-30, LH residues 34-37, hCG residues 34-37, and FSH residues 32-35). An alignment of these peptides based on the superposition of these regions (Fig. 4, top) gave a qualitative first approximation of the inherent homologies of this region.[1]

[1]A recent report by Kurosky *et al.* (10) comparing the entire sequence of the B protein of cholera toxin with the β subunits of the glycoproteins has confirmed this region as the only region of significant analogy.

FIGURE 4 Top: Sequences of the amino terminal 41 residues of the B protein of cholera toxin and of portions of the β subunits of TSH, LH, and hCG defined by positions from the amino terminal residue of each molecule. The residue symbols corresponding to the standard residue abbreviations are as follows

A, ala P, pro D, asp T, thr S, ser L, leu I, ile V, val E, glu G, gly H, his C, cys N, asn M, met Q, gln Z, glx B, asx R, arg F, phe Y, tyr K, lys W, trp X, undetermined.

Middle: The sequence of the amino terminal portions of the B protein of cholera toxin and the β subunits of TSH, LH, hCG, and FSH aligned to maximize homologies by the insertion of gaps resulting from potential deletions or insertions in the course of evolution. The alignment was generated by the computer program ALIGN (4). The underlined residues show the half-cystine alignment that results from this analysis. Bottom: The sequences of the amino terminal portions of the α subunits of TSH, LH, hCG, and FSH aligned to show an analogy with a fragment of the A_1 protein of cholera toxin, the cholera toxin subunit known to be responsible for adenylate cyclase activation. The sequence of cholera toxin used included the 42 amino terminal residues of the B protein as reported by Kurosky et al. (21). All other sequences were from the tapes of the "Atlas of Protein Sequence and Structure" (22) updated to December 5, 1975. The computer methods were those previously reported for sequence analysis (4) and involved the use of the programs SEARCH, ALIGN, and RELATE. The computer work was done on the IBM 360/44 computer at the National Biomedical Research Foundation. (Data are taken from reference 4.)

The CAGY region in the TSH-LH-hCG-FSH superfamily represents an important homology that has been highly preserved through the structural and functional differentiation of these hormones. The presence of such highly preserved regions usually implies the existence of loci that are critical for function and poorly susceptible to successful mutation. Cholera toxin, with a region of significant homology occurring at a similar position in its molecular structure (near the amino terminal end), would thus appear to be a structural analog of the process effected by this region of the glycoprotein hormones. Since the homology exists on that portion of the glycoprotein or toxin molecules, the β subunit or the B protein that carries the primary determinants for receptor binding, and since the β subunits of the glycoprotein hormones determine their target organ specificity, this region of sequence

```
                                  Position                                                              Position
CHOLERA TOXIN (partial fragment)   1  T P Q N I T D L C A E Y H N T Q I H T L N D K I F S Y T E S L A G K R E M A I Q T F   42
TSH β CHAIN (bovine)              19  C L T I N T T V C A G Y C M T, R B V B G K L F L P K Y A L S Q D V C T Y R D F M Y K  60
LH β CHAIN (bovine)               26  C I T F T T S I C A G Y C P S M K R V L P V I L P P M P E R V C T Y H E L R F A S V   67
HCG β CHAIN (human)               26  C I T V N T T I C A G Y C P T M T R V L Q G V L P A L P Z L V C N Y R D V R F E S I   67
FSH β CHAIN (human)               24  N T T(W,B, T) E T C A G Y C Y T R D L V Y K D P A K P R I Q K T C T F K E L V Y E T V  65
```

```
RESIDUE NO. FROM AMINO TERMINUS
                  1                   2                   3             4                 5                   6                   7
1 2 3 4 5 6 7 8 9 0 1 2 3 4 5 6 7 8 9 0 1 2 3 4 5 6 7 8 9 0 1  2  3 4 5 6 7  8 9 0 1  2 3 4 5 6 7 8  9  0 1 2 3 4 5 6 7 8 9 0 1 2 3 4 5 6 7 8 9 0 1

CHOLERA TOXIN (partial fragment, B protein)
- - - - - - - - - - - - - - - - - - - - - - - - - - T P Q N I  T  D L - - -  C A E Y  H N T Q I H T  L  N D K I F S Y T E S L A G K R E M A I Q T F

TSH β CHAIN (bovine)
- - - - - - - F C I P T E Y M M H V E R K E C A Y C L T I N T  T  V - - - -  C A G Y  C - - M T R B  V  B G K L F L P K Y A L S Q D V C T Y R D F M

LH β CHAIN (bovine)
S R G P L R P L C Q P I N A T L A A E K E A C P V C I T F T T  S  I - - - -  C A G Y  C P S M K R V  L  P V I L P P M P E R V C T Y H E L R F A S V

HCG β CHAIN (human)
S K Q P L R P R C R P I N A T L A V E K E G C P V C I T V N T  T  I - - - -  C A G Y  C P T M T R V  L  Q G V L P A L P Z L V C N X R D V R F E S I

FSH β CHAIN (human)
- - - - - -(B,S)C E L T N I T I A I E K E E C R.F C I S I N T  T  (T,B,W)E T  C A G Y  C - - Y T R D  L  V Y K D P A K P R.I Q K T C T F K E L V Y E
```

```
                                  Position                                   Position
CHOLERA TOXIN (partial fragment)       Z Z P H I H A P G G  C P B  A P R
TSH α CHAIN (bovine)               1   F P D G E F T M Z G  C P Z  C K L K E N K Y F S K P B   26
LH α CHAIN (bovine)                1   F P B G Z F T M Z G  C P Z  C K L K Z B K Y F S K P B   26
HCG α CHAIN (human)                1           A P B V Z B  C P Z  C T L Z Z B P F F S Z P G   22
FSH α CHAIN (human)                1           A P D V E D  C P E  C T L Q E N P F F S Q P G   22
```

TABLE III Activity of Enzymes Involved in the Biosynthetic Pathway for Ganglioside Synthesis in Normal and Tumor Thyroid Tissue

Acceptor	Product formed	Donor[b]	Enzyme	Enzymatic activity (pmoles product formed/mg/protein/hr)			
				Rat thyroid[a]		Thyroid tumor 1-8R[a]	
				Total homogenate	Crude membrane[c] preparation	Total homogenate	Crude membrane[c] preparation
Lactosylceramide	G_{M3}	CMPAcNeu	Sialyltransferase I	281	598	23	133
G_{M3}	G_{M2}	UDPGalNac	N-acetylgalactosaminyltransferase	21	69	0	9
G_{M2}	G_{M1}	UDPGal	Galactosyltransferase	225	470	103	93
G_{M1}	G_{D1a}	CMPAcNeu	Sialyltransferase II	63	89	1138	1973

[a]Enzymatic activities were assayed as described (6). Homogenates and membrane preparations were in 0.25 M sucrose, 1 mM EDTA, 0.1% 2-mercaptoethanol, and 0.01 M tris-acetate buffer, pH 7.0.

[b]CMPAcNeu, cytidine monophosphate N-acetylneuraminic acid; UDPGalNac, uridine diphospho-N-acetylgalactosamine; UDPGal, uridine diphosphogalactose.

[c]The crude membrane preparation was a [illegible] that had been washed 3 times with the buffer used to

similarity is an obvious candidate for the "active site" concerned with binding to receptors on the cell membrane.

A quantitative measure of general homology confirmed the significance of the homologies between cholera toxin, TSH, LH, and hCG (4), but failed to identify a significant homology between cholera toxin and FSH. The inability of the computer programs to derive a quantitative homology between the β subunit of FSH and the B protein of cholera toxin occurred despite the fact that FSH exhibited high quantitative levels of homology between itself and other members of the glycoprotein superfamily (LH-hCG-TSH) and despite the fact that the CAGY region was obviously present in FSH (Fig. 4, top and middle). In this regard, it was notable that FSH was the only member of the glycoprotein hormone superfamily that had not been found to prevent or chase TSH binding to TSH receptors. The sequence derived by the computer program that gave the best possible alignments (Fig. 4, middle) offered an explanation of this phenomena. The best alignment of the half-cystine residues of the β subunit of FSH with its sister molecules required the existence of a four-residue gap in LH, hCG, and TSH just before the CAGY sequence and a similar three-residue gap in the cholera toxin B protein.[1] It was reasonable to presume then that the small change in the CAGY locus might result in a loss of competitive binding activity with respect to other members of the glycoprotein hormone superfamily and with respect either to its own receptor or the receptors specific for TSH, LH, and hCG.

The possibility existed that the structure-function analogy could extend to other regions of the B protein of cholera toxin, and even to the A protein. The presumption was that the A chain of cholera toxin might in some region represent an evolutionary descendant of a fundamental adenylate cyclase-stimulating protein, and homologies might be detected between it and the α subunits of the glycoprotein hormones. Figure 4 (bottom) demonstrates such a region using a peptide fragment that has been characterized within the A_1 subunit of cholera toxin, i.e., the subunit that enhances the adenylate cyclase activity of pigeon erythrocyte membranes *in vitro* (15). It is of interest that the outlined analogy in the α subunits of TSH and LH is presumed to exist in a highly restricted conformational state defined by intrachain disulfide bonds (24).

[1]A recent study of the FSH sequence does not insert the additional residues in FSH but substitutes a tryptophan for a leucine, isoleucine, or valine in the position immediately before the CAGY locus (23).

Cholera toxin interacts with thyroid plasma membranes and these interactions affect both the ability of TSH to bind to the membranes and to stimulate the adenylate cyclase activity of the membranes. Unlabeled cholera toxin, as well as unlabeled TSH, prevents the binding of TSH to thyroid plasma membranes (3). Although there is an initial enhancement of TSH binding, 40% inhibition occurs at cholera toxin concentrations similar to those of unlabeled TSH. The addition of more unlabeled cholera toxin to the incubation medium causes no further inhibition of TSH binding, whereas the addition of more unlabeled TSH results in complete inhibition of binding. The 40% level of maximal inhibition correlates with a G_{M1} content of approximately 25-30% of the total gangliosides present in these membranes.

An evaluation of the kinetics of this inhibition indicates that the interaction of cholera toxin with thyroid plasma membranes results in either an increase or decrease in the number of receptor sites available to bind TSH at concentrations of cholera toxin that enhance or inhibit binding, respectively. No significant alteration in the affinity of the sites for TSH was detectable in these analyses, and the mechanism by which the cholera toxin displaced TSH from the receptor sites appears to be "mixed," i.e., had both competitive and noncompetitive elements (3).

If G_{M1} were a structural component of the membrane that interacted with both TSH and cholera toxin and if the mode of action of TSH and cholera toxin were similar, G_{M1} should induce in cholera toxin a conformational alteration which would be both analogous to that imposed upon TSH and could be detected by physical techniques such as fluorescence spectroscopy. The fluorescence emission spectrum of cholera toxin shows a maximum centered at 342 nm, which is typical of proteins containing tryptophan. Upon addition of G_{M1}, the peak emission is shifted to 330 nm. The 12-nm "blue shift" reflects a substantial decrease of the polarity of the indole chromophore environment that is likely to result from a conformational change of the protein. The effect of G_{M1} on the conformation of cholera toxin appears to be specific; thus, the gangliosides G_{M2} and G_{M3} at equivalent or up to fivefold higher concentrations have no effect on the emission maximum of the protein (3).

The hypothesis that TSH and cholera toxin might under certain circumstances compete for a common receptor (G_{M1}) suggests that at high concentrations of the two effectors there would be no additive stimulation of adenylate cyclase activity. In the absence of NAD, this prediction is true (3). Recent work by Gill (25) has, however, suggested that NAD is a necessary cofactor for the cholera toxin-induced stimulation of adenylate cyclase in plasma membrane preparations. In accord with Gill's results, NAD significantly enhances the ability of cholera toxin to stimulate adenylate cyclase activity in thyroid plasma membranes. In contrast, NAD inhibits the ability of TSH to stimulate adenylate cyclase activity in these same thyroid membranes (3). This finding suggests that despite the fact that both cholera toxin and TSH appeared to have an analogous mode of interaction with the membrane at a receptor level, they diverge in the molecular mechanism of adenylate cyclase stimulation.

Perhaps the most important experiment concerning the cholera toxin-TSH interaction with thyroid plasma membranes comes with the attempt to show by direct biochemical means that G_{M1} is the cholera toxin receptor on thyroid plasma membranes. The terminal galactosyl residue in the carbohydrate moiety of gangliosides and glycoprotein membrane components is susceptible to galactose oxidase. Thus, thyroid membranes incubated with galactose oxidase, followed by exposure to sodium borotritide, should have tritium incorporated into ganglioside residues if these residues are available to the enzyme. When such an experiment is performed (Fig. 5a), the major labeled ganglioside is G_{M1} (3). If, however, thyroid membranes were pretreated with cholera toxin, the G_{M1} residues are not labeled with tritium (Fig. 5b). Instead, pretreatment with cholera toxin causes an enhancement of the labeling of other glycolipids contained in the membrane that chromatograph in the region of G_{D1b}, G_{T1}, G_{M2}, and G_{M3}.

This last observation is of special interest in the light of the data that show that low concentrations of cholera toxin enhance TSH binding to thyroid plasma membranes. Taken together, there is a clear indication that there is a cooperative relationship between ganglioside or ganglioside-like receptors. This cooperative relationship was confirmed in studies that compared the inhibition of TSH binding to the receptor by unlabeled cholera toxin alone, by unlabeled TSH alone, by unlabeled TSH plus an amount of unlabeled cholera toxin sufficient to yield no significant net inhibition or enhancement, and by unlabeled cholera toxin plus an amount of unlabeled TSH sufficient to cause only 20% inhibition. Prior exposure of the membranes to a low concentration of TSH totally abolished the enhancement effect of cholera toxin on TSH binding. Similarly, prior exposure of the membranes to cholera toxin resulted in an increased ability of unlabeled TSH to inhibit binding.

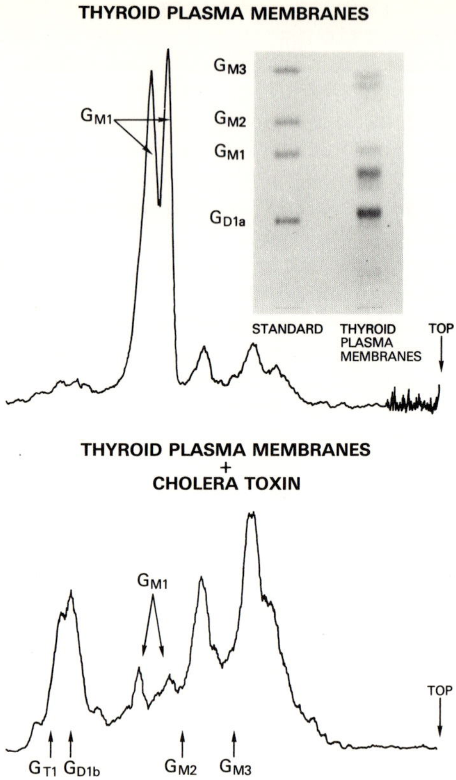

FIGURE 5 (a) Tritium incorporation into the gangliosides of thyroid plasma membranes after exposure to galactose oxidase and subsequent reduction by [^3H] sodium borohydride. The gangliosides were subjected to thin-layer chromatography (photograph), and the plates were scanned using a Vanguard radioscanning instrument (graph). Simultaneous chromatography of authentic G_{M1}, G_{T1}, G_{D1b}, G_{M3}, and G_{M2} allow the definition of the peaks of radioactivity. (b) Tritium incorporation into the gangliosides of thyroid plasma membranes preincubated with cholera toxin prior to galactose oxidase exposure and [^3H] sodium borohydride reduction. The concentration of cholera toxin during the preincubation period was sufficient to give maximal inhibition of ^{125}I-TSH binding. (Data are from reference 3.)

THE ROLE OF GANGLIOSIDE OR GANGLIOSIDE-LIKE COMPOUNDS IN THE STRUCTURE AND FUNCTION OF RECEPTORS FOR OTHER GLYCOPROTEIN HORMONES

The sequence homologies between luteinizing hormone, human chorionic gonadotropin, follicle-stimulating hormone, TSH, and cholera toxin (Fig. 4) raised the obvious possibility that all pituitary glycoprotein hormones would have a similar receptor structure and a similar mechanism of receptor interaction. Each target organ would thus have as its receptor a ganglioside or ganglioside-like structure with a specific and unique carbohydrate sequence, i.e., one that was different from that on other target organs. The interaction of the appropriate hormone with this specific oligosaccharide would result in a unique conformational shift such that the α subunit would be placed in the position favored for membrane translocation and adenylate cyclase activation in that particular target tissue. Interaction with the wrong hormone would result in a different conformation unfavorable position membrane translocation and no transmission of its "message" to the cell membrane.

Studies of human chorionic gonadotropin (hCG) already have validated a portion of this hypothesis (5). Thus, gangliosides inhibit ^{125}I-labeled hCG binding to rat testicular membranes, the efficacy of the inhibition being related to the oligosaccharide structure of the ganglioside (Fig. 6). One of the most effective inhibitors of hCG binding is G_{D1a}, whereas this same ganglioside is the *least* effective inhibitor of TSH binding to thyroid plasma membranes (2) and is an ineffective inhibitor of cholera toxin binding to cell membranes (12-14).

As in the case of the ganglioside inhibition of TSH binding to thyroid plasma membranes, the ganglioside inhibition of hCG binding to testes membranes is the result of the interaction of the ganglioside with the hormone rather than of an interaction with the membranes. Evidence for this conclusion comes from the following experiments. First, there is no significant inhibition of hCG binding when gangliosides are preincubated with the membranes prior to incubation with ^{125}I-hCG. Second, preincubation of ^{125}I-hCG with gangliosides results in a change in the elution pattern of the ^{125}I-hCG when it is subjected to gel filtration chromatography; in contrast, preincubation of ^{125}I-hCG and gangliosides in the presence of a 1000-fold excess of unlabeled hCG prevents the altered elution pattern of the ^{125}I-hCG. Last, fluorescence studies demonstrate that gangliosides interact with hCG (5). This

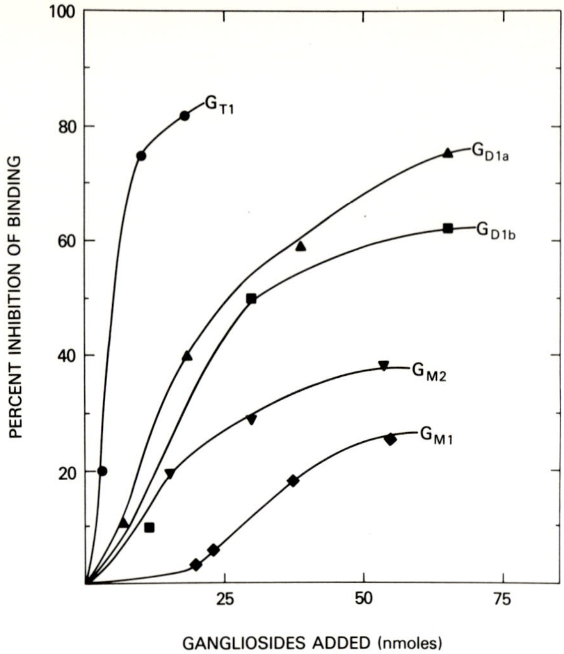

FIGURE 6 Ganglioside inhibition of ^{125}I-hCG binding to rat testes membranes. Binding was for 2 hr at 23° in 0.03 M tris-acetate, pH 6.0. (Data are from reference 5.)

interaction is associated with a large increase of the emission intensity of DNS covalently linked to the hormone (Fig. 7). The magnitude of the fluorescence changes depends on the oligosaccharide structure of the gangliosides tested and reflects rather closely the efficacy with which the ganglioside inhibits binding of ^{125}I-hCG to rat testicular membranes.

These data thus indicate that hCG, like TSH, interacts with a receptor that has a ganglioside or ganglioside-like structure and that, like TSH and cholera toxin, hCG undergoes a change in conformation upon interacting with its receptor structure. In addition, the data suggest that the oligosaccharide moiety of the hCG receptor structure is distinct from the oligosaccharide moiety on the TSH receptor, since the ganglioside inhibition of hCG binding to rat testicular membranes exhibits a different specificity ($G_{T1} > G_{D1a} > G_{D1b} > G_{M2} > G_{M1}$) than the ganglioside inhibition of TSH binding to thyroid plasma membranes ($G_{D1b} > G_{T1} > G_{M1} > G_{M2} > G_{D1a}$). Most dramatic in this regard is the effectiveness of G_{D1a} as an inhibitor of hCG binding to rat testes membranes in con-

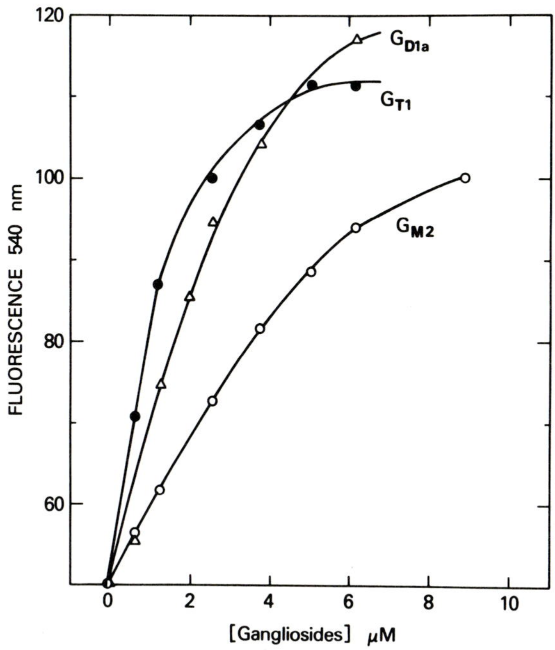

FIGURE 7 Effect of gangliosides on the fluorescence of DNS-hCG in 0.02 M tris-acetate, pH 7.5, at 21°. Excitation was at 340 nm. (Data are from reference 5.)

trast to its ability to inhibit TSH binding to thyroid membranes. Thus, the most effective inhibitors of hCG binding, G_{T1} and G_{D1a}, have a terminal sialic acid on the oligosaccharide chain, wereas the most effective inhibitors of TSH binding, G_{T1} and G_{D1b}, have a disialyl group linked to the internal galactose residue.

Target organ specificity would thus seem to reflect two phenomena: (1) the presence in the membranes of different target organs of a ganglioside or ganglioside-like receptor structure containing oligosaccharide structures different from those in other tissues; and (2) the evolution of primary sequence alterations on the β subunit of the glycoprotein hormones that favor their interaction with the predominant ganglioside or ganglioside-like molecule contained within the receptor structure of that particular target organ.

Studies with LH are consistent with these conclusions at this time.[1]

[1]Manuscript in preparation by G. Lee, S. M. Aloj, and L. D. Kohn.

SIMILARITIES IN THE STRUCTURE AND FUNCTION OF RECEPTORS FOR CHOLERA TOXIN AND THE GLYCOPROTEIN HORMONES EXTEND TO INTERFERON

Evidence has accumulated that suggests that the binding of interferon to specific cell surface receptors is necessary for the development of its antiviral action and that these receptors are also ganglioside or ganglioside-like structures on the outside of the cell membrane containing an oligosaccharide moiety as a critical feature of their structure.

The evidence for this is as follows: *Phaseolus vulgaris* phytohemagglutinin (PHA) blocks interferon action; Sepharose-bound interferon loses its antiviral activity after preincubation with gangliosides; soluble interferon binds to Sepharose-bound gangliosides; interferon binding to gangliosides is inhibited by PHA; and sialyl lactose reverses the ability of gangliosides to inhibit the action of interferon (26-28).

Since the sequence homology on the cholera toxin B protein and the thyrotropin β subunit appears to be a key feature of their structure, enabling them to bind to glycolipid or glycoprotein receptor structures of the cell membrane, it seems reasonable to presume that this same sequence might be important to other agents interacting with analogously structured receptors.

Recent experiments (7, 8) indicate that cholera toxin and TSH do, in fact, inhibit the establishment of the antiviral state by interferon. These experiments were based on the hypothesis that cholera toxin or TSH might either directly or indirectly interfere with the gangliosides or ganglioside-like oligosaccharide structures on L cell membranes that are believed to be important for interferon's interaction with its cell membrane receptors. Direct *in vitro* binding experiments with cholera toxin and TSH (8) already indicate that this notion is valid (Fig. 8). They further suggest that the toxin and TSH do not necessarily interact with the same receptor site as interferon, but that their respective interactions result in a secondary change in the state of the membrane that prevents the interferon from establishing its antiviral effect (8).

Interferons are usually species specific (29). Human interferon does not protect chick cells from viral infection nor does mouse interferon protect human cells. This species specificity does not appear to lie solely at the initial point of receptor-interferon interaction, since mouse interferon can bind to plasma membranes from human KB-3 cells, which are insensitive to both mouse and human interferons (Fig. 9A).

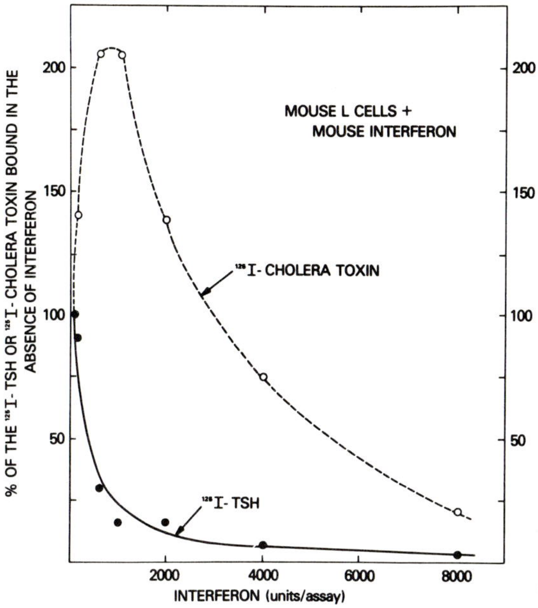

FIGURE 8 The effect of mouse interferon on ^{125}I-labeled TSH or ^{125}I-labeled cholera toxin binding to L cell plasma membranes. The interferon in these experiments was preincubated with the membranes for 1/2 hr at 0° before the ^{125}I-labeled TSH or ^{125}I-labeled cholera toxin was added at 0° for 1 hr. The conditions of the preincubation and binding assay were otherwise standard, 20 mM tris-acetate at pH 6.0 and 0.6% bovine serum albumin in 100 μl. (Data are from reference 8.)

Also, human interferon can interact with plasma membranes from mouse cells (Fig. 9B) to alter the binding of ^{125}I-labeled cholera toxin or ^{125}I-labeled TSH.

Experiments using TSH or cholera toxin as probes of the mehcanism of interferon action thus suggest the following

(1) The antiviral action of interferon is initiated at the plasma membrane.

(2) The initial binding of interferon to a ganglioside or ganglioside-like receptor is insufficient to establish the antiviral state.

(3) Interferon may be analogous to TSH and cholera toxin in having two separate molecular determinants, one for binding to a cell surface receptor and another for the initiation of later membrane events important in establishing its biological activity.

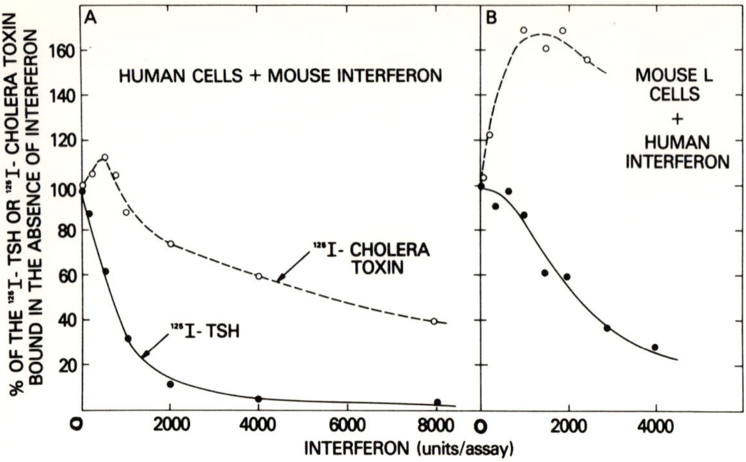

FIGURE 9 (a) The effect of mouse interferon on ^{125}I-labeled TSH (o) or ^{125}I-labeled cholera toxin (o) binding to plasma membranes from human KB-3 cells. (b) The effect of human interferon on the binding of ^{125}I-labeled TSH or ^{125}I-labeled cholera toxin to L cell plasma membranes. In these experiments and those of Fig. 9a, interferon was preincubated with the membranes for 1/2 hr at 0° before addition of the radioactive ligand. Conditions were the same as those in Fig. 8. (Data are from reference 8.)

(4) The common sequence region on the B protein of cholera toxin and the β subunit of TSH may be interesting with regard to the structure of interferon itself. If the hypothesis that this sequence is critical to interactions with ganglioside or ganglioside-like oligosaccharides on the cell membrane is correct, one might anticipate an analogous sequence in the interferon molecule. This possibility is of special interest, given the current problems with structural studies of interferon-both in terms of defining the determinants on the interferon molecule important to receptor interaction and to its antiviral action.

It is possible that the mechanism of action of interferon inhibition by cholera toxin and TSH may be less specific. Interferon could act simply by altering the cell surface and thereby induce changes that result in an intracellular antiviral state with several membrane-associated viral functions altered. This might explain why some reports have suggested that interferon primarily inhibits viral transcription, others, viral translocation, and still others, egress of the virus from the infected cell. Cholera toxin and TSH by binding to sites on the plasma membrane may cause a perturbation of the plasma membrane structure that in turn prevents interferon from inducing specific cell surface alterations that are necessary for the establishment of antiviral activity.

Since the mechanism of action of cholera toxin and TSH are becoming increasingly understood, these findings may help to elucidate the early stages of interferon action.

SIMILARITIES IN THE STRUCTURE AND FUNCTION OF RECEPTORS FOR CHOLERA TOXIN AND THE GLYCOPROTEIN HORMONES CANNOT ONLY BE EXTENDED TO OTHER BACTERIAL TOXINS BUT THIS RELATIONSHIP HAS POTENTIAL PATHOPHYSIOLOGICAL SIGNIFICANCE

The above studies have suggested that gangliosides present on thyroid membranes might be functional components of the cell surface receptor for both TSH and cholera toxin and that both TSH and cholera toxin might have analogous mechanisms by which they transmit their information to the cell. These observations raised the possibility that other bacterial products with a known tropism for gangliosides, such as tetanus toxin (30, 31) and *Escherichia coli* enterotoxin, might exhibit similar interactions with thyroid plasma membranes and have similar influences on TSH or cholera toxin receptor sites on these thyroid membranes, and that these interactions could be of medical significance as pathophysiological effectors of thyroid stimulation. In this regard, tetanus toxin is of particular interest, since the specificity of the ganglioside-tetanus interaction [$G_{D1b} = G_{T1} > G_{M1} > G_{D1a} > G_{M2}$ (30, 31)] is similar to the specificity of the ganglioside-TSH interaction and since the clinical presentation of tetanus includes a clinical state that closely resembles thyroid storm.

The classical symptomology of tetanus involves a syndrome of disinhibition in which tetanus toxin is presumed to abolish the effect of inhibitory pathways in the central nervous system. This leads to muscle rigidity, convulsions, and, without treatment, death, due usually to respiratory failure. In patients treated with tracheostomy, curarization, and positive pressure ventilation, however, a second set of symptoms is often noted consisting of tachycardia, changes in arterial pressure (hypertension or hypotension), increased basal metabolism, and pyrexia (32-34). These symptoms have been attributed to sympathetic overactivity. The symptoms are, however, identical to those seen in thyroid storm (35) where sympathet-

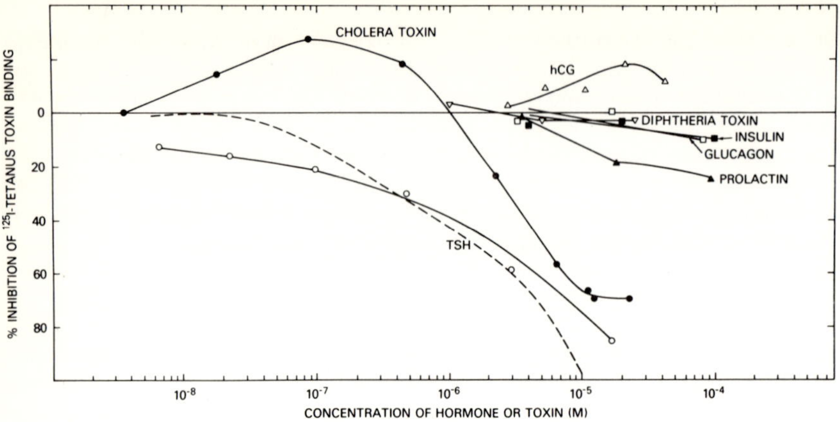

FIGURE 10 The ability of unlabeled TSH (○), glucagon (□), insulin (■), diptheria toxin (∇), cholera toxin (●), and human chorionic gonadotropin (Δ) to prevent ^{125}I-tetanus toxin binding to thyroid plasma membranes compared to the ability of unlabeled TSH to prevent ^{125}I-TSH binding to these membranes (-----). (See reference 11 for details of these data.)

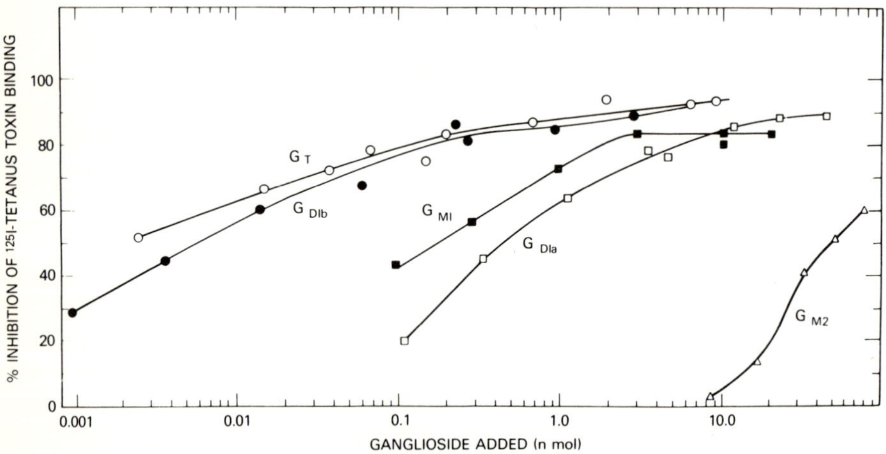

FIGURE 11 The ability of the gangliosides G_{T1} (○), G_{D1b} (●), G_{M1} (■), G_{D1a} (□), and G_{M2} (Δ) to inhibit ^{125}I-tetanus toxin binding to thyroid plasma membranes. Incubation conditions were standard. The ganglioside was preincubated with the toxin under standard incubation conditions for 15 min prior to the membrane addition. (See reference 11 for details of these data.)

ic overactivity is presumed to be related to thyroid hyperfunction.

When recently studied, tetanus toxin was shown to specifically bind to thyroid plasma membranes and that the properties of its interaction with the membranes was shown to closely resemble the TSH interaction with these membranes (11). Thus, the binding of ^{125}I-tetanus toxin exhibited similar time, pH, and salt dependence as previously demonstrated for ^{125}I-TSH binding to TSH receptors on thyroid membranes. In addition, ^{125}I-tetanus toxin binding can be blocked or chased by the addition of either unlabeled tetanus toxin or unlabeled TSH (Fig. 10), and unlabeled cholera toxin has effects on ^{125}I-tetanus toxin binding similar to its effects on ^{125}I-TSH binding. Most intriguing, the different interactions of various gangliosides with tetanus toxin as measured by ^{125}I-tetanus toxin binding to thyroid membranes is effectively the same (Fig. 11) as that measured by van Heyningen using bioassay techniques reflecting the primary disinhibition effects on the central nervous system (30, 31).

The ability of tetanus toxin to interact with thyroid membranes and with TSH receptors on these membranes, as suggested above, has several significant implications and applications. An obvious clinical implication is that in those patients with tetanus whose neurotoxicity is treated with tracheostomy, curarization, and positive pressure ventilation, the syndrome of sympathetic overactivity that can appear is contributed to by the ability of the toxin to induce thyroid hyperfunction. This syndrome would thus be analogous in its etiological relationship to the thyroid as in thyroid storm or as is the hypersensitivity to catecholamines demonstrable in thyrotoxic patients.

A second implication is that tetanus toxin will have a receptor structure and mechanism of effecting cell processes analogous to cholera toxin, the glycoprotein hormones, and interferon. It can thus be predicted that tetanus toxin might well have two different functional subunits, one for binding to the cell membrane and one for membrane translocation. It can be presumed that tetanus toxin will have the common primary sequences within these two subunit or structural components already demonstrated for cholera toxin and the glycoprotein hormones (Fig. 4), that tetanus toxin will undergo a specific conformational change upon interacting with the receptor, and that the tetanus toxin action will also require or induce a change in membrane state or a reordering of membrane components. In short, it seems reasonable to presume that tetanus toxin interactions with the TSH receptor on thyroid plasma membranes are analogous to its interactions with neural tissue membranes and that studies that compare the tetanus toxin effects on thyroid membranes with those of TSH will clarify both the mechanism by which tetanus exerts its neurotoxic effects and TSH its thyroid-stimulating effects.

THE MECHANISM BY WHICH GLYCOPROTEIN HORMONES, INTERFERON, AND THE BACTERIAL TOXINS EFFECT CELL CHANGES

Current views of glycoprotein hormone interactions with cells invoke cyclic AMP as the "second" functional component of message transmission, i.e., the hormone is the "first message" and the cyclic AMP is the "second message." Perhaps the most important implication of the above data that relate the mechanism of action of glycoprotein hormones, toxins, and interferon is that cyclic AMP is not necessarily the "second message" but merely one of several messages responsive to the changes induced by these effectors through the cell membrane. In fact, if there is a common denominator in the action of all these agents and tetanus toxin, it appears to be a change in the membrane, resulting in an alteration of its electrochemical gradient, permeability, and ion transport properties. In tetanus toxin, the result is altered neuron transmission. In cholera toxin, there is a loss of potassium and water from the gut cell. In TSH, there are changes in iodide transport. In interferon, there is the observation that ouabain inhibits the protective effect offered by this agent against viral infection.

As noted above (Fig. 5), cholera toxin has a major effect on the orientation of glycolipids in the cell membrane. This was seen as the exposure of otherwise unaccessible membrane glycolipids to galactose oxidase and borotritide reduction after cholera toxin interacted with the cell. Analogous data have been obtained when TSH interacts with thyroid plasma membranes. Recent experiments suggest that this alteration in the orientation of the membrane glycolipids is associated with effects of TSH on membrane transport before any activation of adenylate cyclase or any change in cyclic AMP levels within the cell.

Thus, TSH causes a rapid uptake of the lipophilic cation, triphenylmethyl phosphonium (TPMP). Since TPMP uptake is a measure of the electrochemical gradient across a cell membrane and since the electrochemical gradient is vital to drive the transport of ions, sugars, amino acids, etc., across that membrane, TSH would appear to have a major effect on membrane transport. This effect precedes adenylate cyclase stimulation in these studies and is independent of cyclase stimulation in studies of thyroid cells in culture (36). These results thus allow one to speculate not only that the common denominator of the action of all these agents is an effect on membrane transport, but that this effect is not mediated by cyclic AMP. This concept is already supported by experimental data, i.e., hCG has been shown to effect changes in adrenal cell ion transport that not only precede adenylate cyclase stimulation, but occur at lower concentrations of hormone effector (37).

It is of interest that these speculative considerations in mammalian cells are compatible with recent studies of Peterkofsky and Gaydar (38) on the mechanism of regulation of adenylate cyclase in *E. coli*. Peterkofsky would suggest that adenylate cyclase activity is regulated by the structural state and catalytic function of enzymes involved in sugar transport, that TSH is to the thyroid cell as glucose is to the bacterial cell, and that the regulation of membrane transport by these two effectors results in changes in the intracellular levels of cyclic nucleotides, i.e., in changes in adenylate and guanylate cyclase activities. These changes in intracellular cyclic nucleotides are thus compensatory mechanisms by which the cell responds to the effector. The physiological significance of the alterations in cyclic AMP levels may thus be that elevated cyclic AMP levels overcome the barrier to transcription of operons important to maintain the viability of the organism, i.e., in the TSH-stimulated human thyroid, the barrier to the transcription of catabolic enzymes and transport systems that result in increased thyroglobulin biosynthesis and thyroid hormone secretion.

With these ideas in mind, it is interesting to conclude with an observation concerning the primary sequence data relating a small area of sequence homology on the A_1 protein of cholera toxin and the α subunit of TSH (Fig. 4). This area is startlingly similar in sequence to the nonapeptide neurohypophyseal hormones, oxytocin and vasopressin (Fig. 12). Thus, six of the nine residues match, either as identical residues (*cys*, *pro*), as analogously charged residues (*gln*, *glx*, *glu*; *arg*, *lys*, *glu*, *glx*), or as similarly hydrophobic residues (*ile*, *val*, *phe*, *met*).

The primary actions of these neurohypophyseal peptide hormones, i.e., muscle or glandular contraction, diuretic changes, and blood pressure effects, would all appear to revolve about changes in the electrochemical gradient across the membrane and in membrane transport. The possibility, therefore, exists that this common sequence on the α subunits of the glycoprotein hormones is not only the site of their membrane active event but also that their modification of membrane changes is similarly induced. In this regard, recent

	Position		Position
		S——————S	
OXYTOCIN	1	Cys-Tyr-Ile-Gln-Asn-Cys-Pro-Leu-Gly	9
ARGININE VASO-TOCIN	1	Cys-Tyr-Ile-Gln-Asn-Cys-Pro-Arg-Gly	9
ARGININE VASO-PRESSIN	1	Cys-Tyr-Phe-Gln-Asn-Cys-Pro-Arg-Gly	9
LYSINE VASO-PRESSIN	1	Cys-Tyr-Phe-Gln-Asn-Cys-Pro-Lys-Gly	9
HCG (human)	1	Ala-Pro-Asx-Val-Glx-Asx-Cys-Pro-Glx-Cys-Thr-Leu	12
FSH (human)	1	Ala-Pro-Asp-Val-Glu-Asp-Cys-Pro-Glu-Cys-Thr-Leu	12
TSH (bovine)	1	Phe-Pro-Asp-Gly-Glu-Phe-Thr-Met-Glx-Gly-Cys-Pro-Glx-Cys-Lys-Leu	16
LH (bovine)	1	Phe-Pro-Asx-Gly-Glx-Phe-Thr-Met-Glx-Gly-Cys-Pro-Glx-Cys-Lys-Leu	16
		S——————S	

FIGURE 12 Comparison of the amino acid sequence of the neurohypophyseal nonapeptide hormones, oxytocin and vasopressin, with the amino acid sequence of a portion of the α subunits of the glycoprotein hormones. Sequence data are derived from the "Atlas of Protein Sequence and Structure" (22). Lines, identical residues; dots, analogously charged residues; and hatched, hydrophobic residues. The glycine, empty area, replacements for cystine have been considered neutral in synthetic peptides that substitute glycine for half-cystine, i.e., there is no loss of function in

studies have reported that a 1400 molecular weight fragment of the A_1 protein of cholera toxin, i.e., a 10- or 11-residue sequence, may be sufficient to cause the adenylate cyclase changes correlated with the action of this agent (39).

ACKNOWLEDGMENTS

We are indebted to Drs. Peter H. Fishman and Roscoe O. Brady of the Developmental and Metabolic Neurology Branch, National Institute of Neurological and Communicative Disorders and Stroke, and Drs. William H. Habig and M. Carolyn Hardegree of the Division of Bacterial Products, Bureau of Biologics, whose collaborative efforts and support in the work presented in this report was the basis of these interpretations. (See references 2-9 and 11).

REFERENCES

1. Fleischer, B., in this volume.
2. Mullin, B. R., Fishman, P. H., Lee, G., Aloj, S. M., Ledley, F. D., Winand, R. J., Kohn, L. D., and Brady, R. O. (1976), *Proc. Natl. Acad. Sci. U.S.A. 73*, 842-846.
3. Mullin, B. R., Aloj, S. M., Fishman, P. H., Lee, G., Kohn, L. D., and Brady, R. O. (1976), *Proc. Natl. Acad. Sci. U.S.A. 73*, 1679-1683.
4. Ledley, F. D., Mullin, B. R., Lee, G., Aloj, S. M., Fishman, P. H., Hunt, L. T., Dayhoff, M. O., and Kohn, L. D. (1976), *Biochem. Biophys. Res. Commun. 69*, 852-859.
5. Lee, G., Aloj, S. M., Brady, R. O., and Kohn, L. D. (1976), *Biochem. Biophys. Res. Commun. 73*, 370-378.
6. Meldolesi, M. F., Fishman, P. H., Aloj, S. M., Kohn, L. D., and Brady, R. O. (1976), *Proc. Natl. Acad. Sci. U.S.A. 73*, 4060-4065.
7. Friedman, R. M., and Kohn, L. D. (1976), *Biochem. Biophys. Res. Commun. 70*, 1078-1083.
8. Kohn, L. D., Friedman, R. M., Holmes, J. M., and Lee, G. (1976), *Proc. Natl. Acad. Sci. U.S.A. 73*, 3695-3699.
9. Kohn, L. D. (1977), *in* "Horizons in Biochemistry and Biophysics," (Quagliariello, E., ed.) Vol. 3, Addison-Wesley Publishing Company, Reading, Massachusetts, pp. 123-154.
10. Kurosky, A., Markel, D. E., Peterson, J. W., and Fitch W. M. (1977), *Science 195*, 299-301.

11. Ledley, F. D., Lee, G., Kohn, L. D., Habig, W. H., and Hardegree, M. C. (1977), *J. Biol. Chem.*, in press.
12. Cuatrecasas, P. (1973), *Biochemistry 12*, 3547-3558.
13. Holmgren, J., Lönnroth, I., and Svennerholmm, L. (1974), *Infect Immun. 8*, 208-214.
14. King, C. A., and van Heyningen, W. E. (1973), *J. Infect. Dis. 127*, 638-647.
15. Gill, D. M., and King, C. A. (1975), *J. Biol. Chem. 250*, 6424-6432.
16. Sahyoun, N., and Cuatrecasas, P. (1975), *Proc. Natl. Acad. Sci. U.S.A. 72*, 3438-3442.
17. Macchia, V., and Meldolesi, M. F. (1974), *in* "Advances in Cytopharmacology," (Ceccarelli, B., Clementi, F., and Meldolesi, J., eds.) Vol. 2, Raven Press, New York, pp. 33-37.
18. Tate, R. L., Holmes, J. M., Kohn, L. D., and Winand, R. J. (1975), *J. Biol. Chem. 250*, 6527-6533.
19. Tate, R. L., Winand, R. J., and Kohn, L. D., (1976) *in* "Thyroid Research: Proceedings of the 7th International Thyroid Conference," Boston, Massachusetts, June 9-11, 1975 (Robbins, J., and Braverman, L., eds.), International Congress Series No. 378, Excerpta Medica, Amsterdam, The Netherlands, pp. 57-60.
20. Kurosky, A., Markel, D., Touchstone, B., and Peterson, J. W. (1975), *Proc. Amer. Soc. Microbiol. 75*, 23-28.
21. Mendez, E., Lai, C. Y., and Wodnar-Filipowicz, A. (1975), *Biochem. Biophys. Res. Commun. 67*, 1435-1443.
22. Dayhoff, M. O. (ed.) (1977), "Atlas of Protein Sequence and Structure," Vol. 5, Supplement 2, National Biomedical Research Foundation, Washington, D. C., in press.
23. Rothman, P. (1976), *J. Biol. Chem. 251*, 993-1005.
24. Liao, T. H., and Pierce, J. G. (1970), *J. Biol. Chem. 245*, 3275-3281.
25. Gill, D. M. (1975), *Proc. Natl. Acad. Sci. U.S.A. 72*, 2064-2068.
26. Besancon, F., and Ankel, H. (1974) *Nature 250*, 784-786.
27. Besancon, F., and Ankel, H. (1974) *Nature 252*, 478-480.
28. Besancon, F., Ankel, H., and Basu, S. (1976), *Nature 259*, 576-578.
29. Tyrell, D. A. J. (1959), *Nature 184*, 452-454.
30. van Heyningen, W. E., and Miller, P. A. (1961), *J. Gen. Microbiol. 24*, 107-119.
31. van Heyningen, W. E. (1974), *Nature 249*, 415-417.
32. Clifton, B. (April 11, 1964), *Lancet*, 785-788.

33. Kerr, J. H., Corbett, J. L., Prys-Roberts, C., Smith, A. C., and Spalding, J. M. K. (August 3, 1968), *Lancet*, 236-241.
34. Corbett, J. L., Kerr, J. H., Prys-Roberts, C., Smith, A. C., and Spalding, J. M. K. (1969), *Anaesthesia 24*, 198-212.
35. Ingbar, S. H. (1966), *N. Engl. J. Med. 274*, 1252-1254.
36. Grollman, E. F., Lee, G., Ambesi-Impiombota, F. S., Meldolesi, M. F., Aloj, S. M., Coon, H., Kaback, H. R., and Kohn, L. D. *Proc. Natl. Acad. Sci. U.S.A.*, in press.
37. Mendolson, C., Dufau, M., and Catt, K. (1976), *J. Biol. Chem. 250*, 8818-8824.
38. Peterkofsky, A., and Gaydar, C. (1975), *Proc. Natl. Acad. Sci. U.S.A. 72*, 2920-2924.
39. Matuo, Y., Wheeler, M. A., and Bitensky, M. W. (1976), *Proc. Natl. Acad. Sci. U.S.A. 73*, 2654-2658.

STRUCTURAL STUDIES ON LECTINS AND LECTIN-SACCHARIDE INTERACTIONS

JOSEPH W. BECKER
GEORGE N. REEKE, JR.
GERALD M. EDELMAN

The Rockefeller University
New York, New York

Concanavalin A (Con A) is one of a large number of plant proteins, called lectins, that bind to specific carbohydrate-containing receptors on cell surfaces. This binding is associated with several biological activities, including mitogenesis, alteration of metabolism, and modulation of cell-surface receptor mobility, and these activities can be inhibited by specific monosaccharides related to D-mannose and D-glucose. We have determined the high-resolution crystal structures of Con A and its complex with a specific monosaccharide. A number of significant, reversible conformational changes are associated with carbohydrate binding, particularly in the metal and carbohydrate binding regions, where some groups move 4-6 Å with respect to the rest of the molecule. The positions of residues associated with both of the large β-structures of the native molecule also appear to change. The saccharide is bound in a shallow depression in the molecular surface 10-13 Å from the essential Mn^{2+} and Ca^{2+} ions. Interactions with the bound saccharide involve residues near positions 14-16, 97-99, 168-169, 207-208, 224-228, and 235-237. Several of these residues are also involved in coordination of the metal ions, consistent with the requirement that the metal ions be bound before saccharide binding can occur.

INTRODUCTION

Carbohydrate-binding proteins from plant and animal sources have been used extensively in studies of both the structure of the cell surface and the role of cell-surface components in fundamental cellular processes such as cell division, cell movement, and cell-cell interaction (1-3). These molecules, termed lectins, interact with a wide variety of cells, presumably through specific binding to the carbohydrate moieties of glycoprotein or glycolipid receptors on the cell surface.

One of the most widely used and best characterized of these proteins is Concanavalin A, the lectin of the jack bean (3). Con A binds to nearly all mammalian cells, and it displays a variety of activities, among which are an insulinlike effect on adipocytes (4), the inhibition of tumor cell migration (5), and the inhibition of phagocytosis by polymorphonuclear leukocytes (6). Con A also can agglutinate many somatic and germ line cell types (7, 8) and, in some cases, cellular transformation and loss of density-dependent growth control are accompanied by an increase in agglutinability (7). Of particular interest are the observations that Con A is mitogenic for lymphocytes (9, 10) and that, when bound to the lymphocyte surface, it can modulate the mobility and distribution of a variety of cell surface molecules (11, 12). All of these activities can be inhibited by specific saccharides related to D-glucose and D-mannose (13, 14) suggesting that they must proceed through binding to receptors containing similar saccharides.

In order to provide the basis for understanding these phenomena at the molecular level, we have carried out a series of related structural and functional studies on Con A. We have determined the complete amino acid sequence of Con A (15-17) as well as the three-dimensional structures of the native active protein (15, 18, 19), the demetallized inactive form (20), and a protein-saccharide complex (20). To correlate the resulting structural data with the activities of the protein, we have also prepared and characterized a series of specific chemical derivatives of Con A and have investigated the effects of these modifications on Con A's interaction with lymphoid cells (21, 22).

STRUCTURE OF CON A

Under physiological conditions, Con A is a tetramer with identical subunits, each of which contains a single polypeptide chain of 237 amino acid residues (23-25, Fig. 1). Each subunit has one specific saccharide-binding site and sites for two metal ions, usually Mn^{2+} and Ca^{2+}, which are required for carbohydrate binding activity (26, 27). The polypeptide chains are folded into compact ellipsoidal domes, approximately 40 × 39 Å in cross section and 42 Å high (Figs. 2, 3). These domes are paired base-to-base to form ellipsoidal dimers and the dimers are paired back-to-back to form tetramers of exact 222 symmetry.

The folding of the polypeptide chain is dominated by two large β-structures, which comprise over half of the residues. A twisted, seven-stranded structure runs through the center of the molecule (Fig. 4a), and a large, flatter six-stranded sheet makes up the entire back of the molecule (Fig. 4b). The two metal ions are bound 5 Å from each other at the top of each dome, between the central pleated sheet and a peptide loop comprising residues 8-24 (Figs. 2, 3, 5). The specific saccharide-binding site is in a shallow depression in the molecular surface adjacent to the metal ions (Figs. 2, 5). There is a large cavity in the molecule near the bottom of each dome and between the two pleated sheets (Fig. 2). This cavity has been shown in crystalline Con A to bind a wide variety of compounds, apparently through nonspecific hydrophobic interactions (18, 28).

DEMETALLIZED CON A

Acid treatment of Con A causes removal of the two metal ions and results in the destruction of the carbohydrate-binding activity of the protein. To investigate the structural features that accompany these phenomena, we have determined the three-dimensional structure of the inactive, demetallized protein for comparison with the native structure (20). Demetallized Con A crystallizes in a different space group from the native protein (29), and therefore the two structures cannot be compared directly. The demetallized structure was solved without the use of heavy atoms by the molecular replacement method (30), using the assumption that the structure would resemble the structure of native Con A, except in the metal-binding region. Such a model was aligned with the unknown demetallized structure using automated rotational (31)

1 10 17
Ala-Asp-Thr-Ile-Val-Ala-Val-Glu-Leu-Asp-Thr-Tyr-Pro-Asn-Thr-Asp-Ile-

18 30 37
Gly-Asp-Pro-Ser-Tyr-Pro-His-Ile-Gly-Ile-Asp-Ile-Lys-Ser-Val-Arg-Ser-Lys-Lys-Thr-

38 50 57
Ala-Lys-Trp-Asn-Met-Gln-Asp-Gly-Lys-Val-Gly-Thr-Ala-His-Ile-Ile-Tyr-Asn-Ser-Val-

58 70 77
Asp-Lys-Arg-Leu-Ser-Ala-Val-Val-Ser-Tyr-Pro-Asn-Ala-Asp-Ala-Thr-Ser-Val-Ser-Tyr-

78 90 97
Asp-Val-Asp-Leu-Asn-Asp-Val-Leu-Pro-Glu-Trp-Val-Arg-Val-Gly-Leu-Ser-Ala-Ser-Thr-

98 110 117
Gly-Leu-Tyr-Lys-Glu-Thr-Asn-Thr-Ile-Leu-Ser-Trp-Ser-Phe-Thr-Ser-Lys-Leu-Lys-Ser-

118 130 137
Asn-Ser-Thr-His-Gln-Thr-Asp-Ala-Leu-His-Phe-Met-Phe-Asn-Gln-Phe-Ser-Lys-Asp-Gln-

138 150 157
Lys-Asp-Leu-Ile-Leu-Gln-Gly-Asp-Ala-Thr-Thr-Gly-Thr-Asp-Gly-Asn-Leu-Glu-Leu-Thr-

158 170 177
Arg-Val-Ser-Ser-Asn-Gly-Ser-Pro-Glu-Gly-Ser-Ser-Val-Gly-Arg-Ala-Leu-Phe-Tyr-Ala-

178 190 197
Pro-Val-His-Ile-Trp-Glu-Ser-Ser-Ala-Thr-Val-Ser-Ala-Phe-Glu-Ala-Thr-Phe-Ala-Phe-

198 210 217
Leu-Ile-Lys-Ser-Pro-Asp-Ser-His-Pro-Ala-Asp-Gly-Ile-Ala-Phe-Phe-Ile-Ser-Asn-Ile-

218 230 237
Asp-Ser-Ser-Ile-Pro-Ser-Gly-Ser-Thr-Gly-Arg-Leu-Leu-Gly-Leu-Phe-Pro-Asp-Ala-Asn

FIGURE 1 The complete amino acid sequence of concanavalin A.

and translational (32) molecular replacement calculations. The resulting model showed excellent agreement with the observed diffraction data, and by comparison of the residual differences between the model and observed data, we have been able to construct a model of the complete demetallized structure.

The demetallized protein has a tetrameric structure strikingly similar to that of the native protein. Significant differences between the two structures are confined to the region of the metal ion and saccharide binding sites (Fig. 5). The

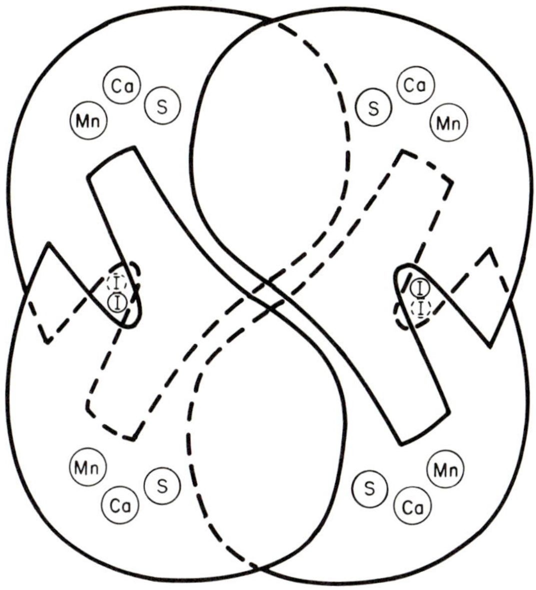

FIGURE 2 Schematic representation of the Con A tetramer. The binding sites for manganese and calcium are indicated by Mn and Ca, respectively; for specific saccharides by S; and for hydrophobic molecules by I.

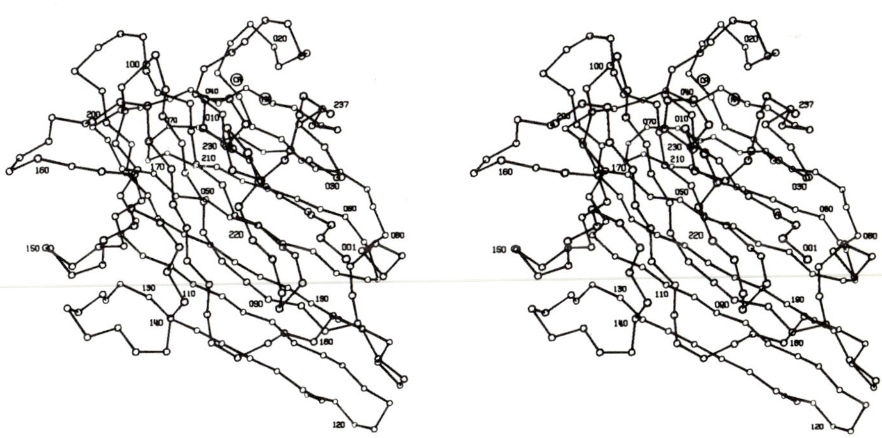

FIGURE 3 Stereoscopic representation of the Con A polypeptide chain. Circles are at the positions of α-carbon atoms. The binding sites for manganese and calcium are labeled Mn and Ca, respectively.

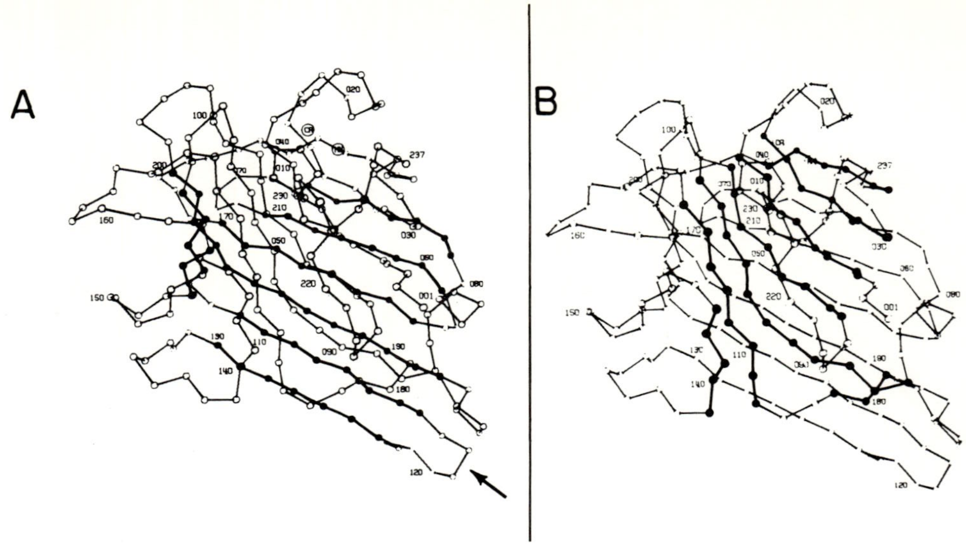

FIGURE 4 *β-Structures in Con A. (A) The large flat β-structure at the back of the protomer is highlighted. (B) The twisted central β-structure is highlighted.*

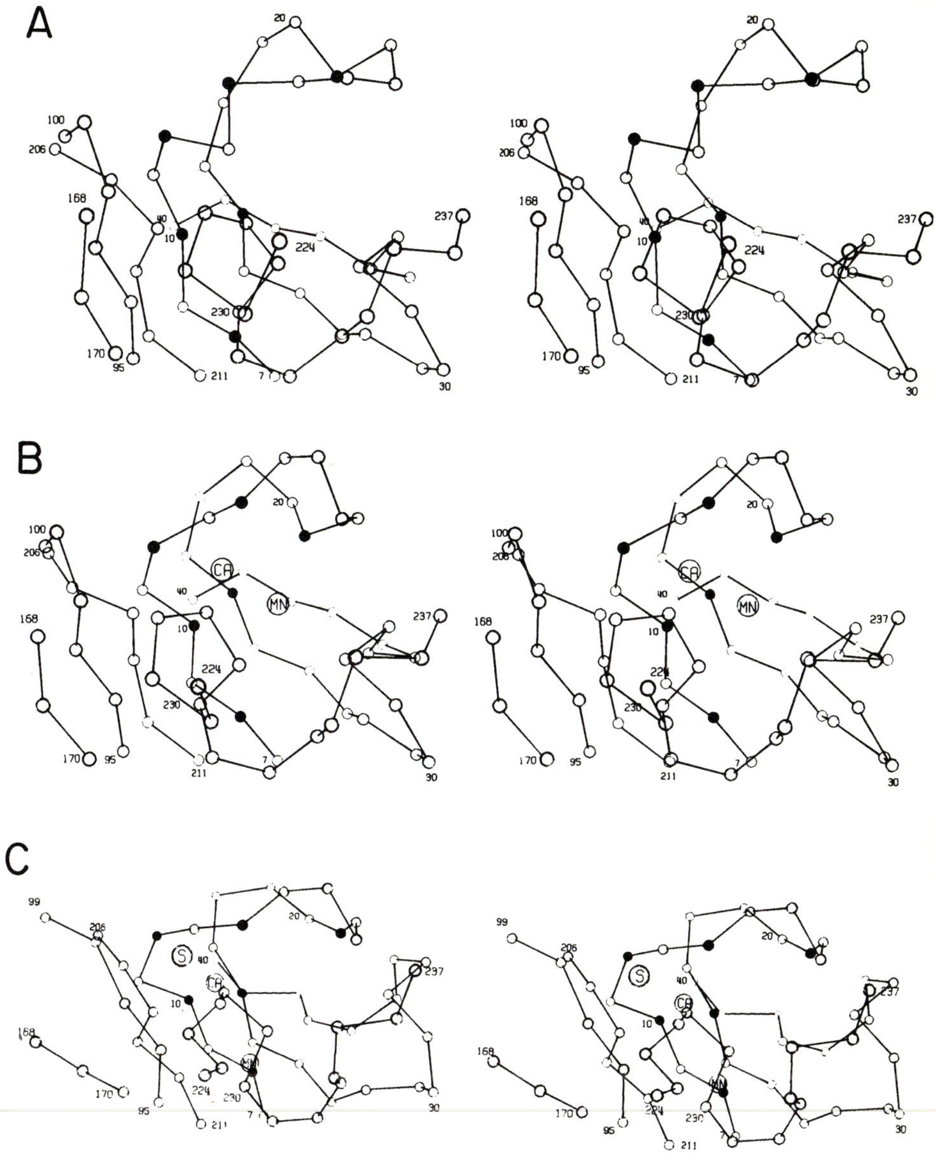

FIGURE 5 Ligand-induced conformational changes in Con A. (A) The potential metal and saccharide binding region of demetallized Con A. Metal ion binding residues are highlighted. (B) The same region in native Con A. (C) The same region in the cross linked saccharide derivative. The position of the bound saccharide is indicated by S.

loop of peptide above the metal ions is raised in the demetallized structure, opening the binding regions to solvent. Several other nearby chains also have different conformations in the two structures, particularly those involving residues 95-100, 206-211, and 224-237. Throughout the rest of the structure, including the regions of intersubunit contact and the hydrophobic binding cavity, the two structures are quite similar, suggesting that any functions of these parts of the molecule should not be sensitive to the presence or absence of the metal ions.

Comparison of the two structures suggests an explanation of the observations that Mn^{2+} binds to its site in Con A more readily than Ca^{2+} does to its site, and that the metal ion-induced conformational change must occur before specific saccharide binding can occur (27, 33, 34). Three of the four protein ligands for Mn^{2+}, Glu 8, Asp 10, and His 24 are in similar positions in both structures and may constitute a preformed Mn^{2+} binding site. On the other hand, three of the Ca^{2+} ligands, Tyr 12, Asn 14, and Asp 19, are in quite different positions, and presumably the binding of Mn^{2+} brings one or more of them into the proper orientation to enable the protein to bind Ca^{2+}. Examination of the Con A saccharide-complex described below likewise indicates that many potential carbohydrate binding residues are also displaced in the demetallized structure.

CON A-SACCHARIDE COMPLEX

Crystallographic observation of Con A-saccharide interactions has long been hampered by the fact that the Con A-saccharide complex is much more soluble than the native protein, causing crystals of the type used in the native structure determination to dissolve when exposed to the appropriate saccharides. In addition, it is now known that the saccharide-binding site is in a region of intermolecular contact, which causes crystals exposed to saccharides to crack or become disordered, even when precautions are taken to prevent their dissolving. To avoid these experimental difficulties, we have taken two independent approaches to a crystallographic examination of the Con A-saccharide complex. First, we have stabilized Con A crystals by extensive cross linking before exposing them to specific saccharides, and, second, we have prepared crystals of the Con A-saccharide complex directly by growing a new crystal form in the presence of inhibitory sugars.

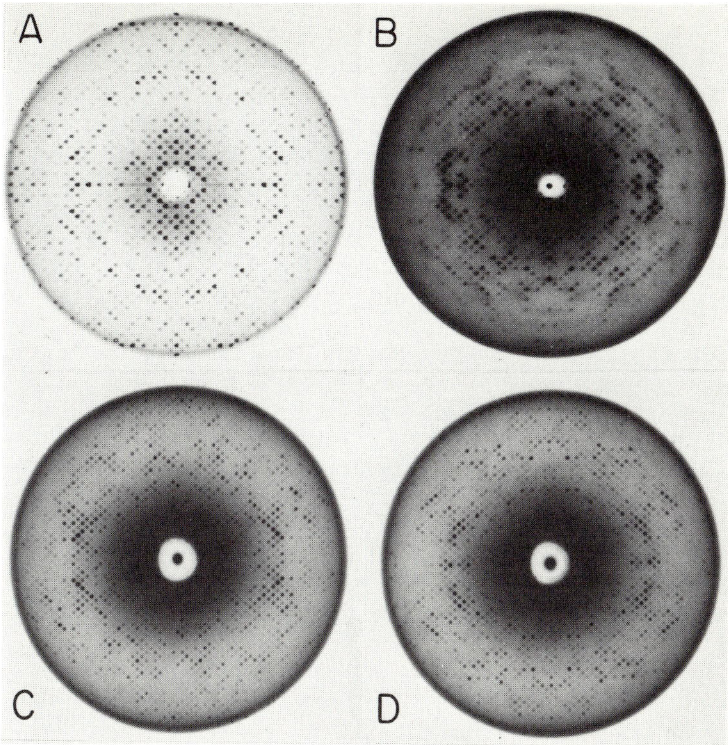

FIGURE 6 Saccharide-induced changes in the diffraction pattern of Con A. (A) hk0 zone of native Con A, showing diffraction from ∞-2.8 Å. Space group I222, a = 89.91, b = 87.23, c = 63.07 Å. (B) hk0 zone of the cross linked saccharide complex. Space group I222 with a = 92.76, b = 86.66, c = 64.99 Å. (C) hk0 zone of Con A crystallized from 1 mM methyl-α-D-glucopyranoside, 20% saturated ammonium sulfate. Space group C1, a = 119.04, b = 103.23, c = 132.09 Å, α = 89° 50', β = 94° 17', γ = 85° 3'. (D) hk0 zone of a Con A crystal grown as in (C) and then treated with 50 mM methyl-α-D-glucopyranoside. Space group $C222_1$, with a = 118.09, b = 103.54, c = 251.8 Å.

Treating a heavily cross linked Con A crystal with an inhibitory saccharide produces some striking changes in the diffraction pattern (Fig. 6). Although the space group does not change and the unit cell dimensions change only slightly, the distribution of intensities is quite different, suggesting a change in the protein's structure. In addition, the protein-saccharide complex does not produce high resolution diffraction spots, suggesting some disordering of the crystal structure. These changes in the diffraction pattern are apparently due to noncovalent binding of specific inhibitory saccharides, since they are induced only by such compounds, and they can be reversed by washing the crystals in saccharide-free buffer.

The conformational changes induced by carbohydrates are sufficiently large that the structure of this complex could not be solved by direct comparison with the native Con A structure. Similarly, our attempts to use molecular replacement methods similar to those that had succeeded in revealing the demetallized structure, also failed to solve this structure. The structure was solved by the standard multiple isomorphous replacement method, using four heavy-atom derivatives similar to those used in the native Con A structure determination, and the resulting map was improved by electron density modification refinement (35). In order to locate the bound saccharide in these low resolution maps, we have used the heavy-atom labeled inhibitory saccharide 2-deoxy-2-iodo-methyl-α-D-mannopyranoside (Fig. 7) in this study.

We have traced the entire polypeptide chain in this complex and, although the interpretation in some regions is still tentative, it is clear that there are several significant conformational changes brought about by saccharide binding. Also, we have located the specifically bound saccharide by comparing the structures of the complexes with the iodinated sugar and its chloro analog. The saccharide-binding site is at the top of the dome-shaped protomers, about 13 Å from the Mn^{2+} ion (Figs. 2 and 5). At the current resolution, only the iodine atom of the bound saccharide can be identified, with the result that we are at present unable to describe the precise geometry and mode of binding of saccharides by Con A. The residues that are in position to be involved in direct saccharide binding are 14-16, 97-99, 168-169, 207-208, 224-228, and 235-237. These residues line a shallow depression at the top of the Con A protomer and many of them are associated with the metal ion-induced conformational change. The fact that the binding region is near a grouping of aromatic side chains, particularly Tyr 12, Tyr 101, and His 205, suggests a possible explanation for the observation that aryl-substituted saccharides are bound more strongly than the unsubstituted analogs (36).

Comparison of this region with the same part of native Con A indicates that the binding of the saccharide has caused the metal-binding region to move considerably with respect to the rest of the molecule. In Fig. 5, these residues have moved generally to the left and out toward the observer. The β-structure on the left of the figure, which is the top of the twisted central pleated sheet (Fig. 4), appears to have been disrupted and the COOH-terminal peptide has moved as well.

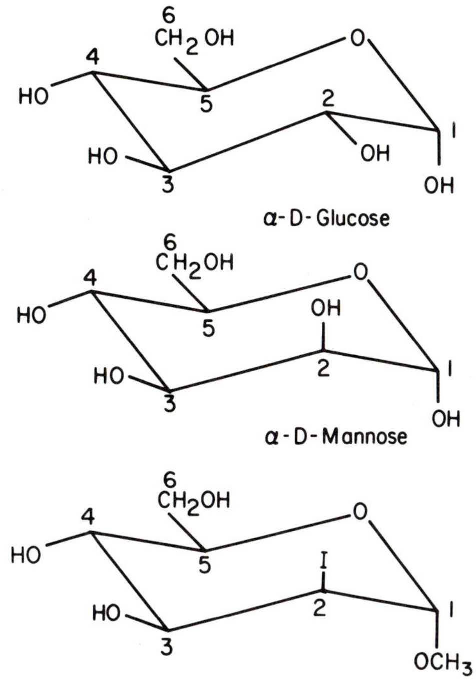

2-Deoxy-2-iodo-methyl-α-D-Mannopyranoside

FIGURE 7 Three monosaccharide inhibitors of Con A.

There are several other conformational changes in other parts of the molecule. For example, the large loop comprising residues 61-77 is thrust out of the back of the molecule, and the loop around residue 160 is pushed away from the body of the molecule.

The ability of the hydrophobic binding site to bind at least small molecules is apparently unaffected by the carbohydrate-induced conformational change, as suggested by the observation that crystals of both Con A and the cross-linked saccharide complex bind *o*-iodophenol with apparently equal strength. The structure of this region of the molecule is relatively unchanged by the binding of sugar, although there are some movements of groups, particularly near the opening to this site. It is possible that these changes might affect potential interactions with more complex molecules.

The carbohydrate-binding region of Con A is also a region of intermolecular contact in this crystal type and this observation suggests that the conformational changes observed here may not all be due to purely saccharide-induced effects. The intermolecular lattice interactions of this crystal form may be sufficiently disrupted by the binding of carbohydrate or by carbohydrate-induced conformational changes that additional conformational changes may be caused simply by packing considerations. To avoid this problem, we have grown a new crystal form of the Con A-saccharide complex (Fig. 6). We have collected diffraction data from the crystals to a 2.8 Å resolution and a structure solution is currently in progress. The availability of these high resolution data suggests that study of this crystal form will lead to knowledge of the atomic-scale details of Con A-saccharide interactions. Also, the availability of structures of two Con A-saccharide complexes should allow saccharide specific conformational changes to be resolved from lattice-dependent ones.

CHEMICAL DERIVATIVES ON CON A

Knowledge of the structure of Con A has facilitated the design of specific chemical modifications, allowing experimental investigation of the relationships between the various structural features of this protein and its biological activities. The addition of Con A to lymphocytes causes several structural and biochemical changes in these cells. Under certain conditions, Con A forms patchy distributions on the cell surface followed by caps (37). At higher doses, Con A displays the opposite effect, producing a diffuse distribution of bound Con A and immobilizing other cell surface receptors (11, 12). Con A also causes lymphocytes to undergo blast transformation and mitosis (9, 10), and this activity also displays dose dependency with the response increasing up to the optimal concentration and decreasing above it.

Extensive succinylation of Con A produces a molecule that is dimeric under physiological conditions, where the native species is the tetramer (21). This loss in valence is apparently due to modification of the amino acid side chains associated with certain intersubunit interactions, and it is unaccompanied by any change in the affinity or specificity of the carbohydrate-binding activity (21). The dimeric derivative, however, is unable to induce or inhibit cap formation in lymphocytes (21). The fact that treatment with succinyl Con A followed by anti-Con A restores surface modulation suggests a critical role in these activities for valence and the ability to cross-link cell-surface receptors. The divalent derivative is, however, equally as potent a mitogen as the native, tetrameric species, and the mitogenic response to succinyl-Con A shows saturation behavior in contrast to the unimodal dose-response curve of native Con A (21). These observations suggest that the reduced valence of the succinyl derivative is still consistent with full mitogenic activity and that the higher valence of the native molecule is associated with a separate ability to inhibit cellular proliferation.

It is clear from these data that a monovalent form of Con A would be an extremely valuable reagent for further testing of the effects of the protein's valence on its mitogenic activity. Such a molecule has proven quite difficult to prepare and characterize, as the monomeric form of Con A apparently does not exist as a stable, long-lived solution species. Recently, two different approaches have been employed to prepare a monovalent form of Con A. Fraser *et al.* (22) partially inactivated succinyl Con A with a photoaffinity label and isolated a monovalent product by affinity chromatography, and Wands *et al.* (38) partially digested Con A with trypsin and performed a similar isolation. The initial experiments indicate that the affinity-labeled preparation is mitogenic, while the trypsinized protein is not. These experiments leave the question of the effect of valence on mitogenicity currently unresolved. Both preparations require more extensive characterization, but the apparent difference in activity may arise from more complex factors. For example, mitogenesis may require interaction with a protease-sensitive region of the Con A molecule other than the carbohydrate binding site or it may involve interaction with another agent, such as macrophages, with which these two species have different interactions. However, the suggestion that a molecule that is monovalent for carbohydrate can possess full mitogenic activity is supported by the observation that carbohydrate-specific immunoglobulins are potent mitogens in both divalent and monovalent forms (39).

CONCLUSIONS

Con A is the first lectin the complete amino acid sequence and three-dimensional structure of which are known. Many of the structural features revealed by the above studies, particularly those relating to the valence and ligand-binding properties, have been associated with the chemical and biological activities of the protein. The structure of the native molecule thus serves as a model of a molecular species capable of recognizing and interacting with specific cell-surface receptors, and the structure of the saccharide complex serves as a model of the actual interaction. In addition, Con A serves as a reference structure against which the structures of other cell-surface probes, particularly other lectins, may be compared, as studies of their structures approach similar levels of refinement.

ACKNOWLEDGMENTS

This work was supported by grants AI-11378, AI-09273, AM-04256, and GM-22663 from the National Institutes of Health. George N. Reeke is an Alfred P. Sloan Research Fellow.

REFERENCES

1. Lis, H., and Sharon, N. (1973). The biochemistry of plant lectins (Phytohemagglutinins), *Ann. Rev. Biochem. 42,* 541-574.
2. Cohen, E. (ed.). (1974). "Biomedical Perspectives of Agglutinins of Invertebrate and Plant Origins," *Ann. N. Y. Acad. Sci. 234.*
3. Chowdhury, T. K., and Weiss, A. K., (eds.). (1975). "Concanavalin A," *Adv. Exp. Med. Biol. 55.*
4. Cuatrecasas, P., and Tell, G. P. E. (1973). Insulin-like activity of concanavalin A and wheat germ agglutinin--Direct interactions with insulin receptors, *Proc. Nat. Acad. Sci. U.S.A. 70,* 485-489.
5. Friberg, S., Cochran, A., and Golub, S. (1971). Concanavalin A inhibits tumor cell migration, *Nature New Biol. 232,* 121-122.
6. Berlin, R. D. (1971). Effect of concanavalin A on phagocytosis, *Nature New Biol. 235,* 44-45.
7. Inbar, M., and Sachs, L. (1969). Interaction of the carbohydrate-binding protein concanavalin A with normal and transformed cells, *Proc. Nat. Acad. Sci. U.S.A. 63,* 1418-1425.
8. Edelman, G. M., and Millette, C. F. (1971). Molecular probes of spermatozoan structure, *Proc. Nat. Acad. Sci. U.S.A. 68,* 2436-2440.
9. Powell, A. E., and Leon, M. A. (1970). Reversible interaction of human lymphocytes with the mitogen concanavalin A, *Exp. Cell Res. 62,* 315-325.

10. Beckert, W. H., and Sharkey, M. M. (1970) Mitogenic activity of the jack bean (*Canavalia ensiformis*) with rabbit peripheral blood lymphocytes, *Int. Arch. Allergy Appl. Immunol. 39,* 337-341.
11. Yahara, I., and Edelman, G. M. (1972). Restriction of the mobility of lymphocyte immunoglobulin receptors by concanavalin A, *Proc. Nat. Acad. Sci, U.S.A. 69,* 608-612.
12. Yahara, I., and Edelman, G. M. (1973). The effects of concanavalin A on the mobility of lymphocyte surface receptors, *Exp. Cell Res. 81,* 143-145.
13. Goldstein, I. J., Hollerman, C. E., and Smith, E. E. (1965). Protein-carbohydrate interaction. II. Inhibition studies on the interaction of concanavalin A with polysaccharides, *Biochemistry 4,* 876-883.
14. Goldstein, I. J., Reichert, C. M., and Misaki, A. (1974), Interaction of concanavalin A with model substrates, *Ann. N.Y. Acad. Sci. 234,* 283-296.
15. Edelman, G. M., Cunningham, B. A., Reeke, G. N., Becker, J. W., Waxdal, M. J., and Wang, J. L. (1972). The covalent and three-dimensional structure of concanavalin A, *Proc. Nat. Acad. Sci. U.S.A. 69,* 2580-2584.
16. Wang, J. L., Cunningham, B. A., Waxdal, M. J., and Edelman, G. M., (1975). The covalent and three-dimensional structure of concanavalin A I. Amino acid sequence of cyanogen bromide fragments F_1 and F_2, *J. Biol. Chem. 250,* 1490-1502.
17. Cunningham, B. A., Wang, J. L., Waxdal, M. J., and Edelman, G. M. (1975). The covalent and three-dimensional structure of concanavalin A. II. Amino acid sequence of cyanogen bromide fragment F_3, *J. Biol. Chem. 250,* 1503-1512.
18. Becker, J. W., Reeke, G. N., Jr., Wang, J. L., Cunningham, B. A., and Edelman, G. M. (1975). The covalent and three-dimensional structure of concanavalin A. III. Structure of the monomer and its interactions with metals and saccharides, *J. Biol. Chem. 250,* 1513-1524.
19. Reeke, G. N., Jr., Becker, J. W., and Edelman, G. M. (1975). The covalent and three-dimensional structure of concanavalin A. IV. Atomic coordinates, hydrogen bonding, and quaternary structure, *J. Biol. Chem. 250,* 1525-1547.
20. Becker, J. W., Reeke, G. N., Jr., Cunningham, B. A., and Edelman, G. M. (1976). New evidence on the location of the saccharide-binding site of concanavalin A, *Nature 259,* 406-409.
21. Gunther, G. R., Wang, J. L., Yahara, I., Cunningham, B. A., and Edelman, G. M. (1973). Concanavalin A derivatives with altered biological activities, *Proc. Nat. Acad. Sci. U.S.A. 70,* 1012-1016.

22. Fraser, A. R., Hemperly, J. J., Wang, J. L., and Edelman, G. M. (1976). Monovalent derivatives of concanavalin A, *Proc. Nat. Acad. Sci. U.S.A. 73,* 790-794.
23. Wang, J. L., Cunningham, B. A., and Edelman, G. M. (1971). Unusual fragments in the subunit structure of concanavalin A, *Proc. Nat. Acad. Sci. U.S.A. 68,* 1130-1134.
24. Edmundson, A. B., Ely, K. R., Sly, D. A., Westholm, F. A., Powers, D. A., and Liener, I. E. (1971). Isolation and characterization of concanavalin A polypeptide chains, *Biochemistry 10,* 3554-3559.
25. Kalb, A. J., and Lustig, A. (1968). The molecular weight of concanavalin A, *Biochim. Biophys. Acta. 168,* 366-367.
26. Sumner, J. B., and Howell, S. F. (1936). The identification of the hemagglutinin of the jack bean with concanavalin A, *J. Bacteriol. 32,* 227-237.
27. Kalb, A. J., and Levitzki, A. (1968). Metal-binding sites of concanavalin A and their role in the binding of α-methyl-D-glucopyranoside, *Biochem. J. 109,* 669-672.
28. Hardman, K. D., and Ainsworth, C. F. (1973). Binding of nonpolar molecules by crystalline concanavalin A, *Biochemistry 12,* 4442-4447.
29. Jack, A., Weinzierl, J., and Kalb, A. J. (1971). An X-ray crystallographic study of demetallized concanavalin A, *J. Mol. Biol. 58,* 389-395.
30. Rossmann, M. G., and Blow, D. M. (1962). The detection of sub-units within the crystallographic asymmetric unit, *Acta. Cryst. 15,* 24-31.
31. Crowther, R. A. (1971) *in* "The Molecular Replacement Method" (Rossmann, M. G., ed.), pp. 173-178, Gordon and Breach, New York.
32. Crowther, R. A., and Blow, D. M. (1967). A method of positioning a known molecule in an unknown crystal structure, *Acta. Cryst. 23,* 544-548.
33. Yariv, J., Kalb, A. J., and Levitzki, A. (1968). The interaction of concanavalin A with methyl-α-D-glucopyranoside, *Biochim. Biophys. Acta. 165,* 303-305.
34. Agrawal, B. B. L., and Goldstein, I. J. (1968). Protein-carbohydrate interaction. XV. The role of bivalent cations in concanavalin A-polysaccharide interactions, *Can. J. Biochem. 46,* 1147-1150.
35. Collins, D. M., Cotton, F. A., Hazen, E. E., Meyer, E. F., and Morimoto, C. N. (1975). Protein crystal structures: Quicker, cheaper approaches, *Science 190,* 1047-1053.
36. Poretz, R. D., and Goldstein, I. J. (1971). Protein-carbohydrate interaction: On the mode of binding of aromatic moieties to concanavalin A, *Biochem. Pharmacol. 20,* 2727-2739.

37. Unanue, E. R., Perkins, W. D., and Karnovsky, M. J. (1972). Ligand-induced movement of lymphocyte membrane molecules. I. Analysis by immunofluorescence and ultrastructural radiography, *J. Exp. Med. 136,* 885-906.
38. Wands, J. R., Podolsky, D. K., and Isselbacher, K. J. (1976). Mechanism of human lymphocyte stimulation by concanavalin A: Role of valence and surface binding sites, *Proc. Nat. Acad. Sci. U.S.A. 73,* 2118-2122.
39. Sela, B. A., Wang, J. L., and Edelman, G. M. (1976). Lymphocyte activation by monovalent fragments of antibodies reactive with cell surface carbohydrates, *J. Exp. Med. 143,* 665-671.

ALTERATIONS OF SURFACE GLYCOCONJUGATES AND CELL MORPHOLOGY INDUCED BY BUTYRIC ACID

PETER H. FISHMAN
ROSCOE O. BRADY

Developmental and Metabolic Neurology Branch

RICHARD C. HENNEBERRY
ERNST FREESE

Laboratory of Molecular Biology

National Institute of Neurological
and Communicative Disorders and Stroke
National Institutes of Health
Bethesda, Maryland

Butyrate (C_4) induces morphological and biochemical differentiation in a variety of mammalian cells. Effects of butyrate (and other short chain fatty acids) have been studied extensively in HeLa cells. HeLa cells extend neuritelike processes in the presence of butyrate (and C_3 and C_5). Morphological changes are preceded by increases in CMP-sialic acid:lactosylceramide sialyltransferase activity (∿ 20-fold) and its product G_{M3} (sialolactosylceramide) (3-5 fold). Induction of shape changes and enzyme activity are inhibited by cycloheximide and Actinomycin D. Only inducers (C_3-C_5) of process formation cause increased sialyltransferase activity; analogs of butyrate are ineffective. Levels of other glycolipids, glycosyltransferases, or glycohydrolases are not substantially altered by the addition of butyrate to HeLa cultures. Surface gangliosides can be labeled by selective oxidation of exposed sialyl residues with periodate and reduction with borotritide. About 5-fold more G_{M3} is labeled in butyrate-treated cells than in control cells.

The growth inhibitory properties of fatty acids are well known; both butyrate and decanoate inhibit the growth of HeLa. There is an enrichment of sialoglycopeptides derived from the surface of HeLa cells treated with butyrate or decanoate. However, decanoate does not cause shape changes. Thus, changes in surface glycoproteins are not related to changes in HeLa cell morphology. Induction of ganglioside biosynthesis appears to be uniquely associated with the morphological differentiation in HeLa cells caused by butyrate.

INTRODUCTION

Cell surface carbohydrates are recognized as antigenic determinants (1) and as receptors for lectins (2), toxins (2-5), and possibly hormones (6-7). A role for surface glycoconjugates has been implicated in various biological processes including cellular adhesiveness, intercellular recognition, malignant transformation, and morphological differentiation (1, 8-12). This last process lends itself readily to experimental evaluation. A number of methods for inducing morphological differentiation in cultured mammalian cells have been described (Table I). Although cyclic AMP (or its butyrylated derivatives) has been implicated in many of these systems, it is now apparent that butyrate itself is a potent inducer of morphological changes. In addition, butyrate can induce biochemical differentiation in a variety of mammalian cells (Table II). Although the mechanism(s) by which butyrate causes these changes is unknown, the specific effects of butyrate on HeLa cell morphology and ganglioside biosynthesis provide an excellent model to investigate the relationship between morphology and cell surface carbohydrates.

EFFECT OF BUTYRATE ON CELL MORPHOLOGY

It was first reported by Ginsburg *et al.* (22) that lipophilic agents inhibited the growth of mammalian cells and that short-chain fatty acids such as butyric caused the cells to change shape. Butyrate induced HeLa cells to become more jagged and to extend neuritelike processes (Fig. 1). Process formation commenced 7-8 hr after exposure to butyrate, was maximal by 14-18 hr, and was inhibited by actinomycin D, cycloheximide, and colchicine (22, 24).

In addition to butyrate, pentanoate and propionate caused similar shape changes in HeLa whereas shorter (acetate) or longer chain fatty acids including decanoate did not (22, 24).

TABLE I Induction of Morphological Differentiation in Mammalian Cells in Culture[a]

Cell line	Treatment	Morphological change	References
Mouse neuroblastoma	Serum-deprivation	Formation and extension of axons	13
	Bu_2cAMP; BucAMP		14
	Prostaglandins		14
	RO20-1724; papaverine		14
	Sodium butyrate		15, 16, 18
Human neuroblastoma	Serum-deprivation	Formation and extension of axons	17
	Bu_2cAMP		17
	Prostaglandins		17
	Papaverine		17
	Sodium butyrate		17
Chinese hamster ovary	Bu_2cAMP	Formation of elongated, fibroblastlike cells growing in parallel arrays	19, 21
	Sodium butyrate		20, 21
HeLa	Sodium butyrate	Formation of jagged shaped cells and extension of neuritelike processes	22

[a]Abbreviations: Bu_2cAMP, dibutyryl cyclic AMP; BucAMP, monobutyryl cyclic AMP; RO 20-1724, 4-(3-butoxy-4-methyoxybenzl)-2-imidazolidinone (a phosphodiesterase inhibitor).

TABLE II *Biochemical Differentiation Induced in Mammalian Cells by Butyrate*[a]

Cell line	Biochemical change	References
HeLa	Increase in G_{M3} synthesis	23, 24
	Increase in alkaline phosphatase	25
	Increase in β-subunit of HCG	26
Mouse neuroblastoma	Increase in tyrosine hydroxylase	14, 17
	Increase in acetylcholinesterase	15
	Increase in choline acetyltransferase	14, 17
	Increase in plasminogen activator	16
Erythroleukemic cells	Increase in hemoglobin synthesis	27, 28

[a]*Abbreviations:* G_{M3}, *N-acetylneuraminylgalactosylglucosylceramide; HCG, human chorionic gonadotropin.*

Cyclic AMP (5 m*M*) and monobutyryl cyclic AMP (1 m*M*) had no effect on HeLa cell morphology (22).

EFFECT OF BUTYRATE ON GANGLIOSIDE METABOLISM IN HELA

Glycosphingolipids were isolated from untreated HeLa cells and from cells exposed to 5 m*M* sodium butyrate for 24 hr at which time maximum morphological changes were induced (24). The amount of G_{M3} in butyrate-treated cells was found to be increased 3.5-fold under these conditions whereas other glycosphingolipid levels were essentially unchanged (Table III). When *N*-[acetyl-^{3}H]-D-mannosamine, a precursor of sialic acid, was added to the culture medium, [^{3}H] incorporation into G_{M3} was increased 6.5-fold in HeLa cells exposed to butyrate and the specific activity of the G_{M3} was twice that from control cells (Table IV). In contrast, [^{3}H] incorporation into G_{D3} was not altered by exposure of HeLa to butyrate.

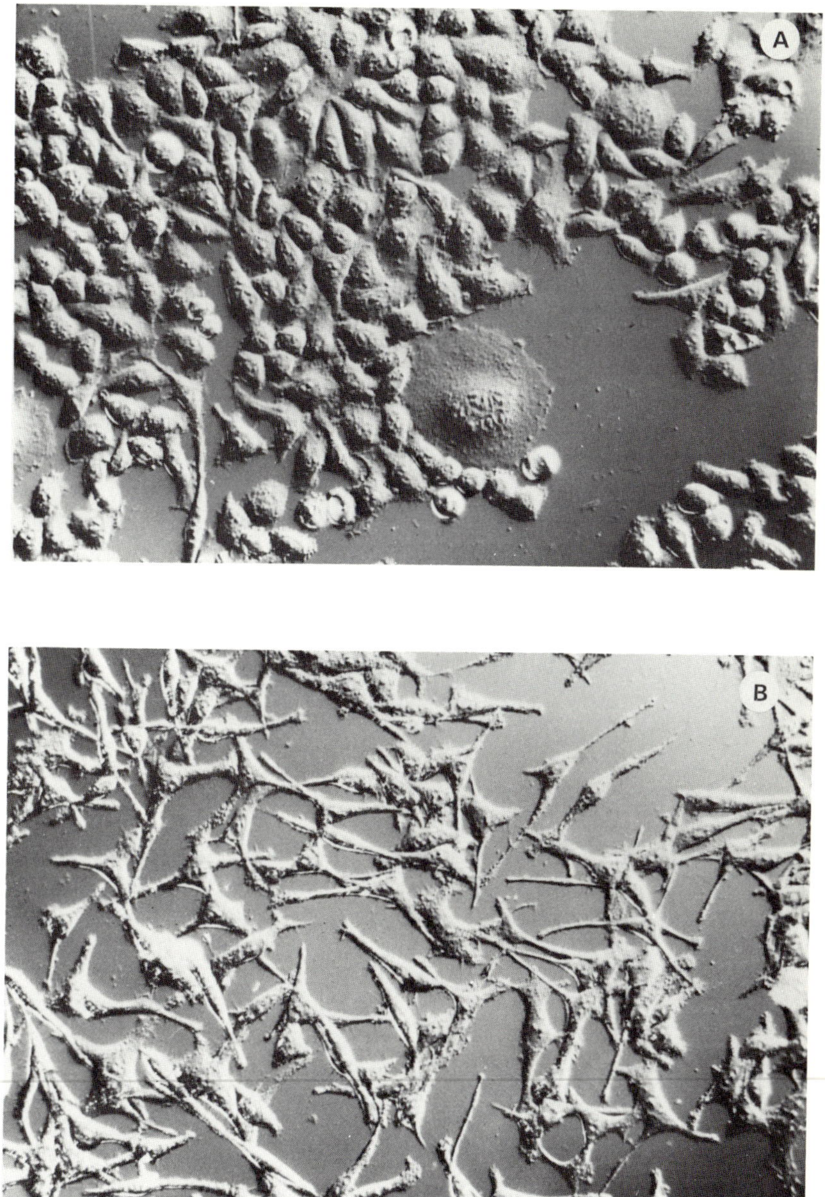

FIGURE 1 Photomicrographs showing effects of butyrate on HeLa cell morphology. HeLa cells were cultured 12 hr in normal medium (A) or medium supplemented with 5 mM sodium butyrate (B). Photomicrographs were taken with Normarski interference optics.

TABLE III Glycolipid Composition of HeLa Cells Grown in Normal and Butyrate Supplemented Media[a]

Glycolipid	Structure[b]	Control	Butyrate
		(nmol/mg cell protein)	
GL-1	cer-glc	0.55	0.95
GL-2	cer-glc-gal	0.41	0.45
GL-3	cer-glc-gal-gal	1.85	2.21
G_{M3}[c]	cer-glc-gal-NAN	0.49	1.72
G_{D3}	cer-glc-gal-NAN-NAN	0.23	0.19

[a]Data from reference (24). HeLa cells were cultured for 24 hr at 37°C in normal and sodium butyrate (5 mM)-supplemented media. Glycolipids were extracted, isolated, separated, and quantified. Results are the average of two experiments.

[b]Abbreviations: cer, ceramide; glc, glucose; gal, galactose; NAN, N-acetylneuraminic acid.

[c]The G_{M3} content of HeLa cells varied from experiment to experiment (29); however, in six separate analyses, the G_{M3} content of butyrate-treated cells was 3.5 to 7 times greater than that of control cells.

TABLE IV Incorporation of N-[Acetyl-^{3}H]-D-Mannosamine into Gangliosides of Control and Butyrate-Treated HeLa Cells[a]

	Control cells		Butyrated-treated cells	
Ganglioside	*cpm/mg protein*	*cpm/nmol*	*cpm/mg protein*	*cpm/nmol*
G_{M3}	7,249	14,794	47,421	27,570
G_{D3}	14,983	32,572[b]	11,646	30,647[b]

[a]*Data from Simmons et al. (24). HeLa cells were cultured at 37°C in medium supplemented with N-[Acetyl-^{3}H]-D-mannosamine (156 µM, 62.5 µCi/ml) and 5 mM sodium butyrate where indicated. After 24 hr, the cells were harvested and the radioactive gangliosides were isolated and analyzed.*

[b]*The higher specific activity of G_{D3} represents the two sialic acid residues per molecule.*

When cells labeled in the presence of butyrate were transferred to fresh medium (minus label and butyrate), there was a decrease in both specific activity and content of G_{M3} in the cells after 24 hr (Table V). By this time, the cells had reverted to the typical HeLa cell shape (as in Fig. 1a). Thus butyrate had a specific and reversible effect on G_{M3} metabolism in HeLa cells.

Gangliosides are presumed to be localized in the plasma membrane with the oligosaccharide chains on the external surface (30). The localization of G_{M3} on untreated and butyrate-treated cells was examined with a modified procedure for externally labeling membrane glycoproteins (31). The cells were incubated with sodium periodate under mild conditions to oxidize surface sialic acid residues to the corresponding seven-carbon aldehyde. The cells, which remained viable, were harvested; the ganglioside fraction was isolated and treated with $NaB[^3H]_4$ to reduce the aldehyde groups to tritiated alcohols.

Incorporation of [^{3}H] into G_{M3} increased over 5-fold in the butyrate-treated cells (Table VI). Since the G_{M3} isolated from control and butyrate-treated cells had similar specific activities, exposure to butyrate appears to increase both surface G_{M3} and total G_{M3} proportionally. There are approximately 200 million molecules of G_{M3} per butyrate-treated cell compared to about 34 million molecules in an untreated cell.

TABLE V Decrease in $[^3H]G_{M3}$ Following Removal of Butyrate from HeLa Cells[a]

	$[^3H]$-G_{M3}	
Hours after removal of butyrate	*cpm/nmol*	*nmol/mg protein*
0	25,619	1.20
24	14,027	0.48

[a]*HeLa cells were cultured at 37°C for 24 hr with medium containing N-[acetyl-^3H]-D-mannosamine (50 µCi/ml) and 5 mM sodium butyrate. The medium was replaced with fresh medium (without label and butyrate) and the cells harvested at the times indicated. Radioactive G_{M3} was isolated from the cells and quantified.*

EFFECT OF BUTYRATE ON ENZYME ACTIVITIES

The increased G_{M3} content in HeLa cells exposed to butyrate could be due either to a reduction in G_{M3} catabolism or to an increase in G_{M3} biosynthesis. The activities of sialidase as well as other glycohydrolases were determined in cell homogenates (Table VII). Enzyme activites of HeLa cells grown in the presence or absence of 5 m*M* sodium butyrate were very similar (33).

The biosynthesis of G_{M3} proceeds by the following reaction

CER-Glc-Gal + CMP-NAN → CER-Glc-Gal-NAN + CMP

that is catalyzed by CMP-NAN:GL-2 sialyltransferase (referred to as sialyltransferase I). The activity of this enzyme increased over 20-fold in HeLa cells exposed to 5 m*M* sodium butyrate (Table VIII). Other glycosyltransferase activities did not increase. These included galactosyltransferase I, which catalyzes the formation of GL-2, the acceptor for sialyltransferase I, and a second sialyltransferase, sialyltransferase II. Competition experiments demonstrated that sialyltransferase I activity was distinct from both sialyltransferase II and the sialyltransferase that forms G_{D3} (34). Thus, butyrate specifically induces the synthesis of G_{M3} in HeLa cells.

TABLE VI Incorporation of ^{3}H into G_{M3} of Control and Butyrate-Treated HeLa Cells by Cell Surface Labeling with Sodium Periodate and $NaB[^{3}H]_4$[a]

	Control	+Butyrate
$[^{3}H]$-G_M (cpm/10^7 cells)	44,300	242,300
G_{M3} (nmol/10^7 cells)	0.56	3.22
$[^{3}H]$-G_{M3} (cpm/nmol)	79,000	75,250
G_{M3} (molecules/cell)	34×10^6	194×10^6

[a]HeLa cells were cultured in control and 5 mM sodium butyrate-supplemented medium. After 24 hr, cells were harvested by brief trypsinization, washed with phosphate-buffered saline (pH 7.4), and resuspended in the same buffer containing 10 mM $NaIO_4$. After 10 min at 25°C, oxidation was terminated by adding 10 mM D-glucose. The cells were then washed in buffer, an aliquot analyzed for cell numbers, and the remaining cells extracted for gangliosides. The isolated gangliosides were reduced with $NaB[^{3}H]_4$ (32), separated by thin-layer chromatography, and analyzed for radioactivity and sialic acid as previously described (24).

INDUCTION OF SIALYLTRANSFERASE I ACTIVITY

Following exposure of HeLa cells to butyrate, sialyltransferase I activity increased rapidly after a 3-4 hr lag (Fig. 2). The effect of butyrate was reversible; enzyme activity returned to control levels upon removal of butyrate (Fig. 3) and the half-life of the enzyme appeared to be 6-8 hr. Addition of actinomycin D or cycloheximide completely blocked the increase in enzyme activity suggesting the induction of messenger RNA coding for sialyltransferase I (Fig. 4). Arresting the cells in the S phase of the cell cycle with thymidine or in the *M* phase with Colcemid, however, did not block enzyme induction by butyrate. In addition, when homogenates of control and butyrate-treated cells were admixed and assayed for enzyme activity, there was no evidence of an inhibitor in the control cells or an activator in the butyrate-treated cells (24). Addition of 5 m*M* butyrate directly to the enzyme assay had no effect on enzyme activity.

TABLE VII *Glycohydrolase Activities in Control and Butyrate-Treated HeLa Cells*[a]

Glycohydrolase	Substrate[b]	Control	+Butyrate
		(nmol/mg protein/hr)	
β-Glucosidase	4-MU-β-Glc	48	54
α-Galactosidase	4-MU-α-Gal	34	50
β-Galactosidase	4-MU-β-Gal	104	139
β-N-Acetylgalactosaminidase	4-MU-β-GalNAc	313	354
β-Glucosidase	CER-[^{14}C]Glc	53	46
β-Galactosidase	CER-Glc-[^{3}H]Gal	1.5	1.8
α-Galactosidase	CER-Glc-Gal-[^{3}H]Gal	3.8	5.7
α-Sialidase	CER-Glc-Gal-[^{14}C]NAN	4.2	3.9
α-Sialidase	CER-glc-Gal-GalNAc-Gal (Gal linked to NAN; terminal Gal linked to [^{14}C]NAN)	6.6	6.8

[a]HeLa cells were cultured for 24 hr in control and 5 mM sodium butyrate supplemented media, then harvested, homogenized, and assayed for glycohydrolase activities with either artificial fluorogenic or natural glycolipid substrates. Data from Tallman et al. (33).

[b]Abbreviations: 4-MU, 4-methylumbelliferyl; Glc, glucose; Gal, galactose; GalNAc, N-acetylgalactosamine; NAN, N-acetylneuraminic acid; CER, ceramide.

TABLE VIII Glycosyltransferase Activities in Control and Butyrate-Treated HeLa Cells[a]

Glycosyltransferase	Reaction[b]	Control (pmol/mg protein/hr)	+Butyrate (pmol/mg protein/hr)
Galactosyltransferase I	GL-1 + UDP-Gal → GL-2 + UDP	233	156
Galactosyltransferase II	GL-2 + UDP-Gal → GL-3 + UDP	154	193
Sialyltransferase I	GL-2 + CMP-NAN → G_{M3} + CMP	12	268
Sialyltransferase II	G_{M1} + CMP-NAN → G_{D1a} + CMP	254	274

[a]Data from references (23, 24, 34). Activities represent formation of products from exogenous acceptors.

[b]Abbreviations: GL-1, glucosylceramide; GL-2, lactosylceramide; GL-3, galactosylgalactosylglucosylceramide; G_{M3}, N-acetylneuraminylgalactosylglucosylceramide; G_{M1}, galactosyl-N-acetylgalactosaminyl-[N-acetylneuraminyl]-galactosylglucosylceramide; G_{D1a}, N-acetylneuraminyl-galactosyl-N-acetylgalactosaminyl-[N-acetylneuraminyl]-galactosylglucosylceramide; Gal, galactose; NAN, N-acetylneuraminic acid.

The increase in sialyltransferase I activity depended on the concentration of sodium butyrate (Fig. 5). Even at 0.5 m*M* butyrate, enzyme activity was induced substantially whereas decanoate even at 0.9 m*M* had no effect. Other fatty acids were tested for their effect on enzyme activity (Table IX). Only propionate and pentanoate produced a substantial increase in sialyltransferase I activity; unsaturated analogs of butyrate had no stimulatory effect.

A ROLE FOR CYCLIC AMP?

Cyclic AMP and monobutyryl cyclic AMP had no effect on HeLa cell morphology (22). However, these and other compounds that elevate intracellular cyclic AMP levels caused morphological changes in other cells (as indicated in Table I). Ex-

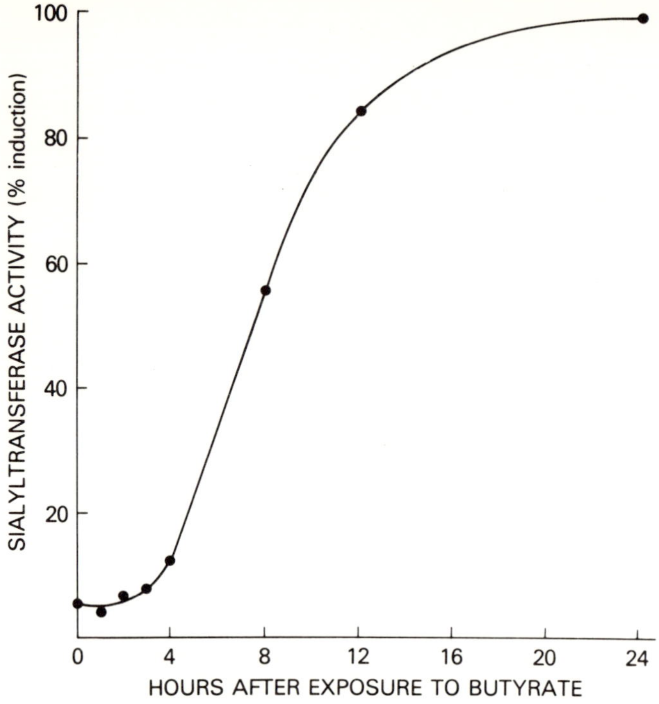

FIGURE 2 Increase of sialyltransferase I activity in HeLa cells after exposure to butyrate. At the designated times, some cultures were harvested and assayed for enzyme activity. Data from Simmons et al. (24).

posure of HeLa cells to compounds of this nature did not induce sialyltransferase I activity (Fig. 6). A slight effect on enzyme activity was seen when cells were exposed to 5 m*M* dibutyryl cyclic AMP. However, HeLa cells can hydrolyze this compound to monobutyryl cyclic AMP and free butyrate (25, 35). As little as 10% breakdown would form sufficient butyrate to induce this level of sialyltransferase I activity (see Fig. 5). The fact that cyclic AMP itself or its monobutyryl derivative did not cause an increase in enzyme activity supports this explanation.

HeLa cells exposed to butyrate showed a small (2- to 3-fold) transient increase in intracellular cyclic AMP levels, which returned to basal values within 24 hr (36). Similar increases in cyclic AMP levels were also observed in HeLa cells exposed to thymidine or decanoate, which also arrests cell growth (36). Since these two compounds induce neither morphological changes nor sialyltransferase I activity, the increase

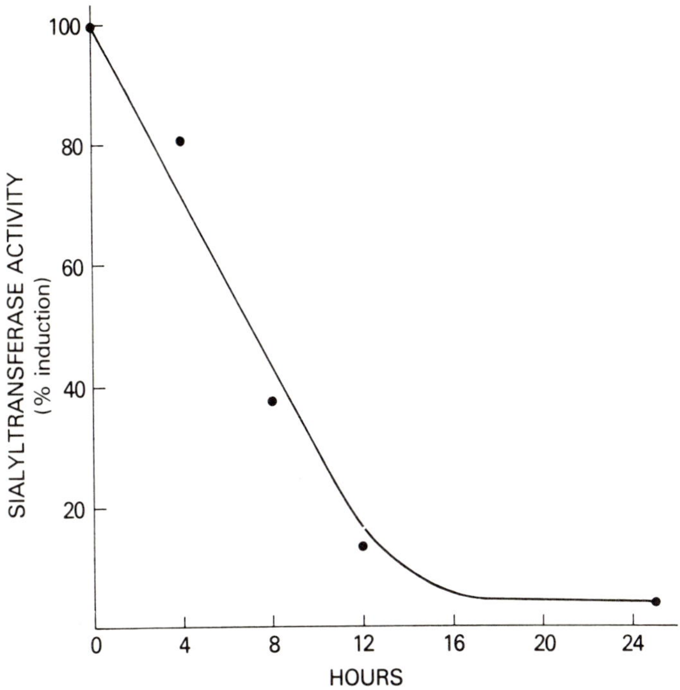

FIGURE 3 Decrease of sialyltransferase I activity in HeLa cells after removal of butyrate. Cells were grown for 12 hr in medium containing 5 mM sodium butyrate. At time 0, the medium was replaced with normal medium and at the designated times, cells were harvested and assayed for enzyme activity. Data from Simmons et al. (24).

of cyclic AMP is not sufficient to explain the specific effects of butyrate.

CORRELATION BETWEEN INDUCTION OF G_{M3} SYNTHESIS AND MORPHOLOGICAL DIFFERENTIATION BY BUTYRATE

The preceding studies indicated a striking correlation between changes in HeLa cell morphology and induction of G_{M3} biosynthesis. In the presence of butyrate, sialyltransferase I activity increased several hours prior to extension of cell processes; upon removal of the inducer, enzyme activity decayed before the cells retracted their neurite-like processes.

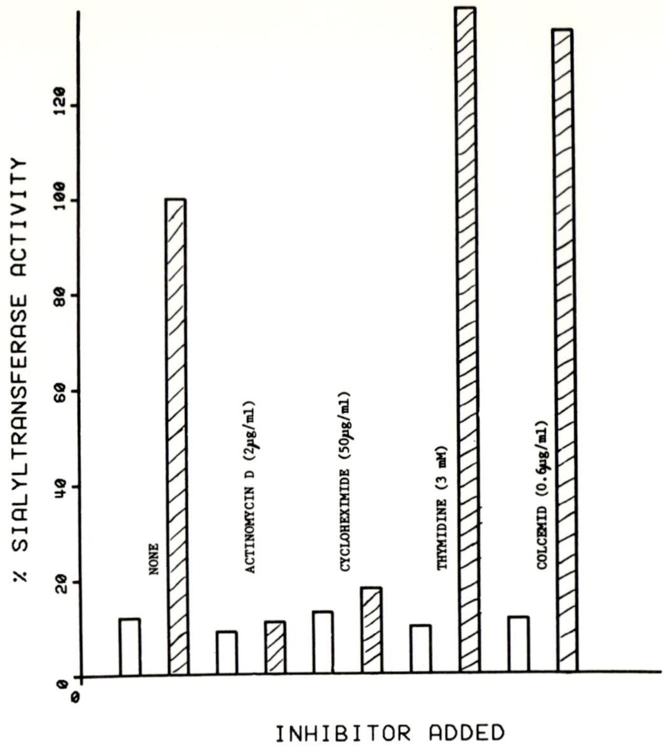

FIGURE 4 Effect of inhibitors of macromolecular synthesis and the cell cycle on induction of sialyltransferase I activity by butyrate. HeLa cells were cultured in the presence of the indicated inhibitor in normal medium (open bars) or medium supplemented with 5 mM sodium butyrate (hatched bars) as follows: for 12 hr with thymidine (3 mM) and Colcemid (0.6 µg/ml) or 16 hr with actinomycin D (2 µg/ml) and cycloheximide (50 µg/ml). Prior to the addition of butyrate, thymidine-treated cells were incubated 24 hr with thymidine (3 mM). Cells were harvested and assayed for enzyme activity. Similar results were obtained with 0.5 µg/ml of actinomycin D and 10 µg/ml of cycloheximide. Data from Simmons et al. (24) and Fishman et al. (23).

Both enzyme induction and shape changes depended on RNA and protein synthesis since both effects of butyrate were blocked by actinomycin D and cycloheximide.

The same three fatty acids, propionate, butyrate, and pentanoate, that induced sialyltransferase I activity caused the cells to form neuritelike processes. Other fatty acids as

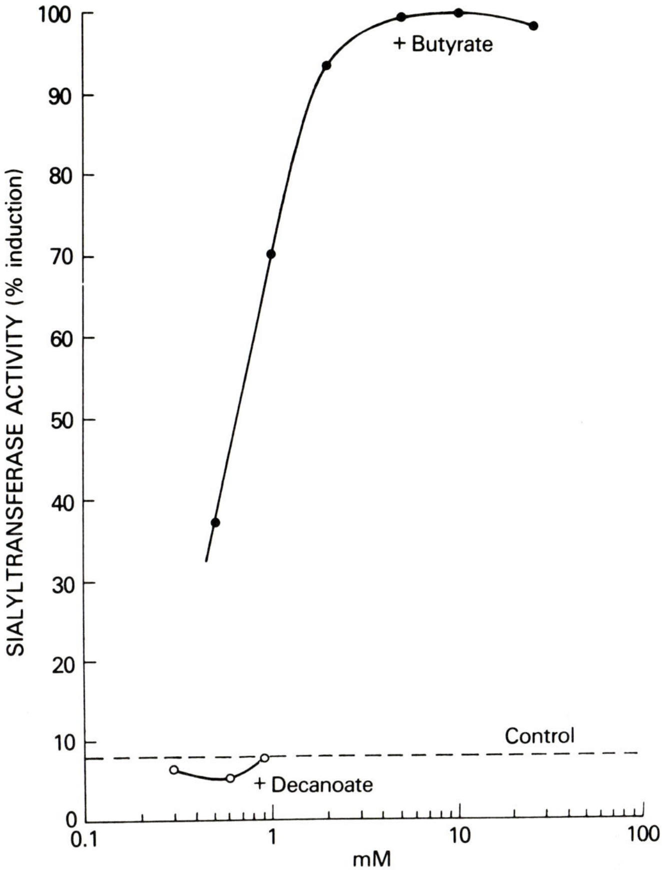

FIGURE 5 Effect of butyrate and decanoate concentrations on sialyltransferase I activity on HeLa cells. Cells were cultured for 12 hr in media containing the indicated concentrations of sodium butyrate (●) or sodium decanoate (o), harvested and assayed for enzyme activity. Activity in untreated cells was also determined (- - -). Data from Simmons et al. (24).

well as many derivatives of butyrate (24) did not cause these changes. Cyclic nucleotides, which can cause shape changes in other cell lines, and agents such as prostaglandins and theophylline, which increase intracellular cyclic AMP levels, did not elevate sialyltransferase I activity or cause morphological changes in HeLa cells.

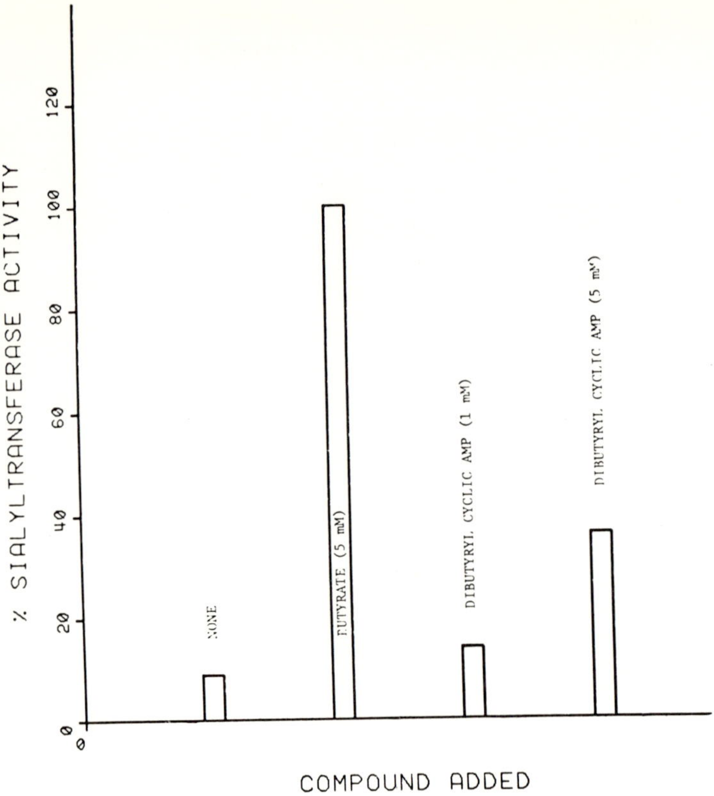

FIGURE 6 Effect of various cyclic AMP agents on sialyltransferase activity in HeLa cells. Cells were cultured for 24 hr in media containing the indicated compound, harvested, and assayed for enzyme activity. Addition of 2 mM cyclic AMP, 0.5 mM N^6-monobutyryl cyclic AMP plus 1 mM theophylline, 10^{-5} M PG_{E1} or 2×10^{-5} M PG_{E2} did not cause an increase in enzyme activity. Data from Tai et al. (36).

In no circumstance did HeLa cells produce the characteristic formation of neuritelike processes observed with butyrate without a corresponding induction of sialyltransferase I activity. Under certain conditions, enzyme activity was induced in the absence of shape changes. Colcemid, which blocks shape changes by preventing microtubule formation, did not inhibit enzyme induction by butyrate. The calcium ionophore A23187 also prevented morphological changes in HeLa at concentrations that did not inhibit sialyltransferase I induction (Table X). Only at higher ionophore concentrations was enzyme

TABLE IX Effect of Fatty Acids on Induction of Sialyltransferase I Activity in HeLa Cells[a]

Fatty acid	Sialyltransferase I activity, relative stimulation[b]	Process formation[c]
None	0	-
Acetate (20 mM)	19	-
Propionate (10 mM)	97	-
Butyrate (1 mM)	67	+
Butyrate (5 mM)	100	+
Crotonate (2 mM)	-12	-
Tetrolate (2 mM)	-7	-
Pentanoate (5 mM)	71	+
Hexanoate (5 mM)	0	-
Octanoate (2 mM)	14	-
Decanoate (0.9 mM)	0	-

[a]Data from Simmons et al. (24). HeLa cells were cultured from 24 hr in medium supplemented with the above fatty acids (as their sodium salts), examined under the microscope, harvested, homogenized, and assayed for enzyme activity.

[b]Relative stimulation = (activity with fatty acid - activity with none)/(activity with butyrate - activity with none).

[c]Morphological changes were determined microscopically as indicated in Fig. 1; (+) process formation, (-) no process formation.

induction prevented. The ionophore presumably causes an increase in intracellular calcium, which in turn prevents the assembly of microtubules and process formation (21). The inhibition of sialyltransferase I induction by higher concentrations of ionophore is unclear but suggests a role for calcium in regulating the induction of this enzyme. At these high ionophore concentrations the cells remained viable, overall protein synthesis was not inhibited, and galactosyltransferase I activity was not reduced (21). The ionophore, which also blocked the morphological changes induced in CHO cells by dibutyryl cyclic AMP, did not inhibit the basal level of sialyltransferase I activity in these cells (21).

Further support for an association between G_{M3} synthesis and shape changes induced in HeLa cells by butyrate is the effect of low concentrations of cycloheximide (29). When butyrate-containing medium was removed and replaced with normal medium containing cycloheximide (<0.5 μg/ml), the cells main-

TABLE X Effect of Butyrate and the Calcium Ionophore A-23187 on HeLa Cell Morphology and Sialyltransferase Activity[a]

Additions to medium			
Butyrate	Ionophore	Sialyltransferase activity (%)	Process formation
-	-	<5	-
3.5 mM	-	100	+
3.5 mM	1.8 μM	100	-
3.5 mM	3.6 μM	18	-
3.5 mM	5.4 μM	<5	-

[a]Data from Henneberry et al. (21). HeLa cells were incubated for 18 hr at 37°C in medium containing the indicated additions. The cells were examined under the microscope for morphological changes, harvested, homogenized, and assayed for sialyltransferase I activity.

tained their neuritelike processes for at least 72 hr. In addition, they maintained their high levels of G_{M3} in the absence of butyrate if cycloheximide was present (29). When cells were labeled with *N*-[acetyl-^{3}H]-D-mannosamine for 24 hr and transferred to fresh medium, cycloheximide prevented any decrease in the specific activity of [^{3}H]-G_{M3} isolated from cells exposed to butyrate during the labeling period (Table XI). However, cycloheximide at these low concentrations did not prevent a decay in induced sialyltransferase I activity nor did it inhibit sialidase activity (Table XII).

The effects of cycloheximide on the maintenance of morphological differentiation and of elevated G_{M3} levels in the absence of butyrate appeared to be specific for this drug as puromycin did not block the cells from reverting to their normal shape or prevent G_{M3} from decaying to control levels (29). These observations strongly implicated an essential role for G_{M3} in the shape changes that HeLa cells undergo in response to butyrate. The mechanism by which cycloheximide exerts its effects is unknown, but it appears to be distinct from its role as an inhibitor of protein synthesis (29). An effect of cycloheximide on calcium fluxes has been reported (37). Experiments with the calcium ionophore suggest that butyrate-treated cells have a lower free intracellular calcium concen-

TABLE XI Effect of Cycloheximide on G_{M3} Metabolism in HeLa Cells[a]

Hours after change to fresh medium	Addition to fresh medium	Specific radioactivity of G_{M3} Control (cpm/nmole)	+Butyrate (cpm/nmole)
0	- - - -	11,800	25,600
24	- - - -	4,500	14,000
24	Cycloheximide	7,650	25,300

[a]Data from Henneberry and Fishman (29). HeLa cells were incubated in medium containing N-[acetyl-^3H]-mannosamine (50 μCi/ml) with or without 5 mM sodium butyrate. After 24 hr, the labeled medium was replaced with fresh medium with or without cycloheximide (0.5 μg/ml). At the indicated times, the cells were harvested, G_{M3} was isolated, and its specific radioactivity was determined.

tration than untreated cells. Cycloheximide may maintain these low calcium levels in the absence of butyrate.

OTHER BIOCHEMICAL CHANGES INDUCED IN HELA CELLS BY BUTYRATE

Butyrate causes in HeLa cells other biochemical changes that do not appear to be directly related to changes in cell morphology. Alkaline phosphatase activity was elevated by butyrate (and pentanoate) and the increase in enzyme activity was inhibited by cycloheximide (25). Ghosh *et al.* also observed changes in cell morphology and correlated them with the increased alkaline phosphatase activity (38). However, a direct correlation is unlikely as the following observations explain. Cortisol and related steroids are potent inducers of HeLa alkaline phosphatase (25, 39) but do not induce the cells to form neuritelike processes (40). When induced by butyrate, alkaline phosphatase activity began to increase at 24 hr at which time the cell processes were fully extended and sialyltransferase I activity was fully induced (Fig. 7). Finally, 11-deoxycorticosterone inhibited the induction of

TABLE XII Effect of Cycloheximide on Sialyltransferase I and G_{M3}-Sialidase Activity in Butyrate-Treated HeLa Cells[a]

Incubation medium			
1st Day	*2nd Day*	*Sialyltransferase I (pmol/mg protein/hr)*	*G_{M3}-sialidase (nmol/mg protein/hr)*
Butyrate	*Butyrate*	*270*	*3.9*
Butyrate	*Control*	*10*	*3.4*
Butyrate	*Cycloheximide*	*26*	*4.5*

[a]Data from Henneberry and Fishman (29). HeLa cells were cultured in control medium or medium containing 5 mM sodium butyrate or 0.5 μg/ml of cycloheximide as indicated. After the 2nd day, the cells were harvested and assayed for enzyme activities. Addition of cycloheximide (0.1-100 μg/ml) directly to the enzyme assays had no significant effect on enzyme activities.

alkaline phosphatase by butyrate, but neither the induction of shape changes nor sialyltransferase I activity (40). A more plausible hypothesis is that an additional effect of butyrate unrelated to changes in cell shape and G_{M3} content is to induce alkaline phosphatase activity by a mechanism analogous to the induction of alkaline phosphatase by steroids.

The effect of butyrate on cell surface glycoproteins of HeLa cells has also been examined (41) by means of established procedures (42, 43). Untreated cells were grown in medium containing [^{14}C]-fucose and butyrate-treated cells in medium containing [^{3}H]-fucose. Surface glycoproteins were removed from the cells with trypsin and digested to glycopeptides with pronase. The glycopeptide digests from the two cell cultures were co-chromatographed on a Bio-Gel P-10 column (Fig. 8). Glycopeptides derived from the surfaces of butyrate-treated cells were enriched in earlier eluting components compared to glycopeptides from untreated cells. When the glycopeptides were digested first with purified neuraminidase and then chromatographed, the elution profiles were virtually identical and all of the radioactivity peaks eluted later from the column (41). This result suggested that the glycopeptides

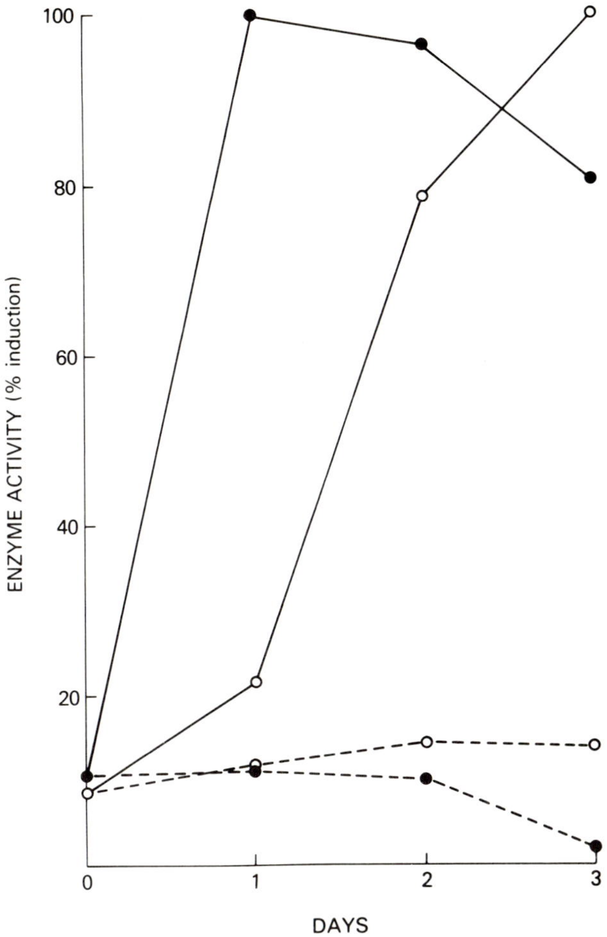

FIGURE 7 Induction of sialyltransferase I and alkaline phosphatase activity in HeLa cells by butyrate. Cells were grown for 24 hr in normal media and at day 0, the media was replaced with normal medium (----) or medium containing 5 mM sodium butyrate (——). At indicated times, cells were harvested and assayed for sialyltransferase I activity (●) and alkaline phosphatase activity (o). Data from Atikkan et al. (40).

contained sialic acid and that the difference between glycopeptides derived from the surfaces of control and butyrate-treated cells resided partially in the amount of sialic acid they contained.

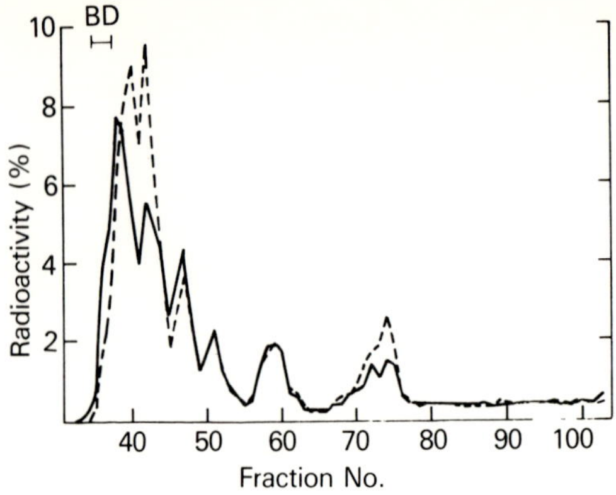

FIGURE 8 Elution profiles of fucose-labeled surface glycopeptides from untreated and butyrate-treated HeLa cells cochromatographed on a Bio-Gel P-10 column. Cells were grown with [^{14}C]-fucose, 1 μCi per ml of medium (---), or [^{3}H]-fucose, 1.5 μCi per ml of medium containing 5 mM sodium butyrate (——). After 24 hr, surface components were removed with trypsin (1 mg/ml for 15 min at 37°C), separated from the cells by centrifuging, and further digested with pronase as described elsewhere (41, 42). The glycopeptide fractions were admixed and chromatographed on a 1 × 100 cm column of Bio-Gel P-10. Fractions of 0.7 ml were collected and analyzed for ^{3}H and ^{14}C; and the data were plotted by computer (41, 42). Data from Fishman and Henneberry (41). BD, blue dextran 2000 containing fractions.

Other investigators have observed differences in the surface glycopeptides derived from normal and transformed cells, growing and nongrowing cells, or differentiated and undifferentiated cells (42-49). These differences in glycopeptide elution profiles were minimized or eliminated by prior digestion of the glycopeptides with neuraminidase (42-46 and 48). Thus, these differences are partially due to differences in the sialic acid content of the glycopeptides (42, 43, 48). However, in contrast to our results, transformed, growing, or less differentiated cells are enriched in the earlier eluting, more sialylated glycopeptides compared to their untransformed, nongrowing, or more differentiated counterparts (42-49).

The differences in glycopeptide elution patterns between control and butyrate-treated HeLa cells appeared to be confined to surface membrane components. When the cells from which the surface material had been removed by trypsin were digested with pronase and the glycopeptides were co-chromato-

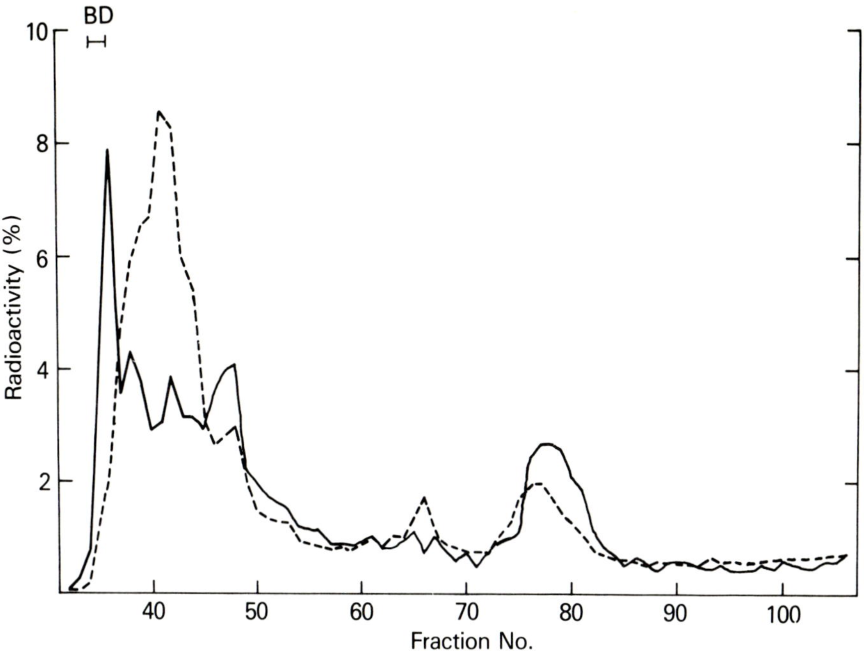

FIGURE 9 Elution profiles of fucose-labeled surface glycopeptides from untreated and decanoate-treated HeLa cells chromatographed as a Bio-Gel P-10 column. Details are the same as in Fig. 8 except HeLa cells were grown for 24 hr with [^{14}C]-fucose (---) or [^{3}H] fucose and 0.5 mM sodium decanoate (——). BD, blue dextran 2000 containing fractions. Data from Fishman and Henneberry (41).

graphed, the elution profiles were very similar (41); the glycopeptides derived from butyrate-treated cells were not enriched in earlier eluting material.

Finally, glycopeptides derived from the surfaces of HeLa cells exposed to decanoate were co-chromatographed with glycopeptides from untreated cells (Fig. 9). Again, the glycopeptides from the decanoate-treated cells were enriched in earlier eluting material. Thus, both butyrate and decanoate caused a similar change in the surface glycoproteins of HeLa cells. However, as we had indicated earlier, decanoate did not induce the cells to form neuritelike processes, whereas butyrate does. Both fatty acids inhibited cell growth and caused a small increase in intracellular cyclic AMP levels. The changes in surface glycoproteins might be a consequence of other inhibitory effects of fatty acids on HeLa cells. If surface glycoproteins should have a role in the morphological differentiation of HeLa, it is secondary to some unique function that butyrate, but not decanoate, induces. The most likely candidate for this unique function is the ganglioside G_{M3}.

ROLE OF G_{M3} IN MORPHOLOGICAL DIFFERENTIATION

Although we can demonstrate an association between the increased G_{M3} synthesis and the shape changes in HeLa cells induced by butyrate, we as yet do not know the nature of this association. One possible mechanism is suggested by the interaction of cholera toxin with its membrane receptor, the ganglioside G_{M1}. Cholera toxin binds to and induces a redistribution (patching and capping) of the G_{M1} receptors (50-52). Redistribution of receptors is inhibited by concanavalin A (50) and colchicine and cytochalasin B, which promote disassembly of microtubules and microfilaments, respectively (50-52). In addition, cholera toxin induces the redistribution and capping of anti-immunoglobulin receptors (52). These observations indicate that gangliosides can move in the lateral plane of the membrane with involvement of microfilaments and microtubules. In addition, gangliosides appear to be linked to other membrane components such as receptors for Con A and anti-immunoglobulin.

This phenomenon is reminiscent of the distribution, mobility, and interaction of membrane components in lymphocytes and erythrocytes (53). Based on studies with Con A, Edelman has proposed that lectin and antibody receptors are linked directly or indirectly to a common assembly composed of colchicine binding proteins (presumably microtubules), which also can interact with microfilament systems (54). Thus, the distribution and mobility of surface receptors is controlled by the transmembrane interaction with microtubule and microfilament systems (53, 54). Nicolson has proposed a similar transmembrane control mechanism in the erythrocyte (53, 55). The erythrocyte contains integral glycoproteins such as glycophorin that span the membrane and peripheral proteins such as spectrin that are associated with the inner surface of the membrane. Membrane-intercalated particles appear to be composed of glycophorin, spectrin, other proteins, and lipids and are connected to actinlike components. A network of spectrin and actin controls the distribution and mobility of the particles. Using cross-linking reagents, Ji demonstrated specific associations between membrane proteins and glycolipids in the erythrocyte (56). A similar association between membrane proteins and ganglioside G_{M1} in lymphocytes could explain the redistribution and capping of receptors by cholera toxin (50).

In HeLa cells, G_{M3} may be part of an analogous system involving microtubule and microfilament networks. G_{M3} may be a component of a "seed" point for the attachment of microtubules (and microfilaments) to the plasma membrane. These attachment sites could be membrane-intercalated particles composed of proteins, lipids, and ganglioside. Newly synthesized G_{M3} could complete and/or stabilize existing particles or induce the formation of new particles from existing membrane components. Butyrate would cause the cells to extend processes by inducing G_{M3} synthesis and lowering free intracellular calcium levels, or by elevating cyclic AMP levels. An interrelationship between Ca and cyclic AMP has been suggested (57, 58) and cyclic AMP may be involved in microtubule assembly (59, 60). The increase in G_{M3} may lead to the formation of more microtubule attachment sites to which microtubule assembly proteins could bind. The reduction in calcium concentration may promote the polymerization of tubulin and attachment of the microtubules to the membrane may result in extension of cell processes.

CONCLUSIONS

Exposure of HeLa cells to millimolar concentrations of sodium butyrate causes numerous morphological and biochemical changes in the cells. The cells extend long neuritelike processes and synthesize new surface glycoconjugates. Process formation depends on the assembly of microtubules and involves low but perhaps varying intracellular calcium levels. Whereas changes in surface glycoproteins can be induced by agents that are not capable of causing morphological changes, the increase in ganglioside G_{M3} synthesis appears to be uniquely associated with the morphological changes induced by butyrate.

The elevated G_{M3} content in butyrate treated cells is due to the specific induction of sialyltransferase I, the biosynthetic enzyme for G_{M3}. Other glycosyltransferases and glycohydrolases, including G_{M3}-sialidase, are not affected by butyrate treatment. Enzyme induction and cell shape changes depend on RNA and protein synthesis and are only caused by butyrate (and propionate and pentanoate). Finally, in the presence of low concentration of cycloheximide, butyrate-treated cells can maintain their processes and high content of G_{M3} in the absence of butyrate.

We can only speculate on the mechanism by which butyrate induce differentiation in HeLa cells and the role that G_{M3} plays in the formation of cellular processes. However, the close association between these two phenomena in HeLa cells provides an excellent model for investigating the function of one class of surface glycoconjugates, the gangliosides.

REFERENCES

1. Winzler, R. J. (1970), *Int. Rev. Cyto. 29*, 77-125.
2. Nicolson, G. L. (1974), *Int. Rev. Cyto. 39*, 89-190.
3. Van Heyningen, W. E. (1974) *Nature 249*, 415-417.
4. Cuatrecasas, P. (1973), *Biochemistry 12*, 3547-3558.
5. Moss, J., Fishman, P. H., Manganiello, V. C., Vaughan, M., and Brady, R. O., (1976), *Proc. Nat. Acad. Sci. U.S.A. 73*, 1034-1037.
6. Winand, R. J., and Kohn, L. D. (1975), *J. Biol. Chem. 250*, 6534-6540.
7. Mullin, B. R., Fishman, P. H., Lee, G., Aloj, S. M., Ledley, F. D., Winand, R. J., Kohn, L. D., and Brady, R. O. (1976), *Proc. Nat. Acad. Sci. U.S.A. 73*, 842-846.
8. Roseman, S. (1970), *Chem. Phys. Lipids 5*, 270-297.
9. Shur, B. D., and Roth, S. (1975), *Biochim. Biophys. Acta 415*, 473-512.
10. Rosen, S. D., Reitherman, R. W., and Barondes, S. H. (1975), *Exp. Cell Res. 95*, 159-166.
11. Fishman, P. H. (1976), *in* "Cell Membranes and Viral Envelopds," (Blough, H. A., and Tiffany, J. M., eds.), Academic Press, New York.
12. Brown, J. C. (1971), *Exp. Cell Res. 69*, 440-442.
13. Seeds, N. W., Gilman, A. G., Amano, T., and Nirenberg, M. W. (1970), *Proc. Nat. Acad. Sci. U.S.A. 66*, 160-167.
14. Prasad, K. N. (1975), *Biol. Rev. 50*, 129-165.
15. Glazer, R. I., and Schneider, F. H. (1975), *J. Biol. Chem. 250*, 2745-2749.
16. Laug, W. E., Jones, P. A., Nye, C. A., and Benedict, W. F. (1976), *Biochem. Biophys. Res. Commun. 68*, 114-119.
17. Prasad, K. N., and Kumar, S. (1974), *in* "Control of Proliferation in Animal Cells," Cold Spring Harbor Conferences on Cell Proliferation, Vol. 1 (Clarkson, B., and Baserga, R., eds.), pp. 581-594. Cold Spring Harbor Laboratories, Cold Spring Harbor, New York.
18. Sheu, C. W., Salomon, D., Simmons, J. L., Sreevalson, T., and Freese, E. (1975), *Antimicrob. Agents and Chemother. 7*, 349-363.

19. Hsie, A. W., and Puck, T. T. (1971), *Proc. Nat. Acad. Sci. U.S.A. 68*, 358-361.
20. Wright, J. A. (1973), *Exp. Cell Res. 78*, 456-460.
21. Henneberry, R. C., Fishman, P. H., and Freese, E. (1975), *Cell 5*, 1-9.
22. Ginsburg, E., Salomon, D., Sreevalsan, T., and Freese, E. (1973), *Proc. Nat. Acad. Sci. U.S.A. 70*, 2457-2461.
23. Fishman, P. H., Simmons, J. L., Brady, R. O., and Freese, E. (1974), *Biochem. Biophys. Res. Commun. 59*, 292-299.
24. Simmons, J. L., Fishman, P. H., Freese, E., and Brady, R. O. (1975), *J. Cell Biol. 66*, 414-424.
25. Griffin, M. J., Price, G. H., Bazzell, K. L., Cox, R. P., and Ghosh, N. K. (1974), *Arch. Biochem. Biophys. 164*, 619-623.
26. Ghosh, N. K., and Cox, R. P. (1976), *Nature 259*, 416-417.
27. Leder, A., and Leder, P. (1975), *Cell 5*, 319-322.
28. Reuben, R. C., Wife, R. L., Breslow, R., Rifkind, R. A., and Marks, P. A. (1976), *Proc. Nat. Acad. Sci. U.S.A. 73*, 862-866.
29. Henneberry, R. C., and Fishman, P. H. (1976), *Exp. Cell Res., 103*, 55-62.
30. Fishman, P. H., and Brady, R. O. (1976), *Science, 194*, 906-915.
31. Liao, T. H., Gallop, P. M., and Blumenfeld, O. O. (1973), *J. Biol. Chem. 248*, 8247-8253.
32. Suzuki, Y., and Suzuki, K. (1972), *J. Lipid Res. 13*, 687-690.
33. Tallman, J. T., Fishman, P. H., and Henneberry, R. C. (1977), *Arch. Biochem. Biophys.*, in press.
34. Fishman, P. H., Bradley, R. M., and Henneberry, R. C. (1976), *Arch. Biochem. Biophys. 172*, 618-626.
35. Kaukel, E., and Hilz, H. (1972), *Biochem. Biophys. Res. Commun. 46*, 1011-1018.
36. Henneberry, R. C., Tai, C. J., Tallman, J. T., and Fishman, P. H., manuscript in preparation.
37. Luttge, U., Lauchli, A., Ball, E., and Pitman, M. G. (1974), *Experentia 30*, 470-471.
38. Ghosh, N. K., Deutsch, S. I., Griffin, M. J., and Cox, R. P. (1975), *J. Cell Physiol. 86*, 663-672.
39. Griffin, M. J., and Cox, R. P. (1966), *Proc. Nat. Acad. Sci. U.S.A. 56*, 946-953.
40. Atikkan, E., Henneberry, R. C., Fishman, P. H., and Freese, E., manuscript in preparation.
41. Fishman, P. H., and Henneberry, R. C., manuscript in preparation
42. Fishman, P. H., Brady, R. O., and Aaronson, S. A. (1976), *Biochemistry 15*, 201-208.

43. Warren, L., Fuhrer, J. P., Buck, C. A., and Walborg, E. F. (1974), *in* "Membrane Transformations in Neoplasia," Miami Winter Symposia Vol. 8 (Schultz, J., and Block, R., eds.), pp. 1-26, Academic Press, New York.
44. Buck, C. A., Glick, M. C., and Warren, L. (1971), *Biochemistry 10*, 2176-2180.
45. Buck, C. A., Fuhrer, J. P., Soslau, G., and Warren, L. (1974), *J. Biol. Chem. 249*, 1541-1550.
46. Van Beek, W. P., Smets, L. A., and Emmelot, P. (1973), *Cancer Res. 33*, 2913-2922.
47. Muramatsu, T., Atkinson, P. H., Nathenson, S. G., and Ceccarini, C. (1973), *J. Mol. Biol. 80*, 781-799.
48. Ceccarini, C., Muramatsu, T., Tsang, J., and Atkinson, P. H. (1975), *Proc. Nat. Acad. Sci. U.S.A. 72*, 3139-3143.
49. Glick, M. C., Kimhi, Y., and Littauer, U. Z. (1973), *Proc. Nat. Acad. Sci. U.S.A. 70*, 1682-1687.
50. Craig, S. W., and Cuatrecases, P. (1975), *Proc. Nat. Acad. Sci. U.S.A. 72*, 3844-3848.
51. Revesz, T., and Greaves, M. (1975), *Nature 257*, 103-106.
52. Sedlacek, H. H., Stark, J., Seiler, F. R., Zeigler, W., and Wiegandt, H. (1976), *FEBS Lett. 61*, 272-276.
53. Nicolson, G. L. (1976), *Biochim. Biophys. Acta 457*, 57-108.
54. Edelman, G. M. (1974), *in* "Control of Proliferation in Animal Cells," Cold Spring Harbor Conferences on Cell Proliferation Vol. 1 (Clarkson, B., and Baserga, R. eds.), pp. 357-377, Cold Spring Harbor Laboratories, Cold Spring Harbor, New York.
55. Nicolson, G. L., and Painter, R. G. (1973), *J. Cell Biol. 59*, 395-406.
56. Ji, T. H. (1974), *J. Biol. Chem. 249*, 7841-7847.
57. Rasmussen, H., Goodman, D. B., and Tenehouse, A. (1972), *Critical Rev. Biochem. 1*, 95-148.
58. Durham, A. C. H. (1974), *Cell 2*, 123-136.
59. Shelanski, M. L. (1973), *J. Histochem. Cytochem. 6*, 529-539.
60. Brinkley, B. R., Fuller, G. M., and Highfield, D. P. (1975), *Proc. Nat. Acad. Sci. U.S.A. 72*, 4981-4985.

SUGAR COMPOSITION OF MAMMALIAN CELL SURFACE MEMBRANE: A FUNCTION OF CARBON SOURCE

DOMENIC PAOLINI
LINDA HUMPHREY
FEDERICO GONZALEZ
JOSEPH COLOFIORE
PETER ROSSOW
HAROLD AMOS

Department of Microbiology and Molecular Genetics
Harvard Medical School
Boston, Massachusetts

MORRIS J. KARNOVSKY

Department of Pathology
Harvard Medical School
Boston, Massachusetts

INTRODUCTION

Of primary interest to us has been the potential influence of metabolizable sugars upon the structural sugar components of mammalian cells. We have chosen to define as structural sugars the oligosaccharides that are identified with glycolipids, glycoproteins, and glycosaminoglycans.

Evidence from studies on fungi and yeast, including *Neurospora* and Saccharomyces, offers support to the possibility that external sugars can alter the proportions of sugar components of the cell wall and of certain exocellular glycopeptides.

Perhaps the first observations linking central sugar metabolism to the composition of the cell wall were those of Tatum and his collaborators with *Neurospora crassa*. A large series of mutants identified by changes in colony morphology were revealed to have significant alterations in cell wall structure (1-3). Biochemical analysis established the site of virtually all mutants in the series to be the genes coding for three central enzymes in hexose metabolism, glucose-6-phosphate dehydrogenase, phosphoglucomutase, and 6-phosphogluconic acid dehydrogenase (5-7). In all cases the mutation did not totally eliminate the enzyme in question, but altered its physical properties and reduced activity to 10 to 6% of wild type. The reduction of enzyme activity was always accompanied by large shifts, as much as 10- to 15-fold, in the pool sizes of intermediates such as glucose-1-phosphate (G-1-P) and glucose-6-phosphate (G-6-P). Even more directly identified with glycosylation, the uridine diphosphoglucose (UDPG) pools varied as much as twofold in the mutant cells. Other nucleotide sugars were not examined.

The same types of colonial and cell wall variations were obtained when wild type *Neurospora* were grown on medium supplemented with sorbose (4). The addition of sorbose did not alter the G-1-P or G-6-P pools but lowered by nearly 50% the UDPG pool. Glucose and glucosamine are the principal constituents of the *Neurospora* cell wall, accounting for 60% of it. When analysis of the sugar composition of the cell wall was made, major changes were noted in the ratio of glucosamine to glucose in both the mutants and the wild type grown on sorbose-containing medium. Finally, the cell wall proved to be a complex structure of four polysaccharide or glycopeptide laminarlike polymers. The proportions of these polymers were altered consistently in the mutants and in the wild type grown on sorbose (4).

A further case of interest is that reported by Drewes and Gander (8) concerning the synthesis of an exocellular glycopeptide by *Penicillium charlesii*. The glycopeptide in question has a mannan backbone and 8 to 10 side chains of galactofuranosyl residues. Mutants deficient in UDP-D-galactose 4'-epimerase grow normally on glucose, but synthesize a peptidophosphogalactomannan (PPGM) with as little as 10% of the wild type galactose content.

We wish to report changes in the sugar composition of membrane preparations from mammalian cells grown on naturally occurring sugars — D-galactose, D-fructose, D-mannose, and D-xylose — in lieu of glucose. It is our working hypothesis that growth on the alternative sugars causes significant perturbations in the nucleotide sugar pools as well as in other intermediates of sugar metabolism. These changes either directly or indirectly modify glycosylation to produce consistent differences in the sugar composition of glycopeptides and glycolipids of the cell.

Our case is built upon evidence gained from monitoring nucleotide sugar synthesis and from observations on the morphology and certain membrane-associated biochemical properties of chick embryo fibroblasts (CEF), Ehrlich ascites cells (EA), hamster fibroblasts (Nil), and Chinese hamster ovary cells (CHO).

METHODS AND MATERIALS

CELLS AND CULTURE TECHNIQUES

The Nil 1Cl hamster cells were obtained through the courtesy of Dr. P. W. Robbins of MIT. They were grown from small inoculum in Eagle's MEM (GIBCO) containing four times the normal concentration of essential and nonessential amino acids and vitamins and 10% fetal calf serum. The medium contained no glucose or pyruvate. Sugars were added as desired. Cells were normally cultivated in Falcon 100 mm plastic Petri dishes in a CO_2 (5%) incubator at 36 to 38°C. Cells were passaged by trypsinization (0.05% trypsin in 0.2% EDTA) and replated at 10^4 cells/cm^2.

Primary chick embryo (CEF) monolayer cultures were prepared by trypsinization of 10-12 day embryos. Culture vessels, whether bottles or shell vials, were inoculated to obtain initial cell densities of 10^4 to 5×10^4 cells/cm^2. If the monolayers were to be cultivated in a 5% CO_2 atmosphere, the medium contained 25 m*M* $NaHCO_3$, whereas those to be incubated without CO_2 contained HEPES or TES buffer (0.01 *M*) without $NaHCO_3$. Secondary cultures were prepared by trypsinization of primaries and redistribution of the cells into fresh medium at the inoculum densities cited above.

Ehrlich ascites tumor cells, obtained from Dr. E. Kaminskas (Department of Medicine, University of Wisconsin Medical School, Milwaukee) were grown in spinner flasks (Bellco Glass, Inc.) at 37°C. The growth medium consisted of Joklik's modified Eagle's minimum essential medium. The medium, obtained in powder form, was sugar free. Glucose or other sugars, as desired, were added at a concentration of 1.0 mg/ml (approximately 5.5 m*M*). Calf serum or dialyzed calf serum (GIBCO) was provided. The cell concentration was maintained at 5 to 10×10^5 cells/ml by replacing half the medium daily with fresh medium.

CHO cells were obtained from Dr. Alice S. Huang (Harvard Medical School). They were grown as monolayer cultures in Dulbecco's modified Eagle's MEM with 10% dialyzed calf or fetal calf serum and sugars as required.

PROCEDURE FOR PULSE LABELING OF NUCLEOTIDE SUGARS

Monolayer cultures were used exclusively for the estimation of carbon flow into nucleotide sugars using ^{14}C-glucose and ^{14}C-galactose. Cells were grown to near-confluency, washed, and refed with fresh medium containing sugars or other additions as indicated. The cells were reincubated for an additional 24 hr or for a shorter period. The 20-min pulse of radiolabeled sugar was administered after washing the monolayer with warm phosphate-buffered saline (PBS), (0.16 *M* NaCl in 0.005 *M* KPO_4 buffer, pH 7.0). (U-^{14}C)-glucose or galactose 4×10^{-5} *M* 132 Ci/ml was provided in the buffer for 20 min at 37°C. The cultures were then washed with ice-cold PBS and killed with 70% ethanol acidified to pH 3.5 with HCl. The ethanol-soluble cell extracts were then spotted on sheets of Whatman #1 paper and chromatographed descending, for 8 hr, in the Paladini-Leloir solvent system (7.5 vol. 95% EtOH; 3 vol. 1 *M* acetate buffer pH 3.5) (9). The chromatograms were then cut into strips and scanned in a Packard strip-scanner. Protein concentration of the ethanol-insoluble fraction estimated by the method of Lowry *et al.* (10) was considered to reflect the cell density of the monolayer.

COVER SLIP CULTURES FOR SCANNING ELECTRON MICROSCOPY

Nil cells were seeded in glucose-containing MEM into Leighton tubes containing cover slips at 2-4 × 10^5 cells/ml. Glass cover slips (4 cm × 1.1 cm), washed, soaked in 95% ethanol, and liberally rinsed with distilled water, were sterilized by autoclaving in the Leighton tubes. After 24 hr incubation at 37°C (in CO_2 incubator) the monolayers were rinsed in sugar-free Kitos-Waymouth medium and provided fresh medium containing appropriate sugars. Under these conditions of plating, an initial cell density of approximately 10^4 cells/cm^2 was usually achieved. The Leighton tubes were incubated in a CO_2 incubator at 37°C and fixed for scanning microscopy and for staining with hematoxylin and Eosin when the cultures appeared to be about 1/3 confluent. The desired level of confluency was usually achieved within 2, 3, or 4 days.

SCANNING MICROSCOPY

The cover slips were rinsed twice in buffered Hanks solution and then fixed for one hour at room temperature in 2.5% glutaraldehyde buffered to pH 7.2 with 0.1 *M* cacodylate buffer. After washing several times with cacodylate buffer, the cells were dehydrated in graded concentrations of acetone. From absolute acetone they were subjected to critical point drying with CO_2 and were then coated with paladium-gold under vacuum, using a rotary table to ensure even coating.

The specimens were then mounted on stubs and examined in an ETEC Autoscan scanning microscope at an accelerating voltage of 20 kv and an angle of tilt ranging from 25 to 45°.

ISOLATION OF CELL MEMBRANES FROM NIL HAMSTER FIBROBLASTS

Nil cell monolayers near confluency were washed twice with ice-cold PBS, harvested by scraping gently with a rubber policeman, and centrifuged at 1000 rpm for 5'. The cell pellet was resuspended in 15 to 40 ml of 10^{-3} *M* tris buffer at pH 7.8 containing 5×10^{-4} *M* $ZnCl_2$ for 15 min at room temperature and then cooled in ice for an additional 20 min. The cells were then ruptured using a Dounce homogenizer with a tight-fitting (type B) pestle. The homogenization was monitored using phase-contrast microscopy: about 80 to 100 strokes were needed to satisfactorily rupture the cells. Plasma membranes were isolated using the method of Brunette and Till (24).

The homogenate was centrifuged for 15' at 1400 rpm in the 269 head of an International Centrifuge. The pellet from centrifugation was then mixed in a two-phase system that contained equal volumes of top and bottom phases of a two-phase partition system and centrifuged for 10' at 8500 rpm in a Sorvall HB4 (swinging bucket) rotor. The membranes that separated at the interface of the two phase system were drawn off with a pasteur pipette, diluted with 5 volumes of distilled water, and spun at 8500 rpm for 10'. The procedure was repeated twice in order to remove nuclear contamination. The final membrane pellet was suspended in water, lyophylized, and stored at -20^oC.

ANALYSIS OF THE SUGAR COMPOSITION BY GAS-LIQUID CHROMATOGRAPHY

Neutral and amino sugars were determined as trimethylsilyl derivatives according to the method of Reinhold (25).

Lyophilized membranes were placed in a teflon-lined minireaction vessel (Supelco) with arabitol as internal standard and methanolyzed with 1 ml of 1 *M* anhydrous methanolic HCl at

80°C for 20 hr under N_2. The sample was taken to dryness under a stream of nitrogen and the amino sugars were re-*N*-acetylated by suspending the residue in 0.1 ml of pyridine followed by 0.1 ml of acetic anhydride. The sample was again taken to dryness and de-O-acetylation achieved by heating with 0.5 *M* methanolic HCl for 1 hr at 65°C. After solvent evaporation, the samples were dried *in vacuo* over KOH pellets and P_2O_5. The methyl glycosides were converted to their trimethylsilyl ethers using Tri-Sil, a commercially prepared silylating agent (containing pyridine, hexamethyldisilazane, and trimethylchlorosilane). One to two μl aliquots were injected directly into the gas chromatograph.

The sugar components were determined using a Perkin-Elmer model 900 gas chromatograph equipped with flame ionization detectors, and a column of O.V. 17 (Supelco, Inc., Bellefonte, Pennsylvania).

RESULTS

GROWTH OF CELLS ON NATURALLY OCCURRING SUGARS IN LIEU OF GLUCOSE

Several mammalian cell lines have been grown in media containing D-mannose, D-xylose, D-galactose, or D-fructose in lieu of glucose. Of these, CHO cells, Nil hamster fibroblasts, human skin fibroblasts, and mouse primary and secondary fibroblasts grow well on each of the four sugars. The results for Nil cells are presented in Fig. 1. From a small inoculum in Eagle's MEM the cells grow at somewhat different rates and cease growing at different densities. These results might have been anticipated from observations made by Cox and Gesner (11, 12) with several mammalian cell lines. Their experiments were all conducted in a medium containing 27.5 m*M* glucose to which some of the same sugars were added as supplements. They reported that under those conditions significant differences in the density at which cells ceased multiplying and changes in cell morphology as recorded by light microscopy were noted. L-fucose and D-mannose in their hands produced consistent and very marked effects, although most were observed at concentrations of sugar in excess of 50 m*M*.

Our results have been obtained with "physiologic" concentrations of sugars (1-5.5 m*M*) and in the virtual absence of glucose. Whereas D-mannose under these conditions (including 10% dialyzed calf serum) supports very well the replication of cells, the cell density at confluence is always lower than

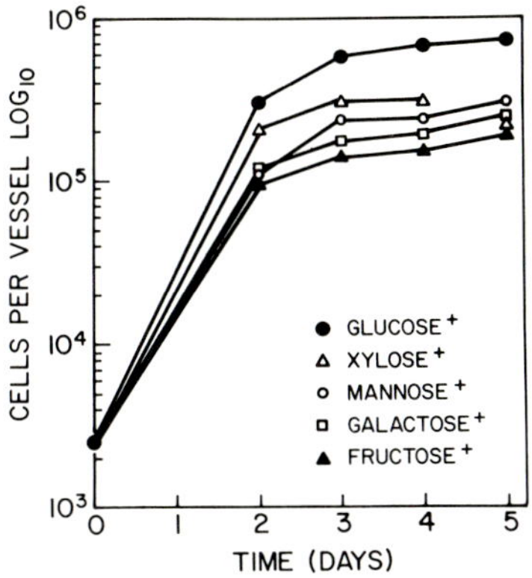

FIGURE 1 Growth of Nil cells on single sugars substituted for glucose. Cells were plated in MEM with 10% dialyzed calf serum and single sugars as indicated. Cells remained in the same medium for the duration of the experiment and were trypsinized and counted in an electronic cell counter. Each point is the average of triplicate cultures.

that achieved with glucose. L-fucose alone is toxic to cells and can be used only in concert with another supportive sugar.

We are particularly interested in the growth characteristics of cells given D-xylose or xylitol as the sole carbon source. Virtually all lines and strains tested grow well on these compounds with little or no lag on subculture from glucose. D-xylose is not transported specifically by mammalian or chick cells in culture but enters the cell by simple diffusion (13). Activity of three enzymes in the xylitol pathway is detectable in extracts of cells grown for generations on glucose, while a three- to six-fold derepression is observed in cells shifted to D-xylose or xylitol (14). L-xylose is not used to support growth probably because the cells do not have an isomerase to convert L-xylose to L-xylulose, which is an intermediate in glucuronic acid metabolism in mammalian cells. L-xylulose has not been available for test purposes.

MORPHOLOGIC DIFFERENCES INDUCED BY SUGARS

As we reported earlier for L-cells (15), cells grown on glucose and shifted while still sparse to media containing other sugars in lieu of glucose, alter their morphology relatively rapidly. If the shift is made to single sugars, the

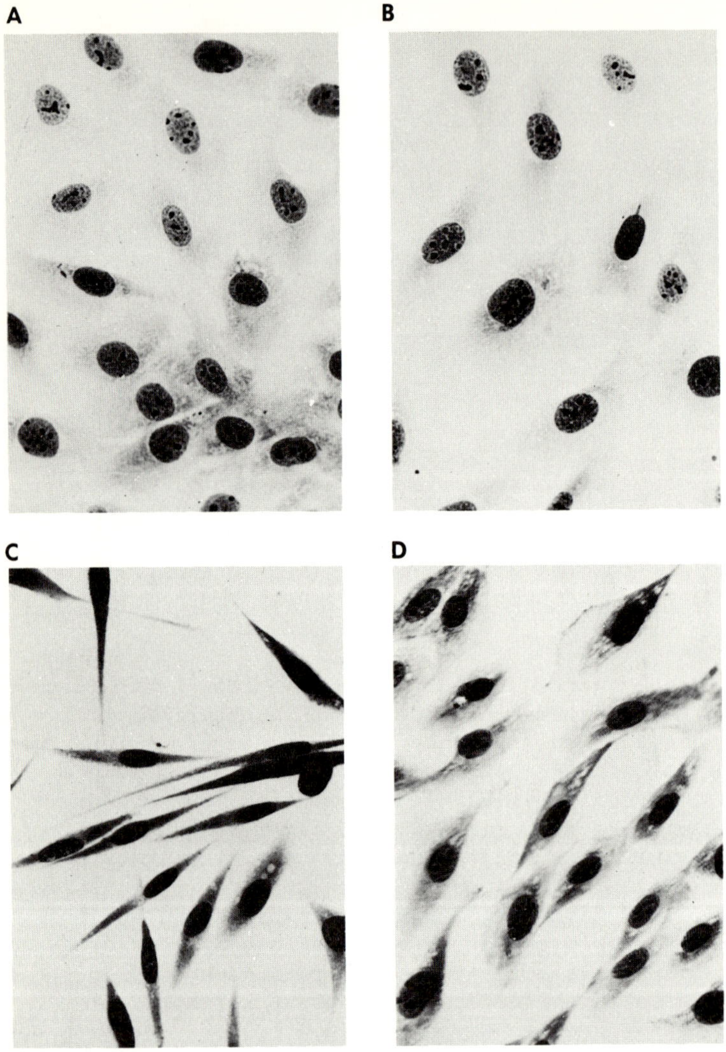

FIGURE 2 Morphology of cells grown on sugars substituted for glucose. Nil cells were plated on coverslips and grown for 48 hr on glucose before transfer to fresh medium with dialyzed calf serum and sugars indicated. After a further 48 hr of incubation at 37°C, the cells were fixed and stained with hematoxylin and eosin stains. 400× magnification by light microscopy. (A) Cells grown on glucose, (B) mannose, (C) galactose, (D) D-xylose.

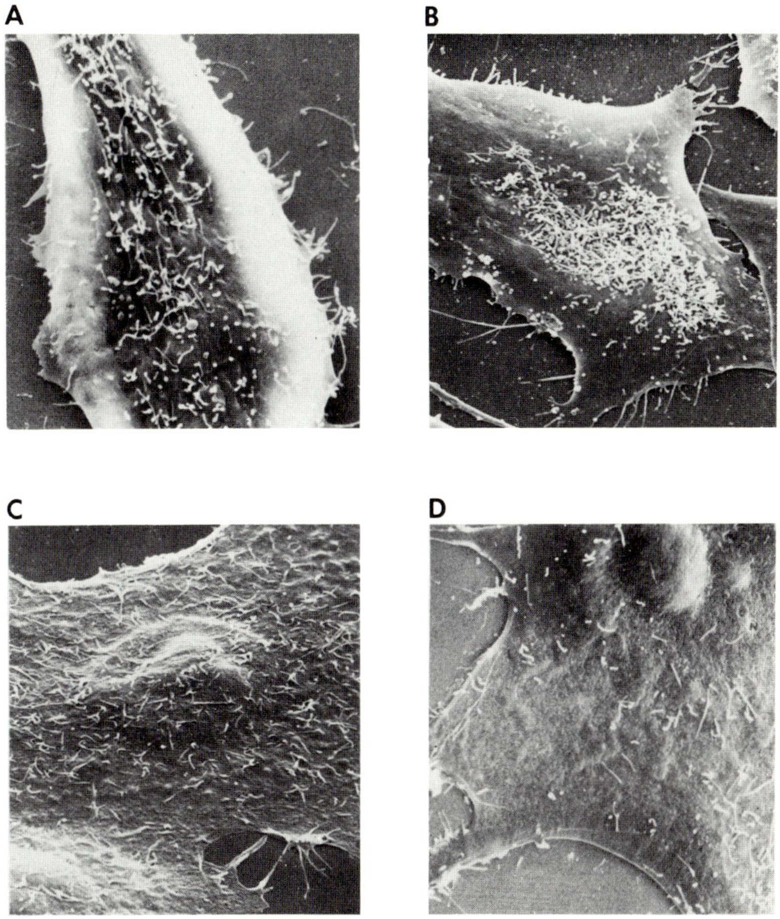

FIGURE 3 Scanning EM micrographs of cells grown on sugars substituted for glucose. Nil cells were plated and grown as described in the legend for Fig. 2. They were fixed and prepared for scanning micrography as described in Section II. (A) Glucose, (B) mannose, (C) galactose; (D) xylose.

changes are detectable by light as well as by EM scanning microscopy within 8 to 12 hr and are well advanced by 24 hr of further incubation at 37°C. If, on the other hand, the new medium contains glucose as well as the alternative sugar at equimolar concentrations, the changes in morphology progress more slowly, requiring 36 instead of 24 hr. When the sugars in question are xylose and galactose, the ultimate modifications are as complete in combination with glucose as on xylose or galactose alone. The photographs of Fig. 2 record the morphology by light microscopy of cells grown on glucose and shifted for 48 hr to mannose, galactose, and xylose. The same cultures viewed by scanning EM (Fig. 3) reveal sharp differences in cell shape, size, apparent nuclear size, prominence of nuclei, and nucleoli. On mannose, glucose, and fructose the cells are raised, the nuclei are not visible from the surface, and there is relatively little spreading over the surface. Both glucose and fructose produce abundant microvilli; mannose less so. Galactose and xylose result in cells spread out and flattened with very few microvilli. The attachment of cells to the surface varies, as might be expected, with the carbon source and the cell line as well as the surface presented to the culture.

EVIDENCE OF BIOCHEMICAL CONCOMITANTS OF MORPHOLOGIC CHANGES

Much of the impetus to pursue the mechanism(s) of morphologic and growth variation associated with specific sugars has come from a series of observations about glucose transport in cells starved for sugar and exposed to sugars other than glucose. Although these findings will not be discussed here, one or two important features of this work are essential to the present studies on nucleotide sugar metabolism. Results obtained with CEF and Nil cells only will be cited.

The initial observation of note was the derepression of glucose uptake when CEF primary or secondary cultures were deprived of sugar in a medium also serum free (16, 17). The derepression was gradual, reaching after 24 hr incubation at 37°C, a rate of entry for glucose some 20 to 25 times higher than for control cells provided with normal concentrations of glucose (5.5 m*M*) during the 24-hr period. The derepression was blocked by inhibitors either of protein or of RNA synthesis. Some naturally occurring sugars and analogs resembled glucose in maintaining low-level transport, while others permitted or stimulated derepression. In the first category at 5.5 m*M* concentration are 2-deoxy-D-glucose, D-mannose, and D-glucosamine; in the second, 6-deoxy- and 3-O-methyl glucose, D-galactose, D-fructose, D-ribose, and D-xylose. Pyruvate and L-lactate accelerate the derepression and result in a higher final rate of glucose transport.

The derepression proved to be of transport, not of subsequent metabolism, since 3-O-methyl D-glucose entry was equally accelerated (18, 19). In fact a high affinity carrier (Km 30-50 μ*M*) was found to be formed during starvation (18). Starvation also increased the Vmax of the constitutive low affinity carrier (Km 1-2 m*M*) (18).

Given the now generally accepted view that the cell surface of mammalian cells is rich in glycoproteins and glycolipids, we were encouraged to speculate that some aspects of the morphologic phenomena as well as of glucose transport changes might at the molecular level be explained by displacement or replacement of structural sugars in oligosaccharides.

NUCLEOTIDE SUGAR METABOLISM

While we were contemplating the potential role of sugars in surface molecules in mid- to late 1972, Dr. Herman Kalckar, whom I want to thank publicly for his generous support of our work in the ruling councils, observed that Nil cells starved of sugar, when pulsed with galactose, accumulated UDP galacturonic acid, whereas those fed normal levels of glucose accumulated principally UDP-galactose (20, 21).

The synthesis of nucleotide sugars and of oligosaccharides involves enzymes of the general classes of pyrophosphorylases (a) and glycosyl transferases (b):

a. Nucleoside-TP + sugar-1-P $\underset{\text{}}{\overset{\text{Pyrophosphorylase}}{\rightleftharpoons}}$ Nucleoside-DP-sugar + PPi

b. Acceptor (X) + UDP/GDP/ADP/CMP Sugar $\xrightarrow[\text{Transferase}]{\text{Glycosyl}}$ X-O-sugar + UMP + Pi

Obviously both the nucleoside triphosphate pool and the sugar-1-phosphate pool are critical to the nucleoside diphospho-sugar levels that determine the availability of substrates for oligosaccharide synthesis. Metabolic perturbations through feedback mechanisms that shift the equilibrium away from sugar-1-phosphate formation in any one of a large set of reactions could have a significant influence upon the nucleotide-sugar pool, hence upon oligosaccharide synthesis.

Ideally, measurements should be made of sugar phosphate and nucleotide sugar pools in the mammalian cells being studied. Such measurement of unlabeled pools would be difficult to accomplish with the limited cell populations of monolayer cultures that can be produced under experimental conditions. To introduce radioactive intermediates either of nucleosides or sugars even at micromolar concentrations would shift equilibria and further defeat the effort to determine steady state levels of sugar phosphates and nucleotide sugars.

We, therefore, decided to use the Kalckar pulse as an indicator of the flow of glucose or galactose to nucleotide sugars and examined the effect of long and short term exposure of cells to sugars, amino sugars, and nucleosides upon the detectable flow. Cells were grown on glucose-containing (5.5 m*M*) medium for 2 or 3 days until they were two thirds confluent, then they were washed and provided with new medium for a 30 min to 24 hr conditioning period (Fig. 4). At the end of the prescribed period ^{14}C-glucose or ^{14}C-galactose (5 μCi/ml; 4×10^{-5} *M*) was introduced, again after appropriate washing, to the monolayer for 5, 10, or 20 min. The acidic (pH 3.5) ethanol-soluble products were extracted from the cells, concentrated, and chromatographed in a Paladini-Leloir solvent system (9). Further enzymic and chromatographic (22, 23) analysis was used to establish the identity of the several peaks (I, II, III, V of Fig. 5; I, II, III of Fig. 6) obtained from labeled glucose and galactose (26; Gonzalez, unpublished results).

The principal findings are illustrated in Figs. 7-9. When starved, Nil cells exhibit a pattern for the flow of ^{14}C-glucose (UL) that shows a major accumulation of UDP-glucuronic acid (peak I), little UDP glucose (peak II), and glucose-6-phosphate (peak III). When pulsed with ^{14}C-galactose, the same cells accumulate high levels of galactose-1-phosphate and little UDP galactose. The counterpart of starved cells, those fed glucose, accumulate little UDP-glucuronic acid and equal amounts of glucose-6-phosphate from a pulse of glucose and considerable UDP galactose, but little galactose-1-phosphate when pulsed with galactose. These experiments are highly reproducible and number in the tens of independent experiments in essential agreement. Why starved cells fail to make UDP galactose is unclear, since their ATP levels are not significantly different from those of fed cells after 24 hr and the same cells make and accumulate large amounts of UDP glucose. Either the UDP galactose is utilized immediately or the Gal-1-Puridyl transferase enzyme that catalyzes the reaction

$$\text{UDP-Glc} + \text{Gal-1-P} \rightleftharpoons \text{UDP-Gal} + \text{Glc-1-P}$$

is not formed or is inactivated by starvation.

MANNOSE AND FRUCTOSE

When Nil cells are conditioned by growth for 24 hr on D-mannose, their transport system for glucose remains repressed. Assessment of the flow of glucose and of galactose by a pulse

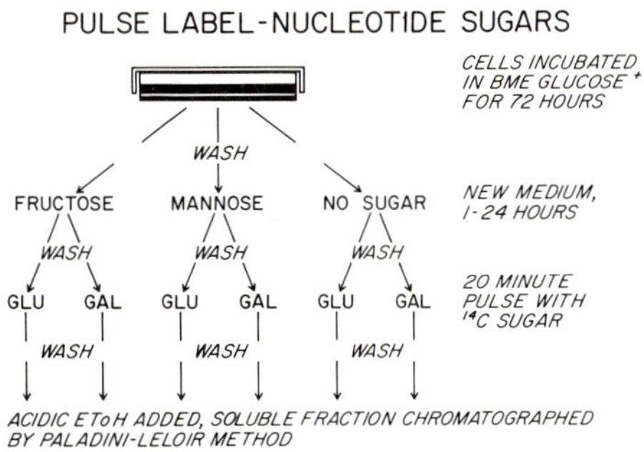

FIGURE 4 Scheme for conditioning cells prior to pulse with ^{14}C-glucose or ^{14}C-galactose to determine nucleotide sugar and sugar phosphate flux.

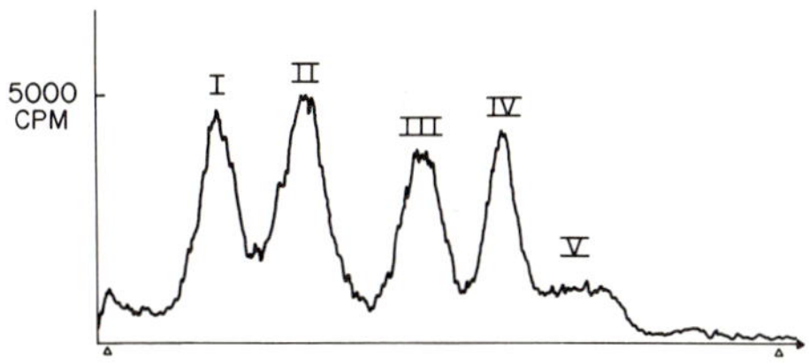

FIGURE 5 Paladini-Leloir chromatogram for acidic (pH 3.5) ethanol soluble fraction obtained after 20 min-pulse of ^{14}C-glucose (described in Section II). Peaks I, UDP hexuronic acid; II, UDP-hexose; III, glucose-6-PO_4+glucose-1-PO_4; IV, ribose; V, glucose. This pattern was obtained from the chromatographing of the acidic alcohol soluble fraction from cells starved of a carbon source for 6 1/2 hr.

of each reveals that mannose-grown cells accumulate UDP-glucuronic acid and relatively little UDP-glucose from labeled glucose, much the pattern of starved cells (Fig. 8). The metabolism of galactose by the same cell population results in UDP-galactose accumulation and little galactose-1-phosphate (Fig. 8), a pattern observed when glucose-fed cells are tested with labeled galactose. Thus cells grown on mannose for 24 hr, even under conditions of low serum such that they do not replicate, offer the paradox of handling glucose as though they were starved and galactose as though they had been metabolizing glucose during the conditioning period. It is clear from these experiments that the delicate equilibrium estab-

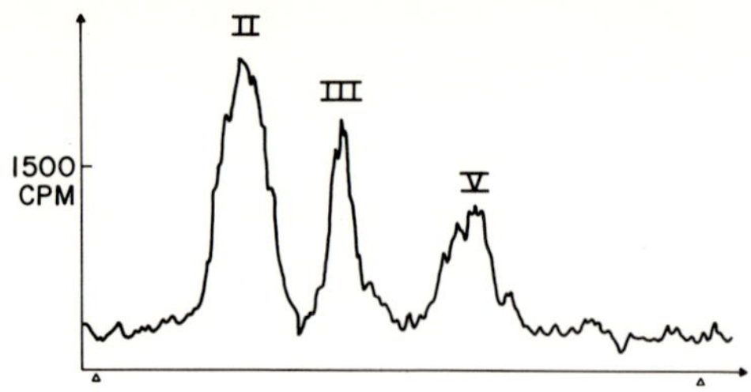

FIGURE 6 Paladini-Leloir chromatogram of acidic (pH 3.5) ethanol soluble fraction obtained after 20 min-pulse of ^{14}C-galactose (described in Section II). Peaks II, UDP hexose; III, galactose-1-PO_4; V, galactose. This chromatogram is a composite of patterns obtained from acidic ethanol extracts of both starved cells and cells provided with glucose.

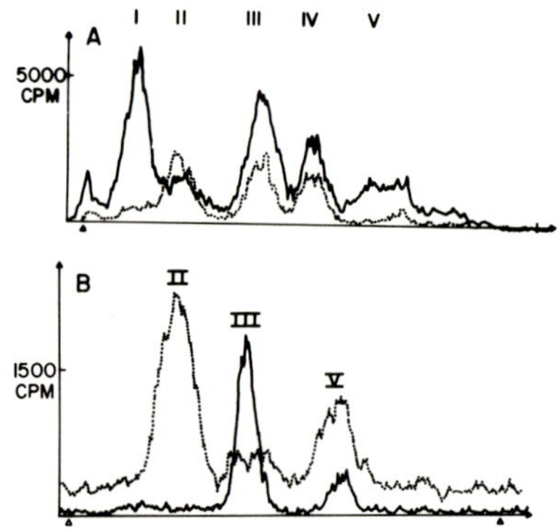

FIGURE 7 Paladini-Leloir chromatogram of acidic (pH 3.5) ethanol soluble fraction from "fed" and starved cells after pulses with ^{14}C-glucose or ^{14}C-galactose. (A) ^{14}C-glucose pulse (20 min), (B) ^{14}C-galactose pulse (20 min). (———) Cells starved of sugar for 24 hr, (-----) cells provided D-glucose (5.5 mM0 for 24 hr.

lished by mannose metabolites is significantly different from that achieved by glucose metabolites and that both differ from the conditions imposed by starvation.

If cells grown or maintained on fructose for 24 hr or more are subjected to the same 20-min pulse of ^{14}C-D-glucose, little labeled UDP-glucuronic acid is detected, while a sizable accumulation of UDP-glucose is observed, essentially the pattern characteristic of cells condition on glucose (Fig. 9).

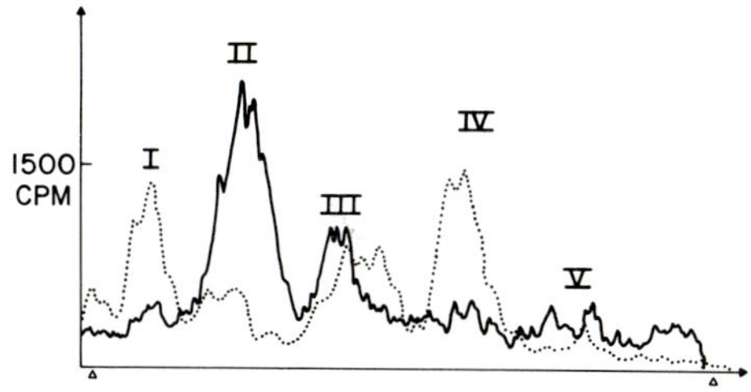

FIGURE 8 Paladini-Leloir chromatogram of acidic (pH 3.5) ethanol soluble fraction from cells grown on D-mannose for 24 hr. (———) Cells pulsed with ^{14}C-glucose for 20 min, (-----) cells pulsed with ^{14}C-galactose for 20 min.

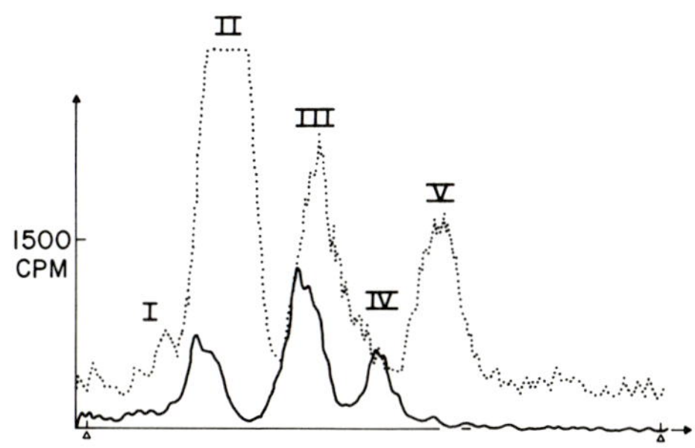

FIGURE 9 Paladini-Leloir chromatograms of acidic (pH 3.5) ethanol soluble fraction from cells grown on D-fructose for 24 hr. (———) Cells pulsed with ^{14}C-glucose for 20 min, (-----) cells pulsed with ^{14}C-galactose for 20 min.

The metabolites accumulating from labeled galactose, on the other hand, present a unique pattern, accumulating a large pool of UDP-galactose and a significant pool of galactose-1-phosphate. Although the mechanisms by which the fluxes, here indicated by the accumulation of labeled reaction products of glucose or galactose metabolism, are controlled remains unknown, the striking differences encountered justify investigation of the enzymes potentially involved and the properties of those enzymes as a function of the carbon source on which the cells grow or are maintained. Thus far hexokinase is the only enzyme that has been examined. Its activity and distribution between the bound and free form have been determined. In starved and glucose-fed cells (18, 19, 27) no significant difference has been detected either in activity or the ratio of free to membrane bound activity. The effect on hexokinase of the substitution of other sugars for glucose has not been examined.

SUGAR COMPOSITION OF MEMBRANES

Nil cells were grown both in roller bottles for some determinations and in standing cultures for others. The values reported (Table I) were obtained from cultures one half to three quarters confluent. For each set of determinations a minimum of 5×10^9 cells, 15-50 roller bottles of cells grown on each sugar were utilized. The sugars D-galactose, D-fructose, D-mannose, and D-xylose contained as received between 0.1 and 0.3% glucose as a contaminant by glucose oxidase assay (data not shown). Dialyzed calf serum (10%) was used to reduce the unwanted glucose to a minimum. The Nil cells grow well from small inoculum (10^4 cells/30 cm^2 plastic vessel) on all five sugars as "sole" carbon source. Indeed, when glucose oxidase was used to rid the four sugar preparations of contaminating glucose (28), little change was observed in the growth of the cells on any of the treated sugars (data not shown) compared to that on the untreated counterparts.

Several reservations must be expressed concerning the potential selection of portions of membrane by the Brunette-Till procedure for isolation of membranes. The cells grown on the various sugars display different levels of resistance to the disrupting procedure. The level of disruption was monitored by examining the homogenate for membrane vesicles with trypan blue stain and light microscopy at 100- to 400-fold magnification. When the cells were 90-95% disrupted by this visualization, the lysis was considered complete. For cells grown on some sugars only 40-50 strokes of the Dounce homogenizer were required, while others consistently resisted until 80-90 strokes had been applied. It is possible that specific patterns of disruption could result in different distribution of membrane segments in certain cell populations. We have assumed random segmentation as the basis for comparing sugar composition of the membrane yield from each lot of cells.

GAS-LIQUID CHROMATOGRAPHIC ANALYSIS OF SUGARS

An identification of mannose, galactose, glucose, inositol, *N*-acetylglucosamine, *N*-acetylgalactosamine, and sialic acid can be made by determination of retention times of standard sugar (Fig. 10). Quantitation was performed using arabitol as an internal standard. Membrane preparations containing

TABLE I Membrane Sugars: Molar Ratios with D-Mannose as Reference Sugar[a]

Grown on	Man	Gal	Glc	GNac	GalNac	Sialic
Glc	1.0	1.2	1.4	4.5	0.2	0.8
Gal	1.0	0.9	1.7	4.7	0.3	0.9
Man	1.0	0.9	1.3	5.6	0.4	1.7
Fruc	1.0	0.4	0.5	4.5	0.3	0.7
Xyl	1.0	0.5	2.8	9.1	0.7	2.4

[a]Approximately 5×10^9 Nil cells grown in roller bottles on single sugars as indicated were harvested by scraping. Membranes were prepared by the Brunette-Till method as described in Section II. The values presented represent two separate lots of cells and duplicate runs by whole procedure on each lot of cells. Absolute sialic acid values were measured chemically (Table II). In this table the mannose values are assumed to be a unit. Ratios of the other sugars were computed with mannose content as the denominator.

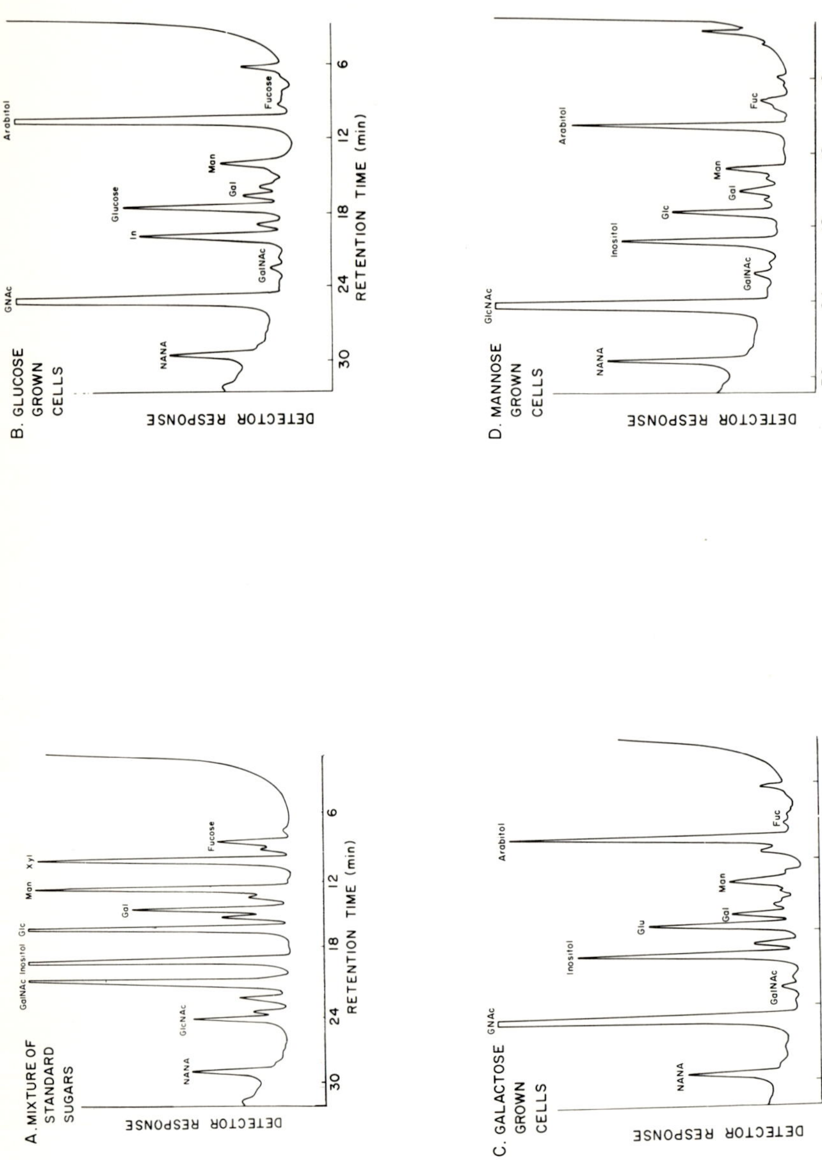
A. MIXTURE OF STANDARD SUGARS
DETECTOR RESPONSE
NANA
GlcNAc
GalNAc
Inositol
Glc
Gal
Man
Xyl
Fucose
30
24
18
12
6
RETENTION TIME (min)
B. GLUCOSE GROWN CELLS
DETECTOR RESPONSE
NANA
GNAc
In
Glucose
Gal
Man
Arabitol
GalNAc
Fucose
30
24
18
12
6
RETENTION TIME (min)
C. GALACTOSE GROWN CELLS
DETECTOR RESPONSE
NANA
GNAc
Inositol
Glu
Gal
Man
Arabitol
GalNAc
Fuc
30
24
18
12
6
RETENTION TIME (min)
D. MANNOSE GROWN CELLS
DETECTOR RESPONSE
NANA
GlcNAc
Inositol
Glc
Gal
Man
Arabitol
GalNAc
Fuc
30
24
18
12
6
RETENTION TIME (min)

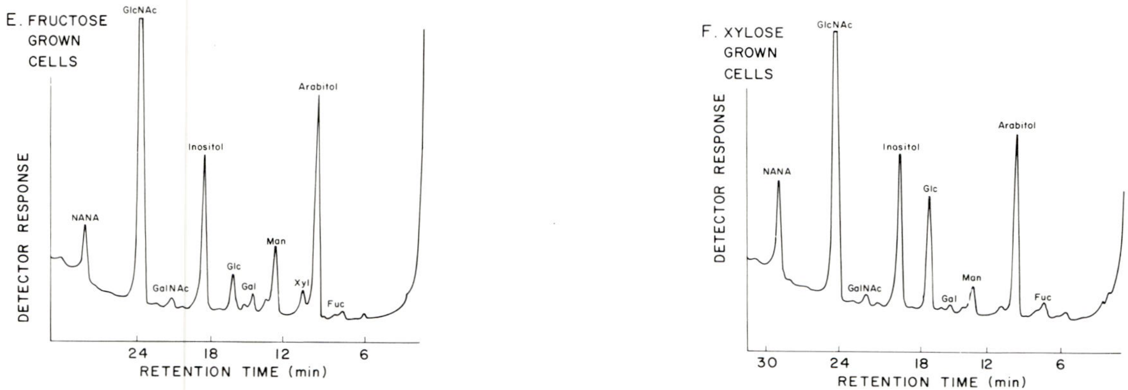

FIGURE 10 *Gas-liquid chromatography of the carbohydrate components of the membrane preparations of Nil cells grown for several generations in one of several sugars as "sole" carbon source. (A) Analysis of a standard mixture of sugars; membranes from Nil cells grown in roller bottles with (B) D-glucose, (C) D-galactose, (D) D-mannose, (E) D-fructose, or (F) D-xylose as carbon and energy source (Arabitol was used as an internal standard).*

1.5 to 2 mg of membrane were methanolyzed and processed (Section II) for gas chromatography.

Simple visual comparison of the relative quantities of sialic acid glucose, galactose, and mannose without efforts at quantitation establish some significant differences. As an example, membranes from glucose (Fig. 10B) and galactose-grown cells (Fig. 10C) yield quite similar patterns, while membranes from xylose-grown cells (Fig. 10F) contain much less galactose and mannose relative to glucose than do membranes from glucose-grown cells. Membranes from fructose-grown cells (Fig. 10E) on the other hand have much less glucose than mannose when compared to membranes from glucose-grown cells.

COMPARATIVE SUGAR CONTENT

The quantity of sugar in each sample was determined by measuring the areas under the peaks with an electronic planimeter (Numanics Corporation model 237 graphics calculator). If one assumes arbitrarily for mannose a unit value and compares the quantity of other sugars to it, the relative values for mannose, galactose, glucose, *N*-acetylglucosamine, *N*-acetylgalactosamine, and sialic acid are as presented in Table I. Relative to mannose, galactose, glucose, and sialic acid vary 3-6 fold as a function of the sugar on which the cells were cultivated. Chemical determination of total sialic acid and of the sialic acids accessible to sialidase revealed that the sialic acid content of the cells remained relatively constant no matter what the carbon source (Table II).

If sialic acid is accepted as the comparison sugar the molar ratios are as presented in Table III. Since the sialic acid values per milligram of cell protein showed little change as a function of carbon source, sialic acid serves as a useful denominator for comparison.

DISCUSSION

The question posed here is whether mammalian cells grown in culture undergo significant changes in membrane sugar composition to account for alterations in phenotype that have been observed. The underlying assumption is that much of the cells' social behavior, in which the plasma membrane plays an important role, is determined by the sugar composition of antigens such as the ABO blood groups (29) or putative glycoproteins such as lectin receptors or the insulin receptor. Gangliosides have been identified as the receptor or one com-

TABLE II Sialic Acid Content — Total Sialidase Releasable

Cells grown on[a]	Total sialic[b] (μg/mg protein)	Released sialic[c] (μg/mg protein)
Glucose	6.3	2.2
Mannose	4.2	2.3
Xylose	6.4	2.6
Galactose	6.1	2.5
Fructose	5.7	4.4

[a]Cells were grown on 150 mM Petri dishes and harvested by scraping when hear confluence.

[b]Total sialic values are an average of three separate experiments, each in duplicate. The values expressed are micrograms of sialic acid per milligram of cell protein of cells, taken immediately after harvest.

[c]Cells of the same cultures after washing were treated with neuraminidase (125 units in 5 ml) for 30 min at 37°C. The values presented are an average of two experiments, each in duplicate. Sialic acid was determined by the thiobarbituric acid procedure (31).

ponent of a receptor for cholera toxin on mammalian cell surfaces (30).

The presence of oligosaccharides on the outer surfaces of mammalian cells has been well established experimentally. Indeed, mild chemical attack upon the sugar components of some surface oligosaccharides accessible to enzymes from the outside is mitogenic. This is true both of periodate treatment and of galactose oxidase oxidation of sugar residues on the outer surface of lymphocytes (31).

We have been able to demonstrate that cells grown on naturally occurring sugars in lieu of glucose or as supplements to glucose induce changes in cell size, shape, and density at confluence. Moreover, the alignment of cells with respect to each other at confluence shows patterns consistent for a particular sugar and a given cell strain or line. More detailed examination of the cell surface by scanning electron microscopy revealed marked differences in prominence of nuclei and nucleoli- size and flatness of the cells, abundance and distribution of microvilli, as well as the form and abundance of filopodia.

TABLE III Membrane Sugars: Molar Ratios[a]

Grown on	Man	Gal	Glc	GNac
Glc	1.3	1.5	1.8	5.6
Gal	1.1	1.0	1.9	5.2
Man	0.6	0.5	0.8	3.3
Fruc	1.4	0.6	0.7	6.4
Xyl	0.4	0.2	1.2	3.8

[a]*Same data as in Table I, computed with sialic acid as denominator for each lot of cells. Sialic μg/mg protein = 1.*

Finally, sharp differences in sugar composition of membranes obtained from cells grown on each of the five sugars were detected. The largest differences were observed in mannose, galactose, and glucose content, whereas the sialic acid content remained virtually the same no matter which of the five sugars was employed for growth.

Although indications point to nucleotide sugar metabolism as a contributor to the altered glycosylation patterns, the specific links between nucleotide sugar pools and oligosaccharide composition have yet to be explored. A study of the glycosyl transferases and their properties in the cells under altered growth conditions may provide a clue to the mechanism(s) that account for the changes in sugar content observed.

REFERENCES

1. Lindegren, C. C., and Lindegren, G., *J. Heredity, 32*, 435 (1941).
2. De Terra, N., and Tatum, E. L., *J. Amer. Bot. 50*, 669 (1963).
3. Mahadevan, P. R., and Tatum, E. L., *Bact. 90*, 1073 (1965).
4. Tatum, E. L., Barratt, R. W., and Cutter, V. M., *Science 109*, 509 (1949).
5. Brody, S., and Tatum, E. L., *Proc. Nat. Acad. Sci. USA 56*, 1290 (1966).
6. Brody, S., and Tatum, E. L., *Proc. Nat. Acad. Sci. USA 58*, 923 (1967).

7. Scott, W. A., and Tatum, E. L., *Proc. Nat. Acad. Sci. USA 66*, 515 (1970).
8. Drewes, L. R., and Gander, J. E., *J. Bact. 121*, 675 (1975).
9. Paladini, C., and Leloir, L. G., *Biochem. J. 51*, 426 (1952).
10. Lowry, O. H., Rosebrough, N. J., Farr, L., and Randall, R. J., *J. Biol. Chem. 193*, 265 (1951).
11. Cox, R. P., and Gesner, B. M., *Proc. Nat. Acad. Sci. USA 54*, 1571 (1965).
12. Cox, R. P., and Gesner, B. M., *Cancer Res. 27*, 974 (1967). (1967).
13. Demetrakopoulos, G. E., Gonzalez, F., Colofiore, J., and Amos, H., *Exp. Cell Res.* (in press, 1977).
14. Demetrakopoulos, G. E., and Amos, H., *Biochem. Biophys. Res. Comm. 72*, 1169 (1976).
15. Amos, H., Leventhal, M., Chu, L., and Karnovsky, M. J., *Cell 7*, 97 (1976).
16. Martineau, R., Kohlbacher, M., Shaw, S., and Amos, H., *Proc. Nat. Acad. Sci. USA 69*, 3407 (1972).
17. Shaw, S. N., and Amos, H., *Biochem. Biophys. Res. Comm. 53*, 357 (1973).
18. Christopher, C. W., Kohlbacher, M., and Amos, H., *Biochem. J. 158*, 439 (1976).
19. Musliner, T., Chrousos, G., and Amos, H., *J. Cell. Physiol.* (1976).
20. Kalckar, H. M., and Ullrey, D., *Proc. Nat. Acad. Sci. USA 70*, 2502 (1973).
21. Ullrey, D., Gamon, M. T., and Kalckar, *Arch. Biochem. Biophys. 167*, 410 (1975).
22. Ginsburg, V., *J. Biol. Chem. 241*, 3750 (1966).
23. Carminatti, H., Passeron, S., Dankert, M., and Recondo, E., *J. Chromatog. 18*, 342 (1965).
24. Brunette, D. M., and Till, J. E., *J. Membrane Biol. 5*, 215 (1971).
25. Reinhold, V. N., *in* "Methods in Enzymology," Hirs, C. H. W., and Timasheff, S. N., eds, Academic Press, New York, Vol. 25, Part B, p. 244 (1972).
26. Rossow, P. W., Radeos, M. S., and Amos, H., *Arch. Biochem. Biophys. 168*, 520 (1975).
27. Kletzien, R. F., and Perdue, J. F., *J. Biol. Chem. 250*, 593 (1975).
28. Burns, R. L., Rosenberger, P. G., and Klebe, R. J., *J. Cell. Physiol. 88*, 307 (1976).
29. Watkins, W. M., *Biochem. Soc. Symp. 40*, 125 (1974).
30. van Heyningen, W. E., *Nature 249*, 415 (1974).
31. O'Brien, R. L., and Parker, J. W., *Cell 7*, 13 (1976).
32. Codington, J. F., Linsley, K. B., and Silber, C., "Carbohydrate Chemistry," Whistler, ed, Vol. 7 (1976).

Note added in proof: The failure of starved cells to make UDP-galactose (p.192 , para. 2, line 13) when pulsed with ^{14}C-galactose can be explained by the requirement for UDP-glucose as substrate for the enzyme, galactose-1-phosphate uridyl transferase, the only pathway for formation of UDP-galactose in mammals (Stryer, L., "Biochemistry," p. 400, Freeman, San Francisco, 1975). The cells starved for 24 hours lack UDP-glucose; hence they cannot make UDP-galactose from galactose-1-phosphate, which they accumulate in large quantity. Traces of glucose (10^{-5} *M*) added to the cells restore the synthesis of UDP-galactose (D. Sens and H. Amos, unpublished data).

INITIAL STUDIES ON THE MECHANISM OF SUBSTRATE ADHESION OF NORMAL AND VIRUS-TRANSFORMED CELLS

LLOYD A. CULP
JOSEFINA F. BUNIEL

Department of Microbiology

JONATHAN J. ROSEN

Department of Biomedical Engineering
School of Medicine
Case Western Reserve University
Cleveland, Ohio

The substrate-attached material (SAM), left tightly bound to the tissue-culture substrate after removal of "normal," virus-transformed, and revertant mammalian cells by EGTA treatment, has been studied by a variety of techniques to determine its prospective role in substrate adhesion. The polysaccharide in SAM was identified as principally hyaluronic acid "bound" to protein as proteoglycan complexes that have been resolved into three size classes by slab SDS-PAGE. A limited number of cell "surface" proteins have also been identified in SAM: C_o (the LETS glycoprotein), C_a (a myosinlike protein), C_1 (MW 56,000, which does not appear to be tubulin), C_2 (actin), C_3-C_5 (histones) artifactually leached from the cell during EGTA treatment and probable binding to the large mass of negatively charged polysaccharide, and several other cell proteins that have been only partially characterized. There are quantitative differences in the amounts of several of these components between normal and transformed cells. The proteoglycans and proteins (except histones) were deposited in a coordinated fashion in well-defined proportions during a variety of cell attachment and growth conditions.

The topographical distribution and metabolic behavior of the polysaccharide and protein components have also been examined. Scanning electron microscopic evidence indicates that

EGTA treatment results in retraction of the cell body away from the substrate-bound footpads by which the cell adheres to the substrate; shearing forces generated during shaking of cultures then break the delicate retracting membrane leaving membrane-enclosed footpads on the substrate as SAM. Our studies indicate that substrate adhesion may be mediated by a cell surface complex containing hyaluronate proteoglycans, the LETS glycoprotein, and subsurface actin-containing microfilaments with several associated proteins.

INTRODUCTION

Considerable evidence has accumulated over the years to indicate that the radically different social behavior of malignant cells, such as altered adhesiveness and reduced control over directional movement and cell division, is the result of changes in the properties of the surface membrane of the cell. Since the surface of the cell is enriched in carbohydrate-containing moieties when compared to most other cell compartments, interest has focused on identifying chemical differences (quantitative and/or qualitative) in the complex polysaccharides and glycoproteins of surface membranes of normal and malignant (in most cases, virus-transformed) cells.

These analyses are complicated by the tremendous heterogeneity in numbers and amounts of cell surface components that may obscure subtle differences in one or more components whose topographical distribution is critically important [see Nicolson (1)]. It therefore became imperative to identify functionally important subsets of surface components in order to ask and hopefully answer more biochemically specific questions.

Interaction of the cell surface with the tissue culture substrate appeared to us to be a unique opportunity for studying a functionally important process biochemically. It has been well established that normal mammalian cells display a stringent requirement for anchorage to this substrate to survive and grow *in vitro*. Malignant cells, on the other hand, can frequently be grown independently of any artificial substrate, an alteration that no doubt reflects significant biochemical differences in the surface components that mediate cell-substrate adhesion. Since the tissue culture substrate is an essentially inert entity, there was considerable hope for enriching for that portion of the cell surface that mediates adhesion for appropriate structure-function studies. Finally, a considerable amount of work had analyzed the cell-substrate adhesion process from a phenomenological point of view (2-4).

IDENTIFICATION OF SUBSTRATE-ATTACHED MATERIAL (SAM)[1]

In order to initiate studies on the structure-function relationships of cell surface components that mediate adhesion, it was necessary to (a) avoid the use of trypsin for detaching cells to minimize damage of interesting surface proteins and glycoproteins and (b) detach cells with a reagent that would enrich for substrate-bound macromolecules. Culp and Black (5) found that treatment of BALB/c 3T3 mouse cells, their SV40-transformed counterpart (SVT2 cells), and concanavalin A-selected revertant (6) cells with the Ca^{2+}-specific chelating agent EGTA resulted in effective cell detachment from the substrate and minimal leaching of polysaccharide-enriched components off the surface of the cell. More importantly, they identified a sizable pool of cellular protein (generally 1-4% of the cell's total content) and polysaccharide (generally 3-9%) that remained dish-bound and required treatment with alkali or SDS for quantitative removal; there was very little RNA or DNA in this material, indicating considerable enrichment of membrane-containing components. Normal and revertant cells that are more "strongly" adhesive to the substrate than SVT2 cells left 3-4 times more of this material behind. Interest then focused on determining the specificity in cellular origin of this substrate-attached material (SAM) and its similarity or dissimilarity to the so-called "microexudates" that had been detected by ellipsometric (7, 8) and electron microscopic (9, 10) techniques.

[1]Abbreviations: Con A, concanavalin A; CSP, cell surface protein as defined in reference 28; EDTA, (ethylenedinitrilo) tetraacetic acid; EGTA, ethylenebis (oxyethylenenitrilo) tetraacetic acid; EM, electron microscopic; GAG, acidic glycosaminoglycans (formerly referred to as mucopolysaccharides); GAP, glycosaminoglycan-associated protein; LETS, large-external-transformation-sensitive glycoprotein as defined in reference 27; MSV, murine sarcoma virus; MW, molecular weight; PAGE, polyacrylamide gel electrophoresis; PBS, phosphate-buffered saline; SDS, sodium dodecyl sulfate; SV40, Simian virus 40.

TOPOGRAPHICAL DISTRIBUTION

When normal or virus-transformed cells were grown as well-isolated colonies in medium containing radioactive leucine or glucosamine to radiolabel the protein or polysaccharide moieties of the cell, respectively, the location of radioactive SAM detected autoradiographically after EGTA-mediated removal of colonies was identical to the original location of the colonies (11). No radioactive material was observed on the substrate at colony-free locations, indicating that deposition of SAM required cellular contact with the substrate and was not the result of secretion of proteins and polysaccharides into the medium of the culture followed by random binding to the substrate. Kinetic experiments (12) analyzing incorporation of radioactive precursors into SAM, cell-associated, and medium-secreted proteins and polysaccharides also verified that the pool of SAM components was not derived from the pool of medium-secreted components. These experiments suggested a vectorially specific process requiring cell surface contact with the substrate for the generation of substrate-attached material.

High resolution autoradiography was then used to show that radioactive SAM protein and polysaccharide was not evenly spread on the substrate but was present in focal pools (11). The density of these focal pools was identical (a) for leucine- or glucosamine-radiolabeled components, (b) for normal or virus-transformed cells, (c) for newly attaching, preradiolabeled cells or long-term-cultured cells, and (d) to the density of subsurface footpads (13) by which the cell adheres to the substrate. The cell apparently adheres to the culture substrate at highly differentiated regions of its surface membrane (see discussion later).

The patterns of SAM deposited by attaching cells that were allowed to spread and initiate movement on the substrate over a 24-hr period were uniquely interesting (11). The area of SAM deposition was considerably larger than the area occupied by the cell itself, indicating that cells may leave "footprints" of SAM behind as they move across the substrate. Therefore, SAM probably results from two different pools of cellular material: (a) as footprint material, which is no longer cell-associated but has been left behind on the substrate in the cell's microenvironment as a result of normal cell movement (probably footpad material that is pinched off the cell surface at the posterior adhesion sites), and (b) as "footpad" material, which is artificially pinched off the cell surface (see scanning electron microscopic evidence later) as a result of the EGTA treatment. Some evidence has been obtained for biochemical similarities and differences between these two classes of SAM components.

BIOCHEMICAL COMPOSITION

Since substrate-attached material was easily radiolabeled by long-term growth of cells in medium containing precursor glucosamine, which is incorporated into cellular polysaccharides at a very low specific radioactivity, the mass of polysaccharide was obviously very large. Pronase digestion of this material to analyze for glycopeptide size heterogeneity by methods similar to those used by Buck *et al.* (14) generated polysaccharide chains that were very large and resolvable into three size classes by Sepharose 6B chromatography (15). Terry and Culp (15) performed a number of analyses to show that the vast majority of the aminosugar-radiolabeled polysaccharide (probably greater than 95% of the polysaccharide accumulated over a 72-hr growth period) was hyaluronic acid, an acidic glycosaminoglycan (GAG) copolymer composed of glucuronic acid and N-acetylglucosamine. This was an interesting observation in light of Defendi and Gasic's (16) discovery that normal and virus-transformed cells synthesize different amounts of GAG and Kraemer's (17) suggestion that GAG's were important components in regulating cell surface phenomena, including adhesion.

Since minor GAG components were not resolvable by glucosamine radiolabeling, radioactive sulfate incorporation was used to demonstrate that SAM also contained a small pool of sulfated GAG (15, 18) that has been incompletely characterized. This sulfated GAG displays unique metabolic properties that will be discussed below.

When alkali- or SDS-extracted SAM was chromatographed on Sepharose 6B columns, most of the protein co-chromatographed with the GAG polysaccharide (15). To further resolve the complexity of the protein components in SAM, we utilized high resolution slab SDS-PAGE gels (19).[1] Glucosamine-radiolabeled hyaluronate from intact components in SAM that had been boiled in SDS- and mercaptoethanol-containing buffers before electrophoresis was resolved into three bands in the 5% stacking gel. Approximately 50% of the polysaccharide was in the largest component called GAG-1 while 25% of the polysaccharide ap-

[1]Unless indicated otherwise, the SAM being analyzed was derived from plastic-grown cells.

peared in the two smaller bands called GAG-2 and GAG-3. Normal, SV40-transformed, and revertant mouse cell SAM's displayed the same distribution, as well as Swiss or BALB/c mouse fibroblast SAM's.

In contrast to these findings using mouse cell lines, hamster Nil-B fibroblasts, which also display density-dependent inhibition of growth, and SV40-transformed Nil-B cells contained a different distribution of polysaccharide—a preponderance of GAG-1 and a much smaller amount of GAG-3 relative to GAG-2 (19). The polysaccharide from these hamster cells was also found to be primarily hyaluronic acid.

When sulfate-radiolabeled GAG in SAM was electrophoresed, a distribution pattern different from that of the glucosamine-radiolabeled GAG was observed. Ninety percent of the radioactive material co-electrophoresed with GAG-1 and the remaining 10% with GAG-3. It will be interesting to determine if these fibroblast GAG's are organized into supramolecular complexes similar to those identified for cartilaginous proteoglycans (43, 44).

When leucine-radiolabeled SAM was electrophoresed, three bands of protein appeared in the stacking gel and co-electrophoresed with the three bands of GAG. Therefore, these bands are called "glycosaminoglycan-associated proteins" (GAP-1, 2, -3) corresponding to the three bands of GAG (19). A number of vigorous treatments of SAM have failed to dissociate most of the protein from the polysaccharide, which suggests that these components are proteoglycan complexes. However, some of the protein at the top of the gel in the GAP-1 band (approximately 75%) can be dispersed by adding carrier histone to the mixtures before boiling and electrophoresis; the remainder of the GAP-1 protein was resistant to dissociation, as were the GAP-2 and GAP-3 bands and the distribution of GAG polysaccharides. This protein appears to be a heterogeneous mixture that may result from some protein denaturation during the boiling of SAM preparations and subsequent trapping at the top of the gel. The ratio of protein:polysaccharide in the proteoglycans was highest in the GAG-3 band and lowest in the GAG-1 band. Approximately 50% of the protein in SAM was present in the GAP bands.

Several cellular SAM proteins were separable from the GAG and appeared as distinct bands in the 20% Ortec slab gels[1] (for resolution of proteins between MW 8000 and approximately 125,000). All SAM proteins are listed in Table I with their partial identities. Protein C_1 does not appear to be tubulin,

[1]High resolution gel techniques were developed by Ortec, Inc., Oak Ridge, Tennessee and were described in their Application Note AN32a.

TABLE I Substrate-Attached Cell Proteins[a]

Cell protein	Apparent mol wt[b]	Properties
GAP-1, -2, -3	Large (bound to GAG)	Not collagenous
C_o	220,000	LETS glycoprotein (rapid turnover and transformation-sensitive)
C_a	200,000	Myosinlike
C_x	175,000	Glycoprotein only found in glass-bound SAM
C_y	145,000	Glycoprotein only found in glass-bound SAM
C_b	85,000	May not be α-actinin[c]
C_c	67,000	?
C_1	56,000	May not be tubulin (reference 19)
C_d'	49,000	Protein only found in glass-bound SAM
C_d	48,000	Rapid turnover
C_2	45,000	Actin
C_e	37,000	?
C_f	27,000	?
C_3	14,000	Histone
C_4	13,000	Histone
C_5	11,000	Histone

[a]These are the components resolved by slab SDS-PAGE whose properties are described in the text and references 19 and 29. Major proteins were originally denoted with subscript numbers and minor proteins with subscript letters.

[b]Apparent molecular weights were determined by slab SDS-PAGE using bovine serum albumin, actin, myoglobin, myosin, and *E. coli* β-galactosidase as markers from plots of the log of the molecular weight vs migration distance (19).

[c]Protein C_b migrated very differently from carrier porcine skeletal α-actinin during two-dimensional analysis on O'Farrell gels (42) (proteins were isoelectric-focused in a disk gel in the first dimension, which was followed by slab SDS-PAGE as the second dimension). The porcine skeletal α-actinin was kindly provided by Dr. Richard Robson.

while protein C_2 has been shown to be actin (19). The identification of substrate-bound actin was extremely interesting in light of the presence of actin-containing microfilaments (75 Å diameter) at surface membrane sites involved in cell-to-cell and cell-to-substrate adhesion and the more highly organized matrices of membrane-associated microfilaments in normal and revertant cells when compared to transformed cells (20-23).

Proteins C_3-C_5 were shown to be histones. These proteins appeared in SVT2 SAM in much greater quantity than in 3T3 or revertant SAM's, varied in amount from preparation to preparation (while the other SAM components were found in similar relative proportions), were not observed in SAM left after scraping cells from the substrate (scraped cells leak very little intracellular protein), and were not found in surface membrane preparations (24). Evidence therefore indicates that the histones are leached from the cell during the EGTA treatment, resulting in artifactual binding to SAM, probably to the large mass of negatively-charged GAG.

A number of minor proteins, C_a-C_f, were also observed in SAM, and little information has been obtained on the nature of C_b-C_f (Table I). Eight percent Ortec gels were used to better resolve protein C_a from GAP-3. Several pieces of evidence (19) indicated that C_a is a myosinlike protein: (a) it co-electrophoresed with rabbit skeletal muscle myosin heavy chain during SDS-PAGE; (b) it was not glycosylated; (c) it was iodinatable as a membrane component using lactoperoxidase, consistent with the recent surface localization of a myosinlike protein (25, 26); and (d) it was specifically immunoprecipitable from Triton-solubilized membrane preparations using an antibody prepared against highly purified mouse L cell myosin.[1] It is reasonable to speculate that this protein is one of a number of microfilament-associated proteins that appear in SAM and may be identical to the protease-labile 205,000 MW protein identified by Teng and Chen (45), which is also iodinatable using lactoperoxidase.

Another high molecular weight protein was also identified in 8% Ortec gels and has been called C_o (Table I). This component appears to be the so-called LETS (27) or CSP (28) glycoprotein because it (a) co-electrophoreses with authentic hamster LETS,[2] (b) is glycosylated, (c) is easily iodinated as a surface membrane component using lactoperoxidase, and (d) is very trypsin-labile as a membrane component (19, 24, 29).

The SAM's from BALB/c 3T3, SVT2, and Con A revertant cells contained similar amounts of these components when grown for 48-72 hr in radioactive leucine-containing medium, except for a slightly elevated amount of GAP protein and larger amounts

[1]Antibodies were kindly provided by Dr. Ira Pastan.

[2]The hamster LETS preparation was kindly provided by Dr. Richard Hynes.

TABLE II Substrate-Attached Serum Proteins[a]

Serum protein	*Apparent mol wt*
S_3	175,000
S_3'	140,000
S_4	85,000
S_5	73,000
S_6	65,000
S_7	49,000
S_8	29,000
S_9	27,000
S_{10}	23,000
S_{11}	11,000

[a]*These are the approximate molecular weights determined by SDS-PAGE analysis as described in the footnote to Table I (references 19 and 30). Proteins S_1 and S_2, which co-electrophoresed with GAP-1 and GAP-3, respectively, were too large to obtain reliable sizing information.*

of C_3-C_5 in SVT2 SAM (19, 29). The sizable amounts of LETS (C_0) in SVT2 SAM was surprising in light of the greatly diminished amount of this component in the surface membranes of a number of virus-transformed mammalian cells (27). However, SVT2 has been found to be a unique transformant in that its surface membrane contains as much LETS (identified by a number of criteria) as BALB/c 3T3 cells (24). On the other hand, MSV-transformed BALB/c 3T3 and SV40-transformed Swiss 3T3 cells were found to contain much less LETS in their SAM's; these transformants do not spread as extensively across the substrate as SVT2 cells do, consistent with the quantitative importance of this component in the substrate adhesion process (19, 29).

These cell-derived SAM proteins were distinguishable in slab SDS-PAGE gels from twelve distinct bands of Coomassie blue-stainable serum proteins that were tightly bound to plastic substrates but were SDS-extractable (30) (Table II). Considerable evidence indicates that one or more of these proteins are critical mediators of "proper" cell adhesion, spreading, and movement across the substrate (10, 30-32). Cells probably do not interact with plastic or glass substrates directly but bind to specific serum components that rapidly saturate the substrate (30). It is interesting to note that serum proteins S_1 and S_2 co-electrophorese with

TABLE III *Growth and Attachment Conditions for Various SAM Analyses*[a]

Experimental conditions	Altered distribution[b] of: Polysaccharide	Protein
(1) Long-term radiolabeling	Normal	Normal
(2) Long-term radiolabeling on glass substrate	Same as (1)	Decreased amount of C_l; new components C_d', C_x, and C_y
(3) Long-term radiolabeing on SAM-coated substrate[c]	ND[d]	Increased amount of GAP-3 and reduced amount of GAP-1
(4) During first hour of attachment	Increased amount of GAG-3	Increased amounts of C_l, C_o, and C_a
(5) During first hour of attachment to SAM-coated substrate	ND[d]	Increased amounts of C_d and GAP-1 [otherwise same as (4)]
(6) Chasing of long-term radiolabeled cells	ND[d]	Same as (1)
(7) Short-term pulse radiolabeling	ND[d]	Much higher levels of C_d and C_o
(8) Chasing of (7)	ND[d]	Most of C_o and C_d disappear; partial loss of other proteins
(9) Detachment with EDTA	Same as (1)	Same as (1)
(10) Detachment by scraping	Same as (1)	No histone

[a]Cells were grown or attached on plastic substrate, unless indicated otherwise, as defined in the first column after growth in medium containing radioactive leucine to identify proteins or radioactive glucosamine to identify GAG polysaccharides. Cells were removed by EGTA treatment, unless noted otherwise, and SAM was removed by SDS treatment for PAGE analysis as described in

reference 19 and 29.

[b]*The patterns are compared to the pattern generated by long-term growth of cells in radioactive precursor. Qualitative and quantitative differences are noted.*

[c]*See Culp (41) for protocols for growth analyses on SAM-coated substrates.*

[d]*Not done.*

cellular GAG-1 and GAG-3, respectively, and may also be high molecular weight proteoglycans.

Cellular SAM components have been analyzed from cells growing and attaching under a wide variety of conditions (19, 29, and Table III). We have consistently found the same proteins and polysaccharides in similar relative proportions, except for elevated levels of one or two components under some conditions. Thus, there appears to be a *coordinate deposition* of these components. This consistency suggests that these components are present in some sort of cell surface complex involving the hyaluronate proteoglycans, the LETS glycoprotein, and microfilament-associated proteins. Consistent with this concept is evidence (Cathcart and Culp, unpublished data) that nonionic detergents are ineffective at removing the SAM components and that removal by stronger reagents results in *coordinate* release from the substrate of most of these components. The molecular organization of this adhesion complex is currently being studied by a number of approaches.

There are two conditions of cell manipulation where elevations in amounts of specific SAM components are uniquely interesting. When growing cells are pulsed with radioactive leucine-containing medium for only two hours, the SAM isolated at the end of the pulsing period has a tenfold higher level of the LETS glycoprotein and a vastly elevated amount of protein C_d (19, 29). This SAM is no doubt enriched in the footpad-derived class of material, rather than the footprint-derived class of SAM, which would be enriched by long-term growth in radioactive precursor. This would indicate that cellular footpads that are actively involved in cell-substrate adhesion sites functionally require much more LETS and protein C_d than the footprint material, which the cell discards during active movement across the substrate. Virtually nothing is known about the "natural" mechanism by which footprint material is generated. Secondly, attachment of preradiolabeled cells to fresh serum-coated substrate (after detachment by EGTA treatment) results in (a) an elevated amount of GAG-3 relative to GAG-1 and GAG-2, (b) higher levels of protein C_1, and (c) an elevated amount of the LETS glycoprotein, consistent with the quantitative importance of this component in footpad-mediated adhesions that form during the first hour or so of attachment.

Culp *et al.* (12) have identified two metabolically different pools of SAM protein—the protein accumulated over a long period of time was stably bound to the substrate, while a small pool identified during short radiolabeling periods with leucine turns over with a half-life of 2-4 hr. Slab SDS-PAGE analysis (19, 29) has indicated that the qualitative and quantitative distribution of SAM components is unaffected by chasing cells for 24 hr with nonradioactive leucine after 48-72 hr of growth in medium containing radioactive leucine.

Chasing of cells that have been radiolabeled for only 2 hr, on the other hand, demonstrated sizable turnover (half-life of 2-4 hr) of the elevated levels of the C_o glycoprotein (LETS) and the C_d protein, such that their final stable concentrations after twenty-four hr of chasing was similar to those found in long-term-growth preparations of SAM (29). The small pool of sulfated GAG was also found to turn over with a half-life of 2-4 hr. Thus, the very similar lability of the LETS glycoprotein, the C_d protein, and the sulfated GAG may suggest a common functional and metabolic role as footpads are formed at the leading lamellae (46, 47) of the moving cell and their eventual separation from the cell as "footprints" at the posterior end of the cell.

SCANNING ELECTRON MICROSCOPIC STUDY

The persistence of microfilament-associated actin, myosin, and perhaps several other proteins in SAM suggested that EGTA treatment was not simply dissociating outer surface membrane components with resultant detachment of the intact membrane of the cell from the substrate. Transmission EM studies (20-22) had already established that microfilaments were associated with the surface of the cell at points of cell-to-substrate and cell-to-cell adhesion. Chemical evidence (33) for the copurification of actin during surface membrane isolation was then obtained.

In order to clarify the cellular origin and morphology of the substrate-attached material, which appeared to be a complex of several surface-localized components, we resorted to scanning electron microscopy (34). As demonstrated previously by Revel *et al.* (13) and Rajaraman *et al.* (35), mouse fibroblasts were bound to the substrate principally by footpads of focally flattened surface membrane (see arrows in Fig. 1a) that could be easily observed during reattachment of EGTA-subcultured cells. Footpads were also observed in highly spread cells during the EGTA treatment, when the cell contracts into a round central cell body with thin membranous retraction fibers extending down to each footpad; the morphology of the footpads was minimally affected by the EGTA treatment.

The second step involved in cell detachment by EGTA treatment required shearing forces in the shaking culture with resultant breakage of the stressed retraction fibers (see arrows in Fig. 1b), repair of the membrane at the cell body and footpad ends, and liberation of the cell body into suspension. Again, the substrate-bound footpad appeared to be unaffected morphologically. The resultant SAM possessed the size, morphology, and density on the substrate of the original cellular footpads with a "nub" of membrane protruding from where it had apparently broken away from the cell body (Fig. 1c). Thus, EGTA treatment (34) results in breakage of the cell away from the substrate in a manner very analogous to that found for trypsin treatment by Revel *et al.* (13).

These studies indicate that EGTA treatment may have little effect on the organization of chemical components directly involved in the footpad adhesion site via the postulated cell surface complex of hyaluronate proteoglycans, the LETS glycoprotein, and actin-containing microfilaments. The most reasonable interpretation of all these studies on the mechanism of EGTA-mediated cell detachment is that the chelating agent disorganizes the supramolecular organization of the subsurface cytoskeleton involving microfilaments and perhaps microtubules, leaving a small pool of membrane-bound microfilaments. The organization of these cytoskeletal elements may be critically important for maintaining a somewhat rigid and organized lattice of surface membrane. Thus, cell-substrate adhesion probably involves two different levels of cell surface organization—interaction of outer surface components with substrate-bound serum components in a specific way, followed by reorganization of the subsurface cytoskeletal elements to form a more rigid surface membrane matrix around the adhesion site.

Some experimental evidence for a two-step adhesion mechanism has indeed been found by Mapstone and Culp (36). When EGTA-detached cells that had been preradiolabeled with leucine or glucosamine precursors were allowed to attach to fresh serum-adsorbed substrate, SAM deposition[1] occurred very rapidly and was more firmly bound to the substrate earlier in the attachment process than stable cell binding. In other words, it took more time for the cell to become firmly substrate-bound than for SAM to be deposited via surface membrane contact with the substrate. This slower step in adhesion may require the reorganization of cellular cytoskeletal elements to provide a stronger surface membrane lattice to resist abusive treatment in the environment of the cell. Definitive experimental evidence for such a two-step adhesion mechanism is still lacking.

[1]SDS-PAGE analysis has demonstrated that the usual SAM proteins and polysaccharides were deposited during the first hour of attachment of cells (19); however, higher concentrations of GAG-3, the LETS glycoprotein, and C_1 were found in attachment SAM's.

CONCLUSIONS

SDS-PAGE analysis of substrate-attached material resistant to EGTA treatment has resulted in the identification of several prominent components: the hyaluronate proteoglycans, the LETS glycoprotein, the myosin-like protein, and actin. Several other proteins remain to be identified, some of which may be associated with the actin-containing microfilaments. The (a) persistence of these components in SAM's deposited during a variety of cell attachment and growth conditions, (b) morphological identification of SAM as virtually intact membrane-enclosed footpads, and (c) sensitivity or resistance of SAM components to treatment with several reagents indicate that the components in SAM exist as a well-organized, tightly associated cell surface complex with *coordinate deposition* of these components during the formation of cell-substrate adhesion sites. The association of these components must not be stringently dependent upon Ca^{2+} and/or Mg^{2+} ions, since they persist during cell detachment by EGTA or EDTA treatments.

All the evidence taken together suggest two possible molecular mechanisms for cell-substrate adhesion dependent upon the role played by the hyaluronate proteoglycans; neither of these mechanisms has been proven by direct experimental evidence as yet. In the first mechanism (29), adhesion is mediated by five "layers" of chemical components. The naked plastic substrate (first layer) is covered by a layer of adsorbed serum proteins (30) (second layer), some of which interact specifically with the hyaluronate proteoglycans (third layer) perhaps in association with sulfated GAG in the glycocalyx of the cell surface. The proteoglycans might then interact with intrinsic transmembrane components (fourth layer), such as the LETS glycoprotein (39) and/or the membrane-associated myosin component, which in turn forms the anchorage point for the actin-containing microfilaments (fifth layer).

An alternative (29) to the mechanism described above involves hyaluronate proteoglycans that are not directly associated with the adhesion site but act as "modulators" due to the negatively charged and helically organized polysaccharide chains (37, 38) at sites adjacent to the true binding sites. The transmembranous LETS glycoprotein may then interact with specific serum components to form a four-layered structure. Some evidence for a modulating role for hyaluronic acid has recently been found by Atherly *et al.* (40), who isolated trypsin-resistant variants of CHO cells and found that they synthesize very little hyaluronic acid, but bind to the substrate much more "strongly" than parental cells.

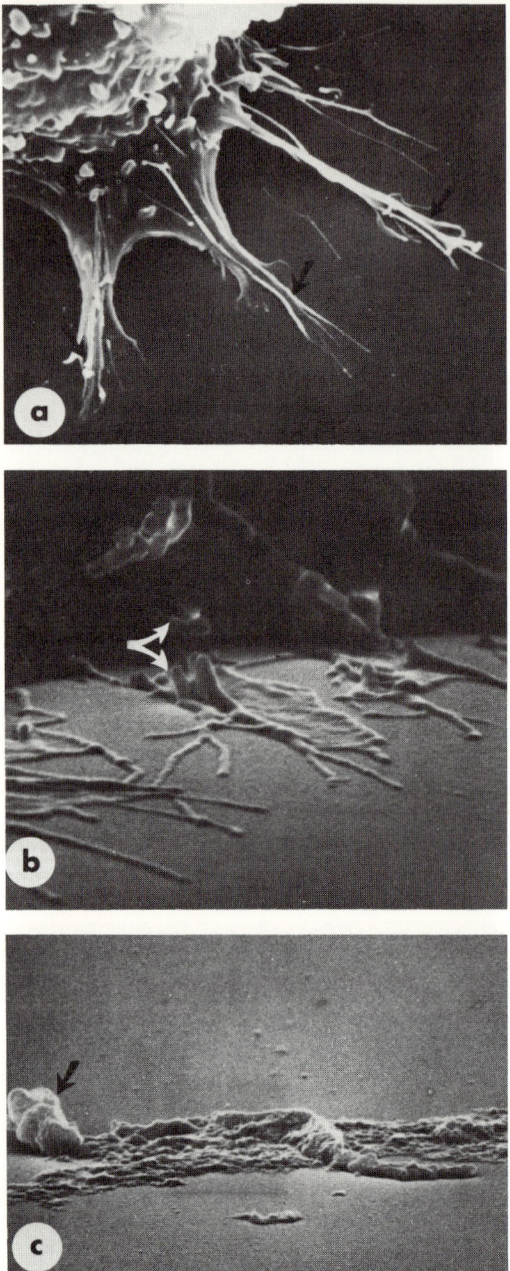
a
b
c

FIGURE 1 Cellular origin of substrate-attached material. Swiss 3T3 cells were detached by EGTA treatment after growth to confluence and then allowed to reattach to fresh glass substrate in medium containing 10% serum. Specimens were processed for scanning electron microscopic observation by fixation in buffered glutaraldehyde, dehydration with graded alcohols and Freon, and critical point drying. Samples were sputter-coated with gold-palladium before SEM analysis. (a) A cell 2 hr after attachment, showing well-developed footpads at the arrows (original magnification 6500×; 20° tilt); (b) a cell detaching after 10 min of treatment in 0.5 mM EGTA in PBS with gentle shaking, showing shearing of the labile retraction fiber (at the arrows) that connects the cell body to the footpad (original magnification 13,500×; 80° tilt); (c) morphology of SAM after removal of cells that had been grown for 24 hr, showing the remnant of the retraction fiber at the arrow orig-(original magnification 13,500×; 80° tilt).

The prominence of the LETS glycoprotein and actin-containing microfilaments in SAM may form an important molecular basis for the altered adhesiveness of malignant cells. Many virus-transformed cells have reduced concentrations of LETS in their surface membranes (27, 28); we (19) have similarly observed reduced concentrations of SAM-containing LETS in MSV-transformed BALB/c 3T3 and SV40-transformed Swiss 3T3 cells (the SVT2 transformant appears to be an exception). Virus-transformed cells also have poorly organized surface-associated microfilaments (20-23) when compared to normal and revertant cells, and this may generate a weaker surface lattice for organization around the footpad adhesion site. If the myosinlike protein proves to be a transmembrane protein (26) forming the anchorage site for the actin-containing microfilaments, its lability to proteases (45) may form the basis for the greater sensitivity of transformed cells because of their poorly organized microfilament arrays.

Virus-transformed cells have also been found to differ from normal and revertant cells in at least three other aspects of interaction with substrate-attached material. (a) Transformed cells deposit less SAM protein and polysaccharide per milligram of protein or per cell (5, 15) than normal or revertant cells. (b) When transformed cells are grown on SAM-coated substrates, their morphology becomes much more epithelioid; their movement over the substrate is reduced; and they resist crawling over (or under) adjacent cells (41). And (c) the presence of SAM coatings specifically inhibits the deposition of new SAM by freshly attaching transformed cells (which had been detached by EGTA treatment) and not by normal or revertant cells (36).

All of this evidence suggests that we may soon have a clearer concept of the molecular basis for differences in substrate adhesion displayed by normal and malignant cells. Several central areas of research command attention. How are the molecular components organized in a three-dimensional sense to form the footpad adhesion site? What are the identities and prospective functions of the unidentified components in SAM? How do these components become organized at the cell surface to generate a prospective "adhesion" complex? And finally, what is the mechanism by which footprint material is generated? Do footprints of SAM play a critical role in regulating the social behavior of substrate-bound cells? Much more biochemical evidence will be required to determine if there are subtle chemical differences in these SAM components between normal and malignant cells. The existence of a cell "surface" complex mediating adhesion may provide considerable versatility in altering the cell's social behavior during the malignant conversion process.

ACKNOWLEDGMENTS

The authors acknowledge funding from the National Cancer Institute (research grant 5-RO1-CA-13513), from the American Cancer Society (research grant BC-217 with partial support from the Ohio division), and from the National Institutes of Health (training grant 5-T01-GM-00171 to the Department of Microbiology) to L.A.C. J.R.R. is a postdoctoral fellow of the National Institutes of Health (fellowship 1F 32-6M-0524401). L.A.C. is a Harry H. Pinney Fellow in Cancer Research at Case Western Reserve University and a Career Development Awardee of the National Cancer Institute (1-K04-CA 70709).

REFERENCES

1. Nicolson, G. L. (1974), *Int. Rev. Cytol. 39*, 89.
2. Taylor, A. C. (1961), *Exp. Cell Res., Suppl. 8*, 154.
3. Weiss, L. (1962), *J. Theor. Biol. 2*, 236.
4. Curtis, A. S. G. (1973), *Prog. Biophys. Mol. Biol. 27*, 317.
5. Culp, L. A., and Black, P. H. (1972), *Biochemistry 11*, 2161.
6. Culp, L. A., and Black, P. H. (1972), *J. Virol. 9*, 611.

7. Rosenberg, M. (1960), *Biophys. J. 1*, 137.
8. Poste, G., Greenham, L. W., Mallucci, L., Reeve, P., and Alexander, D. J. (1973), *Exp. Cell Res. 78*, 303.
9. Yaoi, Y., and Kanaseki, T. (1972), *Nature (London) 237*, 283.
10. Revel, J. P., and Wolken, K. (1973), *Exp. Cell Res. 78*, 1.
11. Culp, L. A. (1975), *Exp. Cell Res. 92*, 467.
12. Culp, L. A., Terry, A. H., and Buniel, J. F. (1975), *Biochemistry 14*, 406.
13. Revel, J. P., Hoch, P., and Ho, D. (1974), *Exp. Cell Res. 84*, 207.
14. Buck, C. A., Glick, M. C., and Warren, L. (1970), *Biochemistry 9*, 4567.
15. Terry, A. H., and Culp, L. A. (1974), *Biochemistry 13*, 414.
16. Defendi, V., and Gasic, G. (1963), *J. Cell. Comp. Physiol. 62*, 23.
17. Kraemer, P. M. (1971), Complex Carbohydrates of Animal Cells: Biochemistry and Physiology of the Cell Periphery, *in* "Biomembranes," Vol. I, p. 67. Manson, L., ed., New York, Plenum.
18. Roblin, R., Albert, S. O., Gelb, N. A., and Black, P. H. (1975), *Biochemistry 14*, 347.
19. Culp, L. A. (1975), *Biochemistry 15*, 4094.
20. McNutt, N. S., Culp, L. A., and Black, P. H. (1971), *J. Cell Biol. 50*, 691.
21. McNutt, N. S., Culp, L. A., and Black, P. H. (1973), *J. Cell Biol. 56*, 412.
22. Perdue, J. F. (1973), *J. Cell Biol. 58*, 265.
23. Pollack, R., Osborn, M., and Weber, K. (1975), *Proc. Natl. Acad. Sci. U.S.A. 72*, 994.
24. Vessey, A. R., and Culp, L. A., submitted for publication.
25. Painter, R. G., Sheetz, M., and Singer, S. J. (1975), *Proc. Natl. Acad. Sci. U.S.A. 72*, 1359.
26. Olden, K., Willingham, M., and Pastan, I. (1976), *Cell 8*, 383.
27. Hynes, R. O., and Bye, J. M. (1974), *Cell 3*, 113.
28. Yamada, K. M., and Weston, J. A. (1975), *Cell 5*, 75.
29. Culp, L. A. (1976), *J. Supramol. Structure and Proceedings of the ICN-UCLA Conference on Cell Shape and Surface Architecture,* Squaw Valley, California, March 7-12, 1976, *5*, 239.
30. Culp, L. A., and Buniel, J. F. (1976), *J. Cell Physiol. 88*, 89.
31. Grinnell, F. (1974), *Arch. Biochem. Biophys. 160*, 304.
32. Grinnell, F. (1976), *Exp. Cell Res. 97*, 265.

33. Gruenstein, E., Rich, A., and Weihing, R. R. (1975), *J. Cell Biol. 64*, 223.
34. Rosen, J. R., and Culp, L. A. (1977), *Exp. Cell Res.*, in press.
35. Rajaraman, R., Rounds, D. E., Yen, S. P. S., and Rembaum, A. (1974), *Exp. Cell Res. 88*, 327.
36. Mapstone, T. B., and Culp, L. A. (1976), *J. Cell Sci. 20*, 479.
37. Atkins, E. D. T., and Sheehan, J. K. (1973), *Science 179*, 562.
38. Dea, I. C. M., Moorhouse, R., Rees, D. A., Arnott, S., Guss, J. M., and Balazs, E. A. (1973), *Science 179*, 560.
39. Hunt, R. C., and Brown, J. C. (1975), *J. Mol. Biol. 97*, 413.
40. Atherly, A. G., Barnhart, B. J., and Kraemer, P. M. (1976), *J. Cell. Physiol.*, in press.
41. Culp, L. A. (1974), *J. Cell Biol. 63*, 71.
42. O'Farrell, P. H. (1975), *J. Biol. Chem. 250*, 4007.
43. Heinegard, D., and Hascall, V. C. (1974), *J. Biol. Chem. 249*, 4250.
44. Hardingham, T. E., and Muir, H. (1974), *Biochem. J. 139*, 565.
45. Teng, N. N. H., and Chen, L. B. (1976), *Nature 259*, 578.
46. Abercrombie, M., Heaysman, J. E. M., and Pegrum, S. M. (1970), *Exp. Cell Res. 62*, 389.
47. Trinkaus, J. P., Betchaku, T., and Krulikowski, L. S. (1971), *Exp. Cell Res. 64*, 291.

SURFACE GLYCOPROTEINS OF NORMAL AND ABNORMAL PLATELETS

G. A. JAMIESON

The American National Red Cross
Blood Research Laboratory
Bethesda, Maryland

Georgetown University
Department of Biochemistry
Schools of Medicine and Dentistry
Washington, D. C.

INTRODUCTION

There are three major types of cells in the circulating blood; the red cell, which is concerned mainly with oxygen transport, the white cells, which are concerned with immunity and tissue repair, and the platelet or thrombocyte, which is concerned with coagulation and hemostasis. Circulating platelets do not normally interact with the endothelial cells lining the lumen of the blood vessel. However, when a blood vessel is severed, or when the endothelial cells are damaged, the platelets circulating in the blood are exposed to the subendothelial components of the vessel wall. Platelets immediately *adhere*, essentially as a monolayer to certain of these components and this initial adhesion is followed by the *release reaction*, in which certain constituents of the platelet are expelled from its interior, and by the *aggregation* of other platelets to those already adhering to the subendothelium to form clumps or microthrombi that extend into the lumen of the blood vessel. Finally, membrane fusion occurs between adjacent platelets leading to viscous metamorphosis and the retraction of the clot under the influence of contractile proteins in the platelet: this complex sequence of hemostatic events can be divided into six distinct stages (1).

These events are quite rapid. Under optimal conditions *in vitro* the adhesion of platelets to collagen is complete in a few seconds, aggregation and release in about three minutes, and membrane fusion, viscous metamorphosis, and clot retraction in about thirty minutes. Each of these obviously involves the platelet membrane; first, in the recognition by the platelet surface of specific subendothelial elements such as collagen and basement membrane, second, in the transfer of this information to the interior of the platelet leading to the release reaction, third, in the interaction of complementary sites on adjacent platelets during aggregation and, fourth, in the loss of membrane integrity leading to membrane fusion and viscous metamorphosis. Thus, the platelet membrane is of fundamental importance in its rapid reaction to specific stimuli. Emerging evidence suggests that platelet surface glycoproteins play a major role in these reactions, certainly in the early stages of adhesion and aggregation.

For a more full discussion of the role of the platelet in hemostasis, the reader is referred to several monographs in this area (2-5) and to recent reviews (6, 7).

PLATELET SURFACE ULTRASTRUCTURE

The platelet is apparently unique among cells in being formed by the fragmentation of its precursor rather than by the maturation of a more primitive cell; the megakaryocyte divides within the bone marrow, each megakaryocyte producing a total of about 2000 platelets (8). During the maturation of the megakaryocyte, demarcation membranes are formed by invagination of its plasma membrane (9) although an origin from internal structures has been suggested (10). These demarcation membranes are rich in carbohydrate and define the outer membrane of the developing platelet. Once formed, the platelets are discoid in shape, 2-3 μm in diameter, with a life-span of about 10 days and a turnover of about 5×10^{11} platelets/day in man.

The trilaminar structure of the platelet plasma membrane is morphologically indistinguishable from that of other cells. However, the surface membrane of the platelet is characterized by deep invaginations into the interior of the platelet leading to a complex network known as the surface-connected system or canalicular system. The canalicular system gives the platelet a "spongelike" quality that may assist in the transport of clotting factors, such as fibrinogen, and other plasma proteins within the circulation. It certainly gives to the platelet a high surface to volume ratio and brings the interior of the platelet and the exterior milieu into close proximity.

Outside the trilaminar membrane, both on the platelet surface and in the channels of the surface connected system, is an extensive exterior coat or glycocalyx that is 20-50 nm in depth and is thought to be rich in carbohydrate on the basis of staining with specific cytochemical reagents (11-14). Recent evidence, to be discussed later, suggests that components of this exterior coat may be important in platelet function. It can be removed by careful proteolysis without extensive changes in the ultrastructure of the membrane (14). At the present time it is not clear whether the glycocalyx should be regarded as an integral component of the platelet membrane or as being more or less loosely associated with it.

In its normal environment the platelet, together with its exterior coat, is surrounded by an ill-defined nimbus of plasma proteins, which has been called the plasmatic atmosphere; differences between the reactions of platelets in plasma and washed platelets may arise from the loss of components in this compartment (15).

Thus, the platelet surface may be regarded as comprising three different zones; the plasmatic atmosphere, which is retained by platelets only under special conditions of isolation and is required for ADP-induced aggregation, the glycocalyx, which remains with the platelet during most isolation procedures and may be the site of thrombin action and lectin-induced agglutination, and the platelet membrane itself, which contains the structural elements associated with transport, release, and the exchange of information between the interior and exterior of the platelet.

PLATELET CHARGE

Platelets have a high net negative charge with an isoelectric point at pH 3.6 and an electrophoretic mobility of -0.85 μm/sec/V/cm at pH 7 (16). This charge is at least partially due to sialic acid residues since treatment with neuraminidase removes about 10^5 molecules of sialic acid per μm^2 and results in 50% reduction in electrophoretic mobility with a concomitant increase in the isoelectric point to pH 4.2.

Calculations show that total surface sialic acid is about 1.9 $\times 10^6$ molecules of sialic acid per μm^2, which is approximately ten times the surface density on red cells.

The electrophoretic mobility of platelets is affected by aggregating agents (16, 17) by aging *in vitro* (18) and in certain pathological conditions such as Bernard-Soulier disease syndrome of congenital macrothrombocytic thrombopenia (19).

GLYCOPROTEINS OF THE PLATELET SURFACE

As mentioned above, indirect evidence for the presence of glycoproteins on the platelet surface arose from the use of specific cytochemical stains combined with electron microscopy and from studies on platelet electrophoresis and viral-induced agglutination.

Further indirect evidence for surface carbohydrate on the platelet has been provided by studies on the reaction of platelets with lectins having different carbohydrate specificities (20). These studies have shown that platelets are strongly agglutinated by wheat germ agglutinin, which has N-acetylglucosamine or sialic acid as its haptenic determinants, by *Ricinus communis*, which has galactose as its haptenic determinant, and by the lectin of *Phaseolus coccineus*; this last lectin has no known monosaccharide determinant but is inhibited by an extended glycopeptide of the red cell surface (21). Lectins with specificities towards glucose and mannose (concanavalin A, *Lathyrus sativus, Lens culinaris* and *Pisum sativum*) did not cause platelet agglutination or the release of ADP, but did cause the release of serotonin at high concentrations suggesting that different receptor sites may exist on the platelet surface for the release of the two metabolites; another galactose-specific lectin (*Agaricus bisporus*) also fell into this class. Interestingly, although concanavalin A did not cause the agglutination of intact platelets it did cause the agglutination of isolated membranes. This is consistent with the suggestion that this lectin may be a marker for galactosyltransferase in certain cells (22) and with the fact that the galactosyltransferase of the platelet is only manifest in isolated membranes but not in intact platelets (23) suggesting a location on the inner aspect of the membrane or buried within the intact membrane in a form that is revealed following the trauma of isolation.

Soybean lectin, with an N-acetylgalactosamine determinant, was virtually without effect on platelet agglutination or release.

Quantitative ultrastructural studies have also been carried out on the binding of lectins to platelets. The erythroagglutinating lectin (E-PHA) of *Phaseolus vulgarus* and the leukoagglutinating lectin (L-PHA) bind to intact platelets causing platelet aggregation and the inhibition of adenyl cyclase, changes that also occur during aggregation induced by thrombin (25). About 600,000 molecules of E-PHA per platelet bind to the surface with an apparent dissociation constant of 0.5×10^{-7} *M*, while about 300,000 molecules of L-PHA bind with an apparent dissociation constant of 4×10^{-7} *M*. Although the lectin of *L. culinaris* does not cause aggregation it does bind tightly (400,000 molecules per platelet, 1.4×10^{-7} *M*) and there is an approximate doubling of receptor sites for this lectin on prior treatment of the platelet with thrombin or L-PHA. This binding has been used for specific labeling in the identification of platelet membrane fractions following thrombin treatment (24).

The use of ferritin-conjugated PHA and the freeze-fracture technique demonstrated that the morphological changes induced by thrombin involved a decrease in the number of lectin receptor sites on the platelet surface and a marked increase in their number in the canalicular system (26). It was suggested that this may arise from the fusion of granule membranes with the plasma membranes of the canalicular system.

Direct evidence for the presence of surface glycoproteins on the platelet was obtained by the isolation of glycopeptides following treatment with trypsin, pronase and papain (27, 28), plasmin and thrombin (29), and chymotrypsin (30). In each case the glycopeptides obtained were generally similar to those obtained with trypsin, the most extensively studied system (28). Three size classes of glycopeptides were obtained with molecular weights of 120,000 (glycopeptide I), 22,500 (glycopeptide II) and 5000 (glycopeptide III); the highest molecular weight glycopeptide (glycopeptide I) has been termed a macroglycopeptide since its molecular weight appears to be an order of magnitude higher than those of glycopeptides from plasma proteins or red cell stroma. The platelet glycopeptides showed several analytical differences between each class; the macroglycopeptide (glycopeptide I) contained about 70% carbohydrate with galactose as essentially its sole neutral sugar, with equimolar amounts of galactosamine and glucosamine, presumably as their N-acetyl derivaties, and with heterosaccharide residues joined to the polypeptide backbone in O- and N-glycosidic linkages. On the other hand, the glycopeptide of lowest molecular weight (glycopeptide-III) contained equimolar amounts of galactose and mannose, had asparagine as its principal amino acid, and appeared to contain only N-glycosidic linkages; the glycopeptide of intermediate molecular weight (glycopeptide-II) was also intermediate in properties but was not studied extensively because it was present in relatively small amounts.

The presence of mucopolysaccharides on the outer surface of the platelet has frequently been suggested on the basis of ultrastructural studies with specific cytochemical reagents. In the proteolytic studies summarized above, it was found that treatment with enzymes that caused the release reaction (trypsin, pronase, papain, plasmin) resulted in the solubilization of a carbohydrate component (glycopeptide-O) that was eluted from Sephadex G200 at the void volume and had properties consistent with being the chondromucopeptide derived from chondroitin sulfate (29). On the other hand, this component was not obtained following digestion with chymotrypsin, which does not cause the release reaction, or among the tryptic products of isolated membranes. These results suggest that mucopolysaccharides are not components of the platelet membrane, as such, but are probably released from the granules under appropriate circumstances.

Differences between the glycopeptides I, II, and III were also shown in their effect on the agglutination of platelets induced by plant lectins (20). Platelet agglutination induced by wheat germ agglutinin was inhibited by the macroglycopeptide but not by glycopeptide II, while the agglutination induced by *P. coccineus* was inhibited by glycopeptide II but not by the macroglycopeptide; glycopeptide III was without effect in either case. Subsequent results have related these observations to the structure of the parent glycoprotein.

A number of studies have been carried out to identify the glycoprotein precursor of the glycopeptides (23, 31-33). Taken together, these studies have indicated that the isolated platelet plasma membrane contains three components that are positive to the periodate-Schiff reagent and that have been termed glycoproteins I, II, and III with approximate molecular weights of 150,000, 120,000, and 100,000, respectively, all of which were thought to be susceptible to lactoperoxidase catalyzed iodination.

A recent study (34) requires reinterpretation of these previous results. This study has utilized surface probes such as neuraminidase transglutaminase/^{14}C glycine ethyl ester (35), lactoperoxidase-catalyzed iodination (32) and proteolytic digestion with trypsin and chymotrypsin. This study (34) has shown that a fourth component, glycoprotein IV of molecular weight 87,000, can be identified under conditions of improved electrophoretic resolution in addition to the previously identified glycoproteins I, II, and III. A densitometric scan of a PAS stained gel of normal platelet membranes is shown in Fig. 1a. Furthermore, since one half of glycopro-

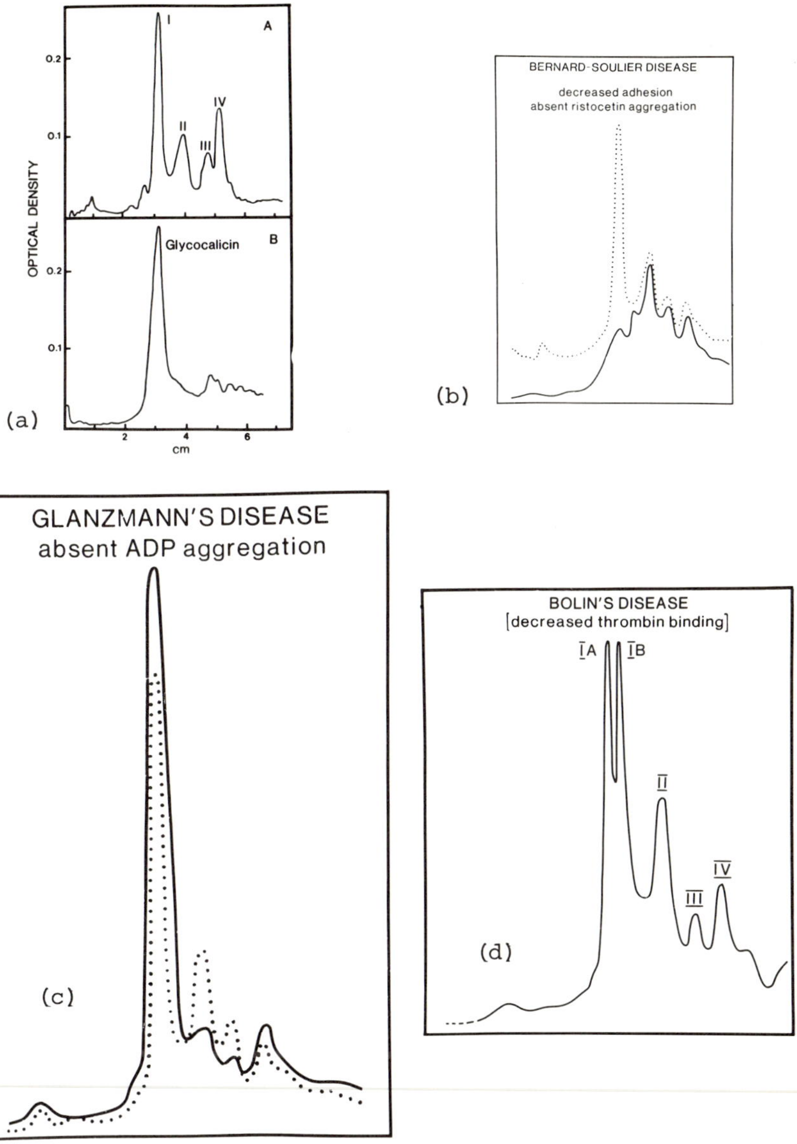

FIGURE 1 Distribution of glycoproteins in plasma membranes from normal and abnormal platelets following SDS-gel electrophoresis and periodate Schiff staining. In the abnormal samples the normal glycoprotein distribution is shown by a dotted line. (a) Normal platelets. The upper portion of the figure (A) is the pattern of membrane bound glycoproteins while the lower portion (B) is the glycoprotein pattern found in the soluble fraction of the platelet homogenate. (b) Bernard-Soulier disease. (c) Glanzmann's thrombasthenia. (d) Bolin's disease.

tein II was degraded by trypsin within one minute but only a further 20% over the following sixty minutes, it appeared likely that two separate components were migrating in the same electrophoretic region. In this study (34) significant labeling with the lactoperoxidase reagent was found only in glycoprotein III while the transglutaminase reagent gave principal labeling in glycoprotein II.

Parenthetically, it may be noted that the terminology that has been introduced in the case of platelet membrane glycoproteins is unfortunate and misleading. The electrophoretic studies in the presence of reducing agents in sodium dodecyl sulfate have not identified glycoproteins, as such, but merely electrophoretic components that react with the periodate-Schiff reagent and may, in fact, be subunits of larger molecules. Moreover, the studies outlined above show that a single "glycoprotein" can be resolved into two components under improved electrophoretic conditions and that there are probably two components in the region of glycoprotein II. Thus, the previously described glycoproteins I, II, and III probably represent at least five components. In this respect, the terminology of Steck (36) with regard to the red cell is much more felicitous in which the various electrophoretic components are described as periodate-Schiff (PAS) 1, PAS-2, etc.

Examination of the soluble fraction of a platelet homogenate shows the presence of a single PAS-positive component that comprises about 10% of the total sialic acid of the intact platelet. This component is apparently released from the platelet surface on membrane disruption since it is accessible to neuraminidase and chymotrypsin in the intact platelet and is radiolabeled with the transglutaminase reagent. This component has been termed glycocalicin because of its origin in the platelet glycocalyx. The electrophoretic pattern is shown in Fig. 1a.

Glycocalicin and the membrane-bound glycoprotein I appear to be similar in terms of electrophoretic mobility in the presence of SDS. In addition, chicken antiglycocalicin antiserum is able to cause the aggregation of both isolated platelet-membranes that glycocalicin and washed platelets. Furthermore, the macroglycopeptides derived from the membrane-bound and soluble glycoproteins give reactions of immunological identity and have similar carbohydrate and amino acid composition.

Four possibilities seem to exist: one is that the soluble and membrane-bound glycoproteins are identical and that the soluble form is released from the membrane by the effects of platelet homogenization and membrane isolation. However, further amounts of the membrane-bound glycoprotein were not solubilized following prolonged sonication of isolated membranes. The second possibility is that the macroglycopeptide portions of the membrane-bound and soluble forms are identical but that the (nonglyco-)peptide "tail" portions differ so that this portion of the soluble glycoprotein is more loosely bound to the membrane and therefore more readily displaced during homogenization and isolation. The third possibility, which is closely related to the second, is that the structural differences in the soluble and membrane-bound forms are sufficiently great that the tail portion is not inserted within the membrane itself but that the soluble glycoprotein is loosely adsorbed to the platelet surface. The fourth possibility is that the soluble and membrane-bound forms are structurally identical but that individual molecules are bound to regions of somehow differing affinity in the membrane perhaps reflecting regions capable of varying degrees of lipophilic interaction.

The two latter possibilities are, perhaps, supported by the fact that the ratio of radiolabeling with ^{14}C-glycine ethyl ester to PAS-stainability is much higher in the soluble glycoprotein than in the equivalent membrane-bound glycoprotein, suggesting that these portions of the molecule are more accessible in the former case. On the other hand, both glycocalicin and glycoprotein I appear to be completely inaccessible to lactoperoxidase-catalyzed iodination in the intact platelet but the former can be readily iodinated in both the peptide and macroglycopeptide portions when isolated in pure form.

This question of accessibility of surface components is particularly important since widespread use of the lactoperoxidase-catalyzed iodination procedure has tended to indicate that only glycoproteins are accessible at the surface of platelets (32) and other cells (36). The application to platelets of the transglutaminase reaction with ^{14}C-glycine ethyl ester, originally developed for analysis of a viral coat protein (35), shows that a number of other components, both glycoprotein and nonglycoprotein in nature, are available at the cell surface. This observation emphasizes that the orientation of individual membrane components cannot be determined by a single labeling procedure and that differences in chemical composition and protein conformation must be taken into account as well as accessibility to the reagent itself.

The fluid mosaic model of membrane structure (37) characterizes membrane surface proteins as intrinsic or extrinsic depending on the degree to which they are bound to the membrane bilayer. Although these proteins are thought to differ in their structural and functional characteristics in the fluid mosaic model, the present findings suggest that closely similar, if not identical, glycoproteins may have the properties of both intrinsic and extrinsic molecules.

A tentative model for the orientation of the various components of the platelet surface may be suggested from this work. The most loosely bound components, which are released from the platelet surface in soluble form following platelet homogenization, include glycocalicin and another component both of which are labeled by the transglutaminase reagent in the intact platelet. However, glycocalicin is not accessible in its entirety to membrane-labeling reagents since differences have been found in the susceptibility to radiolabeling of the glycoprotein before and after solubilization.

Glycoproteins I, II, and III appear to be integral components of the platelet membrane and to be present on its outer surface, based on their accessibility in varying degrees to proteolytic enzymes. Glycoprotein III appears to be the only glycoprotein labeled with the lactoperoxidase reagent. On the other hand, glycoprotein II is most strongly labeled by the transglutaminase method in the intact platelet. Glycoprotein I is only slightly labeled although its analog, glycocalicin, is strongly labeled under the same conditions. Four other membrane-bound proteins also appear to be accessible to this reagent.

Glycoprotein IV is not radiolabeled by either the transglutaminase or lactoperoxidase methods. It is not accessible to trypsin or chymotrypsin in the intact platelet but is susceptible to proteolytic digestion in isolated plasma membranes, suggesting that it is present on the inner aspect of the membrane or is buried within the membrane of the intact platelet but is exposed during isolation procedures. It may be noted that the membrane changes that are associated with the platelet release reaction induced by trypsin are not sufficient to expose glycoprotein IV to proteolytic attack in intact platelets.

Glycocalin itself has been isolated in pure form (38, 39). Tryptic digestion of the purified glycocalicin yields a glycopeptide of molecular weight 118,000 and a peptide of molecular weight 45,000. These data suggest, within the limits of experimental error, that these two components alone make up the intact glycoprotein. The glycopeptide obtained by proteolysis of the glycoprotein is virtually identical in molecular weight and in carbohydrate and amino acid analysis with the purified macroglycopeptide isolated by trypsinization of intact platelets (27).

The presence of similar molar ratios of glucose in the glycoprotein, the tryptic peptide, and the macroglycopeptide is of particular interest (Table I). The presence of glucose in the macroglycopeptide (20), and in glycopeptides isolated

TABLE I Carbohydrate and Amino Acid Analyses of Glycocalicin and Tryptic Products (Moles %)

	Glycocalicin	Macroglycopeptide	Tryptic peptide
NANA	12.5	21.1	1.5
Man	1.2	0.9	0.8
Fuc	1.9	1.3	0.7
Gal	15.1	20.2	1.9
Glc	2.3	1.6	1.5
GlcNAc	7.2	10.9	0.3
GalNAc	5.9	10.0	0.2
Lys	3.3	2.1	4.2
His	1.1	0.6	1.7
Arg	1.1	trace	1.6
Asp	4.7	0.9	8.9
Thr	7.6	7.9	5.5
Ser	4.9	4.8	18.6
Glu	5.8	3.6	14.6
Pro	6.8	6.8	3.3
Gly	2.7	0.7	16.3
Ala	2.3	1.6	8.2
Cys	n.d.	n.d.	n.d.
Val	2.5	0.6	3.3
Met	0.8	trace	trace
Ileu	1.3	1.0	2.2
Leu	6.7	2.3	4.0
Tyr	1.2	trace	1.7
Phe	1.6	1.1	1.6

from other glycoproteins, has previously been interpreted rather cautiously because of the possibility of their contamination with glucose from gel filtration and ion exchange media, sucrose gradients, etc. The similar molar ratios of glucose in glycocalicin and its macroglycopeptide, although isolated at different times and by very different techniques, strongly suggests that glucose is an integral part of this glycoprotein of the platelet surface.

Complementary differences are shown between the amino acid composition of the macroglycopeptide and tryptic peptide isolated from purified glycocalicin (Table I). In particular, the macroglycopeptide is relatively enriched in serine, glycine, and glutamic acid. Overall, this tryptic peptide lacks the preponderance of hydrophobic amino acids present in the tryptic peptide isolated from red cell glycophorin (40, 41). However, Segrest and Feldman (42) have emphasized that hydrophobic sequences as short as ten residues may be sufficient for membrane penetration. Since the tryptic peptide of platelet glycocalicin (M.W. 45,000) is about ten times larger than that from red cell glycophorin (M.W. 4000) it is possible that short hydrophobic domains may be detected in the former when complete amino acid sequence data become available. These authors (42) have also suggested that domains of intermediate hydrophobicity may be present in proteins that have dynamic or reversible binding with membranes. In the light of this, it is interesting to speculate whether there may be differences in the degree or extent of hydrophobic domains in glycocalicin as compared to glycoprotein I, which remains bound to the membrane following platelet homogenization (34).

The immunochemistry of the isolated glycoprotein and its reaction with lectins is also of interest. Extensive studies on platelet immunochemistry have led to the production of monospecific antithrombocyte antisera (43, 44) using the soluble fraction of the platelet homogenate as antigen. The thrombocyte-specific antigen was thought to be a glycoprotein of molecular weight 117,000 and the platelet macroglycopeptide has been found to inhibit the agglutinating activity of the thrombocyte-specific antisera (27). The identical source of glycocalicin and the previous two antigens and the similarity of the immunological data leave little doubt but that glycocalicin is the previously described thrombocyte-specific antigen of platelet homogenates. However, at least three individual determinants are associated with glycocalicin, the macroglycopeptide, and the tryptic peptide. Their combined role in the immunological reactivity of the platelet remains to be elucidated.

In the case of the reaction with lectins, WGA is known to cause an agglutination of platelets and platelet membranes that is inhibited by the platelet macroglycopeptide (20). WGA has been found to give a precipitin reaction with purified glycocalicin. However, purified glycocalicin does not give a precipitin reaction with the lectin of *P. coccineus*, which is also known to cause platelet agglutination but which is inhibited by the intermediate glycopeptide of platelets (glycopeptide II) (20). Thus, glycocalicin is probably a receptor for WGA in effecting agglutination but not for *P. coccineus*. This receptor appears to be affected by the terminal sialic acid of the platelet heterosaccharide chains since the precipitin reaction is lessened following treatment with neuraminidase. Although the lectin of *A. bisporus* gives a precipitin reaction with glycocalicin, its limited activity in the agglutination of platelets may reflect an insufficient number of galactose receptors in the intact platelet since its precipitin activity is increased by treatment of purified glycocalicin with neuraminidase.

Glycocalicin inhibits platelet aggregation induced by thrombin or by ristocetin/factor VIII (45), suggesting that it is the receptor on the platelet surface for these two reagents in inducing platelet aggregation. This observation is consistent with the finding that there is a single receptor for thrombin binding that is located in the glycocalyx rather than on the membrane, as such (46) although the binding of thrombin at the surface shows negative cooperativity (47).

SURFACE GLYCOPROTEIN OF ABNORMAL PLATELETS

Platelets from patients suffering from certain hematological diseases have been found to show abnormal patterns of membrane glycoproteins. Bernard-Soulier disease, or macrothrombocytic thrombopenia, is a congenital platelet defect that is characterized by giant platelets with marked structural abnormalities. Bernard-Soulier platelets are deficient in sialic acid (19), do not aggregate with ristocetin (48, 49), and show decreased adhesion to endothelium (50). Examination of the membrane fraction from Bernard-Soulier platelets has shown the virtual absence of glycoprotein I in this condition although glycoproteins II, III, and IV are relatively unaffected (51) (Fig. 1b); it may be noted that we have found that Bernard-Soulier platelets also lack glycocalicin. Since glycocalicin is the receptor for ristocetin-induced aggregation (45) the absence of it and glycoprotein I suggests the biochemical basis for the functional defects observed in this condition.

Glanzmann's thrombasthenia is characterized by faulty clot retraction, prolonged bleeding time, and the absence of platelet aggregation by ADP or collagen. Examination of platelet membranes from these patients has shown that there is a reduction of about 50% in the amount of glycoprotein II present (33) suggesting that only one of the two glycoproteins present in this region is deleted in this disease and that glycoproteins II, III, and IV are within normal limits (Fig. 1c).

A third type of platelet glycoprotein abnormality has recently been detected by Dr. Robert Bolin of our laboratory. Two unrelated patients were found who had a history of easy bruising and significant bleeding episodes following trauma. In one patient, but not in the other, secondary aggregation with ADP was absent. When platelet membranes from these two patients were examined, the band corresponding to glycoprotein

I was found to exist in two forms: one with the normal electrophoretic mobility, and another that had an electrophoretic mobility corresponding to a molecular weight of 10-12,000 daltons less. These two components were present in equal amounts and their combined amount equaled the amount of glycoprotein I present in normal platelet membranes. Further studies showed that platelets from these patients would bind only about one half as much thrombin as the normal control corresponding to 2700 molecules of thrombin bound per unit of thrombin added as against 6500 for the controls. The pattern is shown in Fig. 1d.

CONCLUSION

The blood platelet is an intensely reactive cell that responds to a variety of stimuli to effect hemostasis. Emerging data suggest that surface glycoproteins are involved in the phenomena of platelet adhesion, the binding of thrombin, and aggregation induced by ristocetin and ADP. At present, three conditions are known to be associated with the absence of specific membrane glycoproteins in different cell types: in platelets, glycoproteins I and II are lacking in Bernard-Soulier and Glanzmann's disease, respectively, while the case described by Bolin appears to be unique since the indication is that both normal and inactive forms of the receptor glycoprotein are present. The third case is the absence of glycophorin in $E_N(a^-)$ individuals although this does not appear to be associated with any changes in red cell function.

REFERENCES

1. Stormorken, H., and Owren, P. A. (1971), *Seminars Haematol.*, *8*: 3.
2. Johnson, S. A. (ed.) (1971), "The Circulating Platelet," Academic Press, New York.
3. Brinkhous, K. M., Shermer, R. W., and Mostofi, F. K. (eds.) (1971), "The Platelet," Williams and Wilkins, Baltimore.
4. Sherry, S., and Scriabine, A. (eds.) (1974), "Platelets and Thrombosis," University Park Press, Baltimore.
5. Baldini, M. G., and Ebbe, S. (eds.) (1974), "Platelets: Production, Trans- and Storage," Grune and Stratton, New York.

6. Weiss, H. J. (1975), *New Eng. J. Med., 293*: 531 and 580.
7. Marcus (1969), *New Eng. J. Med., 280*: 1213, 1278, and 1330.
8. Harker, L. A. (1971), "The Platelet" (Brinkhous, K. M., Shermer, R. W., and Mostofi, F. K., eds.), pp. 13-25, Williams and Wilkins, Baltimore.
9. Behnke, O. (1968), *J. Ultrastruct. Res., 24*: 412.
10. Schulz, H. (1966), *Verh. Deut. Ges. Pathol., 50*: 239.
11. Behnke, O. (1967), *Anat. Rec., 158*: 121.
12. Behnke, O. (1968), *J. Ultrastruct. Res., 24*: 51.
13. Rambourg, A., and Leblond, C. P. (1967), *J. Cell Biol., 32*: 27.
14. Hovig, T. (1968), *Ser. Haemat., 1*: 3.
15. Ardlie, N. G., and Han, P. (1974), *Brit. J. Haemat., 26*: 357.
16. Seaman, G. V. F., and Vassar, P. S. (1966), *Arch. Biochem. Biophys., 117*: 10.
17. Hampton, J. R., and Mitchell, J. R. A. (1966), *Brit. Med. J., 1*: 1074.
18. Mason, R. G., and Shermer, R. W. (1971), *In* "The Platelet" (Brinkhous, K. M., Shermer, R. W., and Mostofi, F. K., eds.), pp. 123-132, Williams and Wilkins, Baltimore.
19. Grottum, K. S., and Solum, N. W., (1969), *Brit. J. Haematol., 16*: 277.
20. Greenberg, J. H., and Jamieson, G. A. (1974), *Biochim. Giophys. Acta, 345*: 231.
21. Kubanek, J., Entlicher, G., and Kocourek, J. (1973), *Biochim. Biophys. Acta, 304*: 93.
22. Lamont, J. T., Perrotto, J. L., Weiser, M. M., and Isselbacher, K. J. (1974), *Proc. Nat. Acad. Sci. USA, 71*: 3726.
23. Barber, A. J., and Jamieson, G. A. (1971), *Biochim. Biophys. Acta, 252*: 546.
24. Rittenhouse-Simmons, S., and Deykin, D. (1976), *Biochim. Biophys. Acta, 426*: 688.
25. Majerus, P. W., and Brodie, G. N. (1972), *J. Biol. Chem., 247*: 4253.
26. Feagler, J. R., Tillack, T. W., Chaplin, D. D., and Majerus, P. W. (1974). *J. Cell Biol., 60*: 541.
27. Pepper, D. S., and Jamieson, G. A. (1969), *Biochemistry, 8*: 3362.
28. Pepper, D. S., and Jamieson, G. A. (1970), *Biochemistry, 9*: 3706.
29. Jamieson, G. A., Fuller, N. A., Barber, A. J., and Lombart, D. (1971), *Ser. Haemat., 4*: 125.
30. Barber, A. J., and Jamieson, G. A. (1971), *Biochemistry, 10*: 4711.

31. Nachman, R. L., and Ferris, B. (1972), *J. Biol. Chem., 247*: 4468.
32. Phillips, D. R. (1972), *Biochemistry, 11*: 4582.
33. Nurden, A. T., and Caen, J. P. (1974), *Brit. J. Haematol., 28*: 253.
34. Okumura, T., and Jamieson, G. A., (1976), *J. Biol. Chem.*, in press.
35. Brewer, G. J., and Singer, S. J. (1974), *Biochemistry, 13*: 3580.
36. Steck, T. L. (1974), *J. Cell Biol., 62*: 1.
37. Singer, S. J., and Nicholson, G. L. (1972), *Science, 175*: 720.
38. Lombart, C., Okumura, T., and Jamieson, G. A. (1974), *FEBS Letters, 41*: 31.
39. Okumura, T., Lombart, C., and Jamieson, G. A. (1976), *J. Biol. Chem.*, in press.
40. Winzler, R. J. (1969), *in*: "Red Cell Membrane-Structure and Function" (Jamieson, G. A., and Greenwalt, T. J., eds.), pp. 157-171, Lippincott, Philadelphia.
41. Marchesi, V. T., Tillack, T. W., Jackson, R. L., Segrest, J. P., and Scott, R. E. (1972), *Proc. Nat. Acad. Sci. USA, 69*: 1445.
42. Segrest, J. P., and Feldman, R. J. (1974), *J. Mol. Biol., 87*: 853.
43. Milgrom, F. F., Campbell, W. A., and Witebsky, E. (1968), *Vox Snag., 15*: 418.
44. Hanna, M., and Nelken, D. (1971), *Immunology, 20*: 533.
45. Okumura, T., and Jamieson, G. A. (1976), *Thromb. Res., 8:* 701.
46. Tollefsen, D. M., and Majerus, P. W. (1976), *Biochemistry, 15*: 2144.
47. Tollefsen, D. M., Feagler, J. R., and Majerus, P. W. (1974), *J. Biol. Chem., 249*: 2646.
48. Howard, M. A., Hutton, R. A., and Hardistry, R. M. (1973), *Brit. J. Med., 4*: 589.
49. Caen, J. P., and Levy-Toledano, S. (1973), *Nature, New Biol., 244*: 149.
50. Weiss, H. J., Tschopp, T. B., and Baumgartner, H. R. (1976), *New Eng. J. Med., 293*: 619.
51. Nurden, A. T., and Caen, J. P. (1975), *Nature, 255*: 720.

MAJOR GLYCOSPHINGOLIPIDS OF BOVINE-ERYTHROCYTE MEMBRANES[1]

J. L. CHIEN
S.-C. LI
Y.-T. LI

Department of Biochemistry
Tulane University
New Orleans, Louisiana

Delta Regional Primate Research Center
Covington, Louisiana

Glycosphingolipids of bovine erythrocyte membranes differ considerably from that of rhesus monkey and human. We have studied the glycolipids of the bovine erythrocyte membrane in great detail. The following neutral glycosphingolipids have been isolated and characterized:

Glcβ1→1Cer
Galβ1→4Glcβ1→1Cer
Galα1→4Galβ1→4Glcβ1→1Cer
Galβ1→4GlcNAcβ1→3Galβ1→4Glcβ1→1Cer
Galα1→3Galβ1→4GlcNAcβ1→3Galβ1→4Glcβ1→1Cer

In addition, two other neutral glycolipids with mobility slower than the pentaglycosylceramide were also detected. The following gangliosides have been isolated and characterized:

NeuAcα2→3Galβ1→4Glcβ1→1Cer
NeuGcα2→3Galβ1→4Glcβ1→1Cer
NeuAcα2→3Galβ1→4GlcNAcβ1→3Galβ1→4Glcβ1→1Cer
NeuGcα2→3Galβ1→4GlcNAcβ1→3Galβ1→4Glcβ1→1Cer
NeuGcα2→3Galβ1→4GlcNAcβ1→3Galβ1→4GlcNAcβ1→3Galβ1→4Glcβ1→1Cer

[1]This investigation was supported by National Science Foundation Grant BMS 74-03500, National Institutes of Health Grants NS 09626 and FR 00164.

INTRODUCTION

Erythrocytes membranes have been used as a model for the elucidation of the structure and function of cell membranes. This is mainly due to the fact that erythrocytes are readily available in large quantities and their plasma membrane can be easily prepared. In order to appreciate the structure and function of the cell membrane, it is imperative to fully elucidate the nature of its constituents. Since the discovery of globoside and hematoside from human erythrocytes by Klenk and Lauenstein (1) and Yamakawa and Suzuki (2), the glycosphingolipids in erythrocytes membranes have been the subject of great interest (3). Besides simple glycosphingolipids such as glucosylceramide, lactosylceramide, and ceramide trihexoside, the glycosphingolipids of mammalian erythrocytes can be classified into two types: (a) those of human, pig, guinea pig, and sheep contain N-acetylgalactosamine (4-7); (b) those of rabbit (8) and bovine (9) contain N-acetylglucosamine. In this paper, we report our results on bovine erythrocyte glycosphingolipids with emphasis on the glucosamine containing neutral glycosphingolipids and gangliosides.

EXPERIMENTAL PROCEDURE

MATERIALS

Bovine blood was obtained fresh from the local slaughter house in Covington, Louisiana. Precoated silica gel plates are products of Analtech, Inc., Newark, Delaware, and Brinkman Instruments, Inc., Westbury, New York. Fatty acid methyl esters, 10% DEGSPS on 80/100 Supelcoport and 3% ECNSS-M on gas chrom Q (100/120) were purchased from Supelco, Inc., Bellefonte, Pennsylvania. The following glycosphingolipids used as references were isolated in this laboratory: galactosylceramide, hematoside (GM3), GM1 and GD1a from human brain, CTH and globoside from human erythrocytes. Forssman hapten was donated by Dr. John M. McKibbin, University of Alabama.

THIN LAYER CHROMATOGRAPHY

The following solvent systems were used: I, chloroform:methanol:2.5 *N* NH_4OH (60:30:6); II, chloroform:menthanol:2.5 *N* NH_4OH (60:35:8); III, chloroform:methanol:H_2O (60:35:8); IV, chloroform:methanol:H_2O (60:40:9); V, chloroform:methanol:H_2O (20:50:10); VI, propanol:H_2O (7:3); VII, chloroform:methanol:H_2O (60:30:6).

EXTRACTION OF GLYCOSPHINGOLIPIDS FROM BOVINE ERYTHROCYTES

Bovine stroma were prepared according to the method of Grochowska and Koscielak (10). Acetone dried bovine stroma, 50 g were swelled in and homogenized with 200 ml of 0.01 *M* KCl solution. To this homogenate, 800 ml of tetrahydrofuran (THF) was added (11). The mixture was stirred overnight at room temperature. After filtering through a Büchner funnel, the residue was re-extracted three times with 400 ml of THF:H_2O (8:1). The combined filtrate was dried in a rotary evaporator. To the same flask, 100 ml of 0.6 *N* NaOH in methanol was added and the mixture was incubated at 37°C for 6 hr. After exhaustive dialysis, the contents were dried completely in a rotary evaporator.

DEAE-Cellulose Column Chromatography

DEAE cellulose, Cl^- form (cellex D) was washed twice with five volumes of 0.1 *N* NaOH. After removing the fine particles and rinsing three times with H_2O, the cellulose was converted into the acetate form by washing twice with 1 *N* acetic acid (12). The sediment was washed three times with H_2O and twice with methanol. After decanting the methanol, the cellulose was suspended in solvent V, packed into a column (2.9 × 30 cm) and equilibrated overnight with the same solvent. The crude lipid extract was dissolved in chloroform:methanol (2:1) and applied to the column. Neutral glycosphingolipids were eluted from the column with 1.2 liters of solvent V. Gangliosides were separated into three fractions by eluting the column with the same solvent containing (A) 0.01 *M*, (B) 0.02 *M*, (C) 0.05 *M* sodium acetate. Effluent from 0.01 *M* sodium acetate (Fraction I) contained GM3 and *N*-acetylglucosamine containing gangliosides. This fraction was further purified by silica gel H column chromatography. Fractions obtained by eluting with 0.02 and 0.05 *M* sodium acetate are currently under study.

SILICA GEL H COLUMN CHROMATOGRAPHY

Silica Gel H was activated overnight at 110° and packed into a column (1.9 × 23 cm) with chloroform. Fraction I from DEAE-cellulose column was dissolved in chloroform:methanol (2:1) and applied to the column. Gangliosides were eluted from the column by 500 ml of solvent I, followed by 500 ml of solvent II. Approximately 5 ml fractions were collected. Aliquots of 50 µl from each fraction were checked for its ganglioside content by thin-layer chromatography (solvent II). Fractions containing certain ganglioside were pooled and further purified by thin-layer chromatography (solvent III or IV). When neutral glycosphingolipids were separated by silica gel H column (1.9 × 26 cm), the column was first washed with chloroform followed by solvent VII.

METHYLATION OF GLYCOSPHINGOLIPIDS

The glycolipid (about 1 mg) was treated with dimethylsulfinyl carbanion base and then methylated with methyl iodide (13). The methylated glycolipid was neutralized and dialyzed according to the method of Stoffel and Hanfland (14). After drying completely in a rotary evaporator, the residue was extracted three times with 500 µl chloroform. The chloroform extract was dried under N_2 and purified by thin-layer chromatography. The purified methylated glycolipid was hydrolyzed with 0.7 *N* H_2SO_4 in 80% aqueous acetic acid (14), then reduced and acetylated according to Björndal *et al.* (15). The partially methylated alditol acetates were analyzed with Finnigan Model 3300 Gas Chromatography - Mass Spectrometer.[1] Neutral sugars were separated isothermally at 180° in a column packed with 3% OV 275, 100/120 mesh Supelcoport. Amino sugars were separated in a column packed with OV 17, 100/120 mesh. The temperature was programmed from 180 to 200° at 2° per min increment.

SEQUENTIAL ARRANGEMENT OF SACCHARIDE UNITS IN GLYCOSPHINGOLIPIDS

For the enzymatic hydrolyses of sialic acid from gangliosides, 30 µg of each ganglioside was dissolved in 150 µl of 0.05 *M* sodium acetate buffer pH 5.0, and incubated overnight with 0.004 units of purified neuraminidase (16) at 37°. One unit of enzyme released 1 µmole of *N*-acetylneuraminic acid from bovine submaxillary mucine per minute. For the hydrolysis of asialogangliosides or neutral glycosphingolipids with various glycosidases, the method described by Li and Li was followed (17).

[1]We are grateful to Dr. Roger Laine of the University of Kentucky for his help in carrying out gas chromatography - mass spectrometry analysis.

RESULTS

GLYCOSPHINGOLIPID PATTERN IN BOVINE, RHESUS MONKEY, AND HUMAN ERYTHROCYTE MEMBRANES

As shown in Fig. 1a, the predominant neutral glycosphingolipids in human erythrocytes is globoside (GalNAcβ1→ 3Galα1→ 4Galβ1→ 4Glcβ1→ 1Cer). The amount of lactosylceramide (Galβ1→ 4Glcβ1→ 1Cer) and triglycosylceramide (Galα1→ 4Galβ1→ 4Glcβ1→ 1Cer) in human erythrocytes is about 1/20 that of globoside. In rhesus monkey, globoside and triglycosylceramide are the major neutral glycosphingolipids. The ratio of triglycosylceramide and globoside is approximately 1:3. In contrast to human and rhesus monkey, no globoside is detected in bovine erythrocytes. The major neutral glycosphingolipid of bovine erythrocytes is pentaglycosylceramide. Lactosylceramide, triglycosylceramide, and lacto *N*-neotetraosylceramide are also detected.

Figure 1b illustrates the ganglioside pattern in the erythrocyte membranes of human, bovine, and rhesus monkey. The predominant ganglioside in human erythrocytes is *N*-acetylneuraminyl lacto-*N*-neotetraosylceramide. A small quantity of hematoside (GM3) and a larger quantity of ganglioside with mobilities close to that of GD1a are also detected. In rhesus monkey erythrocytes, GM3 is the predominant ganglioside. No monosialyl lacto-*N*-neotetraosylceramide can be detected. In bovine erythrocytes, *N*-glycolyneuraminyl lacto-*N*-neotetraosylceramide and monosialyl hexaglycosylceramide are the major gangliosides. *N*-acetyl and *N*-glycolyl GM3 are also present.

NEUTRAL GLYCOSPHINGOLIPIDS IN BOVINE ERYTHROCYTES

Figure 2 shows various purified neutral glycosphingolipids isolated from bovine erythrocytes. On a borate impregnated plate, the monoglycosylceramide (lane A) migrates at the same rate as that of glucosylceramide isolated from bovine spleen. This glycolipid was found to contain only glucose. Diglycosylceramide (lane B) has the same mobility as that of authentic lactosylceramide isolated from human erythrocytes. Incubation of this glycolipid with jack bean β-galactosidase resulted in the production of glucosylceramide. Combined with the results from the permethylation analysis, this glycolipid is identified as lactosylceramide. The migration of triglycosylceramide (lane C) is identical to that of ceramide trihexoside isolated from the kidney of a patient with Fabry's disease. By using sequential enzymic degradation and permethyla-

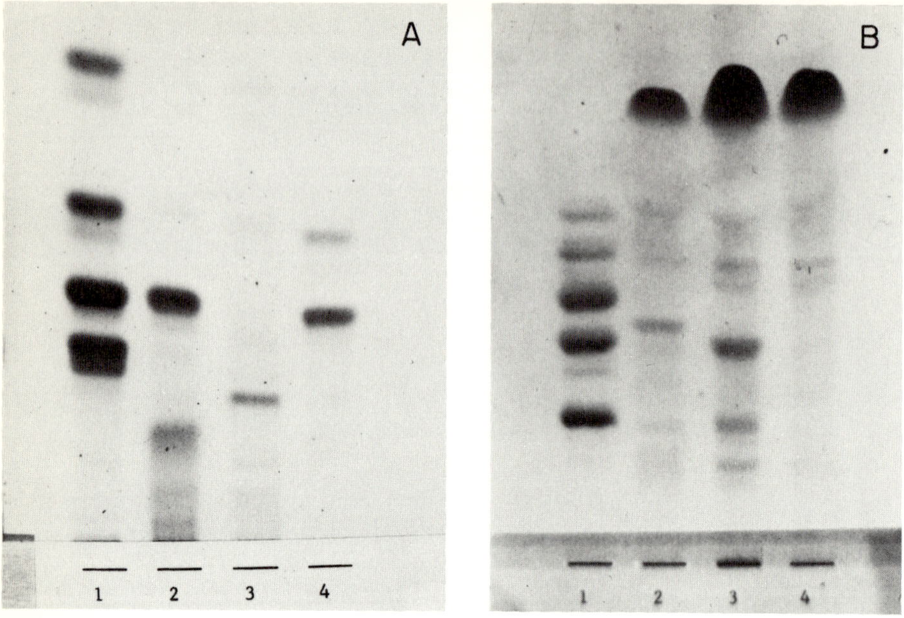

FIGURE 1 (A) Quantitative comparison of neutral glycosphingolipids of human (lane 2), bovine (lane 3), and rhesus monkey (lane 4) erythrocytes. The lipids were extracted and partially purified by DEAE cellulose column (acetate form) and silicic acid column as described in the text. The samples shown on this plate are the acetone:methanol (6:4) effluent from silicic acid column. The plate was developed with solvent III. The standards in lane 1 are lactosylceramide (top), CTH, globoside, and Forssman hapten. (B) Quantitative comparison of gangliosides of human (lane 2), bovine (lane 3), and rhesus monkey (lane 4) erythrocytes. The ganglioside standards in lane 1 are G7 (Top), GM3, GM2, GM1, and GD1a. Solvent II was used.

tion analysis, this glycolipid is identified as Galα1→4Galβ1→4Glc→Cer. Tetraglycosylceramide (lane D) migrates slightly slower than globoside but faster than Forssman hapten. From the sequential enzymic degradation and permethylation analysis, the structure of this glycolipid is determined as Galβ1→4GlcNAcβ1→3Galβ1→4Glc→Cer, which is identical to that isolated from human muscle (18), erythrocytes (19-21) and leukocytes (21). Pentaglycosylceramide (lane E) has the same mobility as that of pentaglycosylceramide isolated from rabbit erythrocytes (2, 22). Fig α-galactosidase converted this glycolipid into lacto-*N*-neotetraosylceramide, which was in turn converted into triglycosylceramide and lactosylceramide successively by

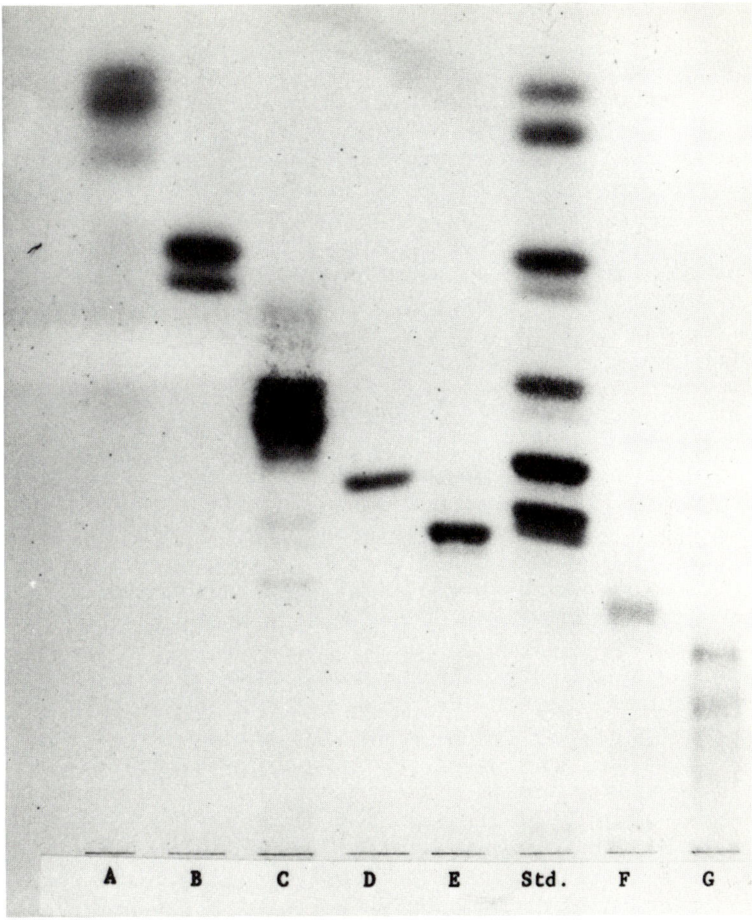

FIGURE 2 Neutral glycosphingolipids purified from bovine erythrocyte membranes (A through G). The standards are galactosylceramide (top), lactosylceramide, CTH, globoside, and Forssman hapten. Solvent III was used.

jack bean β-galactosidase and β-*N*-acetylhexosaminidase. Combined with the results from the permethylation studies, the structure of this pentaglycosylceramide is established to be Galα1→3Galβ1→4GlcNAcβ1→3Galβ1→4Glc→Cer.[1] The characterization of higher neutral glycosphingolipids (lanes F and G) is in progress in our laboratory.

[1]Manuscript in preparation.

GANGLIOSIDES IN BOVINE ERYTHROCYTES

Figure 3 shows the five major gangliosides isolated from bovine erythrocyte membranes. Gangliosides A and B moved identically with authentic *N*-acetyl and *N*-glycolyl neuraminyl GM3, respectively. Gas chromatographic analysis revealed that both gangliosides A and B contain 1 mole of *N*-acetylneuraminic acid and 1 mole of *N*-glycolylneuraminic acid per mole of glycolipid, respectively. Neuraminidase from *Clostridium perfringens* converted both gangliosides into lactosylceramide, which in turn was converted into glucosylceramide by jack bean β-galactosidase. From these results ganglioside A and B are identified as *N*-acetyl GM3 and *N*-glycolyl GM3 respectively. Ganglioside C moved faster than GM1 isolated from human brain, while ganglioside D moved closer to GM1 (Figs. 3 and 4). Gangliosides C and D contain 1 mole of *N*-acetylneuraminic acid and 1 mole of *N*-glycolylneuraminic acid per mole of glycolipid, respectively. Both gangliosides contain glucosamine as the sole amino sugar. Neuraminidase from *C. perfringens* converted these two gangliosides into lacto-*N*-neotetraosylceramide (Fig. 4). Various glycosidases were used to determine the sequential arrangement of saccharide units in the neutral glycolipids derived from gangliosides C and D. As shown in Fig. 5, the neutral glycolipid derived from ganglioside D was converted into triglycosylceramide by jack bean β-galactosidase. The triglycosylceramide was further converted into lactosylceramide and glucosylceramide by β-*N*-acetylhexosaminidase and β-galactosidase. Same results were also obtained for the neutral glycolipid derived from ganglioside C. Both gangliosides C and D were resistant to jack bean β-galactosidase. Permethylation analysis of intact ganglioside resulted in the detection of 1, 3, 5 tri-O-acetyl 2, 4, 6-tri-O-methylgalactital (2, 4, 6-Me_3Gal), 2, 3, 6-Me_3Glc and 3, 6-Me_2GlcNAcMe. Therefore the structures of gangliosides C and D are determined as NeuAcα2→3Galβ1→4GlcNAcβ1→3Galβ1→4Glc→Cer and NeuGcα2→3Galβ1→4GlcNAcβ1→3Galβ1→4Glc→Cer, respectively.

Ganglioside E (Fig. 3) migrates slower than GD1a from human brain. This ganglioside contains Gal, Glc, GlcNAc, sialic acid in the molar ratio 3:1:2:1, respectively. Sialic acid in this ganglioside was found to be *N*-glycolylneuraminic acid. This ganglioside was resistant to β-galactosidase. When treated with neuraminidase, a neutral hexaglycosylceramide was produced. Various exoglycosidases were used to study the carbohydrate sequence of this glycolipid. As shown in Fig. 6, β-galactosidase converted this glycolipid into a compound with mobility slightly slower than the neutral pentaglycosylceramide (Fig. 2). By alternate treatment with β-*N*-acetylhexaminidase and β-galactosidase, this glycolipid was converted stepwise into a tetraglycosylceramide, triglycosyl-

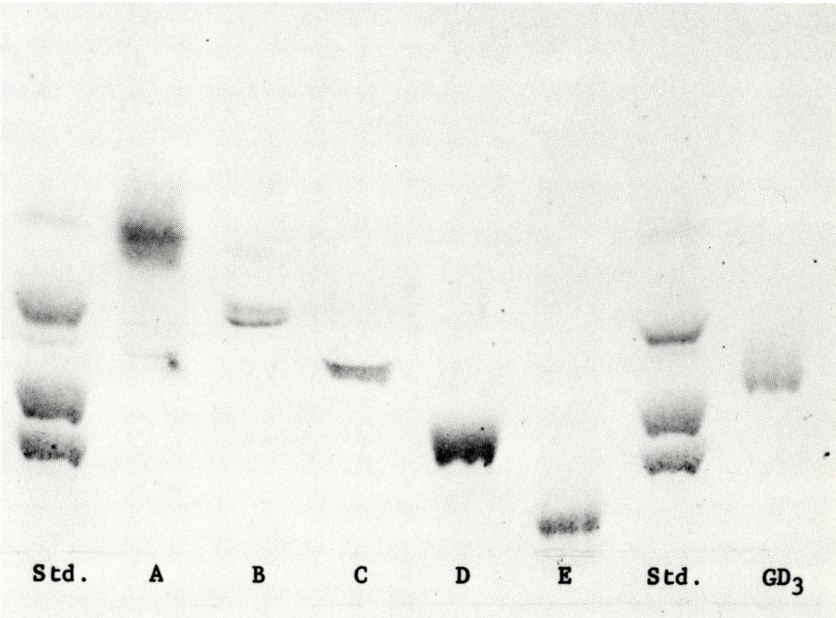

FIGURE 3 Gangliosides (A through E) of bovine erythrocytes. The standards are GM3 (top), GM2, GM1, and GD1a. GD3 standard is at the rightmost lane. Solvent II was used.

ceramide, lactosylceramide, and glucosylceramide. Permethylation of this ganglioside yielded 3, 6-Me_2GlcNAcMe, 2, 4, 6-Me_3Gal, and 2, 3, 6-Me_3Glc. From these results the structure of this ganglioside is determined as NeuGcα2$\rightarrow$3Galβ1$\rightarrow$4GlcNAcβ1$\rightarrow$3Galβ1$\rightarrow$4GlcNAcβ1$\rightarrow$3Galβ1$\rightarrow$4Glc1$\rightarrow$1Cer.

DISCUSSION

Due to the fact that some neutral glycosphingolipids of bovine erythrocytes contain five or more sugar units, the conventional Folch partition cannot give a satisfactory separation between gangliosides and all neutral glycosphingolipids. Therefore DEAE-cellulose column was used to separate these two classes of glycosphingolipids. Subsequently, good separation of individual glycosphingolipids can be achieved by silicic acid column chromatography.

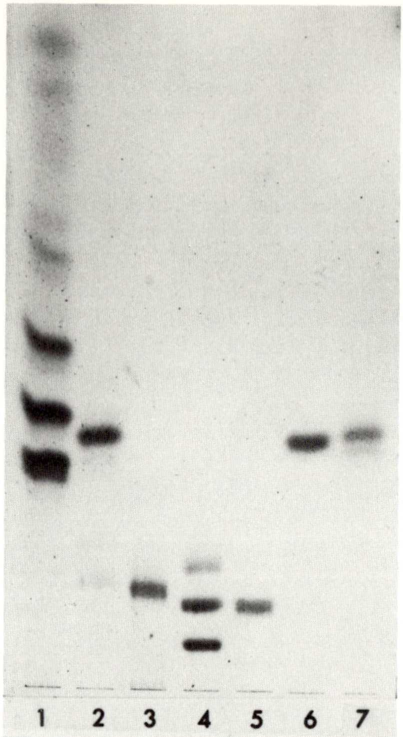

FIGURE 4 Thin-layer chromatography of gangliosides C and D in Fig. 3 before (lanes 3 and 5) and after (lanes 2 and 6) neuraminidase treatment. The standards are in lane 1: lactosylceramide (top), CTH, Fossman hapten; lane 4: GM2, GM1, GD1a; and lane 7: lacto-N-neotetraosylceramide. Solvent III was used.

The present study shows that bovine erythrocyte membranes contain glucosylceramide, lactosylceramide, and triglycosylceramide (CTH) as those found in human, pig, and dog erythrocytes (3). However, the major neutral glycosphingolipid is a pentaglycosylceramide that contains *N*-acetyl glucosamine, whereas in other sources globoside, which contains *N*-acetyl galactosamine, is the most abundant glycolipid. Sequential enzymatic hydrolysis combined with GC-Mass spectrometry established that the structure of this pentaglycosylceramide is Galα1→3Galβ1→4GlcNAcβ1→3Galβ1→4Glc→Cer, which is identical to that of pentaglycosylceramide isolated from human erythrocytes (8, 22). This structure is also similar to that of human

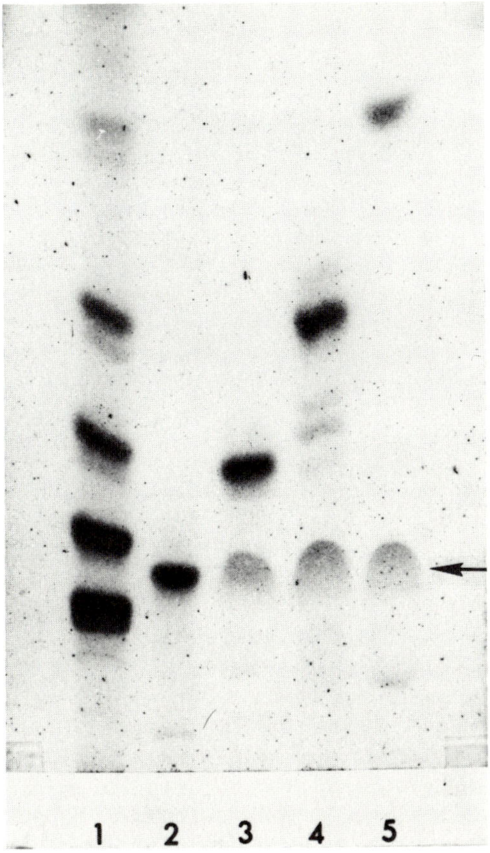

FIGURE 5 Sequential hydrolysis of tetraglycosylceramide obtained from ganglioside D (Fig. 3). Lane 1: glucosylceramide (top), lactosylceramide, CTH, globoside, and Forssman hapten; lane 2, tetraglycosylceramide from ganglioside D; lane 3, 2 + β-galactosidase; lane 4, 3 + β-hexosaminidase; lane 5, 4 + β-galactosidase. Solvent III was used. The spots indicated by arrow are from sodium taurodeoxycholate.

blood group B active glycosphingolipid, Galα1→3(Fucα1→2)Galβ1→4GlcNAcβ1→3Galβ1→4Glc→Cer (23) except that a fucose unit is absent. Besides pentaglycosylceramide, a tetraglycosylceramide with the structure characterized as Galβ1→4GlcNAcβ1→3Galβ1→4Glc→Cer was also present. This tetraglycosylceramide was reported by several investigators to be present in human

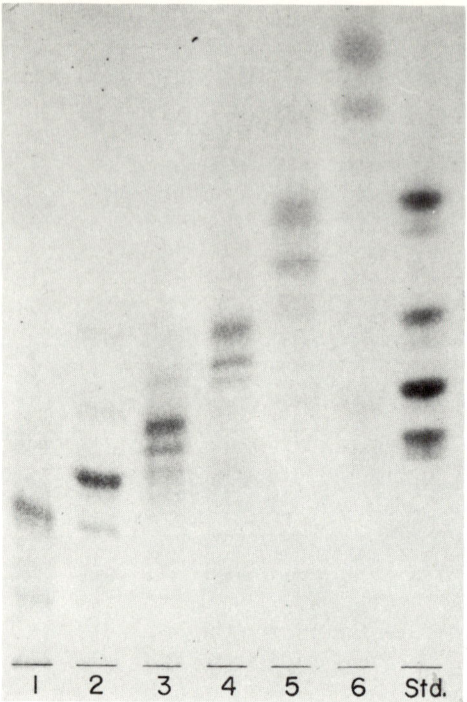

FIGURE 6 Sequential hydrolysis of hexaglycosylceramide derived from ganglioside E (Fig. 3) by neuraminidase treatment. The standards are lactosylceramide (top), CTH, globoside, and Forssman hapten. Lane 1, hexaglycosylceramide; lane 2, 1 + β-galactosidase; lane 3, 2 + β-hexasaminidase; lane 4, 3 + β-galactosidase; lane 5, 4 + β-hexosaminidase; lane 6, 5 + β-galactosidase. Solvent III was used.

erythrocytes (19, 20, 21). The amount of hexaglycosylceramide isolated is not sufficient for a complete structural characterization. However, treatment with jack bean β-galactosidase converted this glycolipid to a pentaglycosylceramide, which in turn was hydrolyzed by β-hexosaminidase to become a tetraglycosylceramide with the mobility identical to that of lacto-*N*-neotetrosylceramide. This glycolipid may have the following structure: Galβ1→4GlcNAcβ1→3Galβ1→4GlcNAcβ1→3Galβ1→4Glc→Cer.

In comparison to neutral glycosphingolipids, very little is known about the gangliosides of bovine erythrocyte membranes, particularly those containing *N*-acetylglucosamine. The glucosamine containing ganglioside was first reported to be present in bovine erythrocyte and spleen (9). However, the intact compound was not isolated until recently (20, 21, 24). The relative mobility of this compound to other known gangliosides on thin-layer plate has not been clearly established. Both *N*-acetyl and *N*-glycolyneuraminyl lacto-*N*-neotetracosylceramide from bovine erythrocytes have been isolated. By enzymatic and permethylation analysis the structures of these two ganglioside were established to be NeuAc and NeuGcα2→3Galβ1→4GlcNAcβ1→3Galβ1→4Glc→Cer, respectively. The behaviors of these two gangliosides in neutral and alkaline solvents are shown in Figs. 3 and 4. In neutral solvent, the ganglioside containing NeuAc moves slightly faster than GM1, while that containing NeuGc has the same mobility as that of GM1. In alkaline solvent the one containing NeuAc has the same mobility as GD3, which moves between GM2 and GM1, and the one with NeuGc moves between GM1 and GD1a. We have also isolated and characterized a novel glucosamine containing ganglioside, which migrates slower than GD1a isolated from human brain. The structure of this ganglioside is established as NeuGcα2→3Galβ1→4GlcNAcβ1→3Galβ1→4GlcNAcβ1→3Galβ1→4Glc→Cer.

Koscielak *et al.* (25) isolated two blood group H active substances from human erythrocytes. These two compounds differ in their chain length of oligosaccharide by adding an extra Galβ1→4GlcNAc unit. Recently, Hanfland (23) also characterized two blood group B active glycosphingolipids from human erythrocytes with the same structural relationship. It is of interest to note that the three gangliosides C, D, and E (Fig. 3) and the two neutral glycosphongolipids (tetraglycosylceramide and hexaglycosylceramide) reported here also have the same relationship. A resorcinol positive band in human erythrocytes that migrate close to bovine monosialylhexaglycosylceramide has also been detected. This ganglioside may very possibly be the same ganglioside present in bovine erythrocytes except that it contains *N*-acetyl neuraminic acid instead of *N*-glycolyl neuraminic acid.

REFERENCES

1. Klenk, E., and Lauenstein, K. (1951), *Z. Physiol. Chem.* *288*, 220.
2. Yamakawa, T., and Suzuki, S. (1951), *J. Biochem.* (Tokyo) *38*, 199.
3. Sweeley, C. G., and Dawson, G. (1969), *in* "Red Cell Membrane Structure and Function," Jamieson, G. A., and Greenwalt, T. J., eds., p. 172, Lippincott, New York.

4. Hakomori, S., Siddiqui, B., Li, Y-T., Li, S-C., and Hellerquist, C. G. (1971), *J. Biol. Chem. 246*, 2271.
5. Miyatake, T., Handa, S., and Yamakawa, T. (1968), *Jap. J. Exp. Med. 38*, 135.
6. Seyama, Y., and Yamakwa, T. (1974), *J. Biochem.* (Tokyo) *75*, 837.
7. Fraser, B. A., and Mallette, M. F. (1974), *Federation Proc. 33*, 1225.
8. Eto, T., Ichikawa, Y., Nishimura, K., Ando, S., Yamakawa, T. (1968), *J. Biochem.* (Tokyo) *64*, 205.
9. Wiegandt, H., and Schulze, B. (1969), *Z. Natuforsch. 24b*, 945.
10. Grochowska, E., and Koscielak, J. (1967), *5th Int. Congr. Biochem. Abstr.*, p. 20.
11. Tettamanti, G., Bonali, F., Marchesini, S., and Zambotti, V. (1973), *Biochim. Biophys. Acta 296*, 160.
12. Hanfland, P., Egli, H. (1975), *Vox Sang. 28*, 438.
13. Hakomori, S. (1964), *J. Biochem. (Tokyo) 55*, 205.
14. Stoffel, W., and Hanfland, P. (1972), *Hoppe-Seyler's Z. Physiol. Chem. 354*, 21.
15. Björndal, H., Hellerquist, C. G., Lindberg, B., and Svensson, S. (1970), *Angew. Chem. Int. Ed. Engl. 9*, 610.
16. Chien, S.-F., Yevich, S. J., Li, S-C., and Li, Y-T. (1975), *Biochem. Biophys. Res. Commun. 65*, 683.
17. Li, Y-T., and Li, S-C. (1973), *CNRS Internat. Sympo. Structure and Methodology of Glycoconjugates 1*, 339, Villeneuve d'Asdq., France.
18. Svennerholm, L., Bruce, A., Manson, J., Rynmark, R.-M., and Vanier, M. T. (1972), *Biochim. Biophys. Acta 280*, 625.
19. Siddiqui, B., and Hakomori, S. (1973), *Biochim. Biophys. Acta 330*, 147.
20. Ando, S., Kon, K., Isobe, M., and Yamakawa, T. (1973), *J. Biochem. 73*, 893.
21. Wherrett, J. R. (1973), *Biochim. Biophys. Acta 326*, 63.
22. Stellner, K., Saito, H., and Hakomori, S. (1973), *Arch. Biochem. Biophys. 155*, 464.
23. Hanfland, P. (1975), *Chem. and Phys. of Lipids 15*, 105.
24. Li, Y-T., Mansson, J.-E., Vanier, M.-T., and Svennerholm, L. (1973), *J. Biol. Chem. 248*, 2634.
25. Koscielak, J., Piasck, A., Gorniak, H., Gardas, A., and Gregor, A. (1973), *Eur. J. Biochem. 37*, 214.

TRANSFERRIN RECEPTOR FROM RABBIT RETICULOCYTE MEMBRANES

DAYLE A. SLY[1]
DIETMAR GROHLICH
ANATOLY BEZKOROVAINY[2]

Department of Biochemistry
Rush-Presbyterian-St. Luke's Medical Center
Chicago, Illinois

INTRODUCTION

Transferrin is a circulatory iron-binding protein that transports iron throughout the mammalian organism. It is a glycoprotein with a molecular weight of near 80,000, and it binds a maximum of two gram atoms of ferric iron per mole of protein. The iron is coordinated with the protein molecule via amino acid side chains. It is believed that each iron atom in transferrin is associated with two to three tyrosyl residues, two histidyl residues, and one bicarbonate ion. When transferrin binds iron, its tertiary structure is apparently altered so that it assumes a more asymmetric shape (Aisen and Brown, 1975; Bezkorovainy and Zschocke, 1974). Under normal physiologic conditions, the iron-transferrin association constant is near 10^{24}, indicating that the association reaction is almost irreversible. However, it is possible to

[1]Present address: Amoco Research Center, P. O. Box 400, Naperville, Illinois 60540.
[2]To whom reprint requests should be directed.

mobilize iron from transferrin to some extent with small chelating agents such as citrate (Aisen and Leibman, 1968).

In addition to its transport function for iron, transferrin is apparently also required for the incorporation of iron into immature red cells. Reticulocytes, either of human or rabbit origin, have been routinely used as convenient model systems representing immature red cells. Thus, if all transferrin is removed from human serum by immunoprecipitation, the serum is no longer capable of stimulating any appreciable iron incorporation into reticulocytes (Eldor *et al.*, 1970). The incorporation of iron into reticulocytes was first studied by Walsh *et al.* (1949), who found that these blood cells, as well as bone marrow aspirates, were capable of absorbing iron from the medium and of incorporating it into hemoglobin. The involvement of transferrin in this process was clarified by Jandl *et al.* (1959) and by Jandl and Katz (1963). These workers discovered that transferrin was able to incorporate its iron into reticulocytes but not erythrocytes, that this was not a simple diffusion process, and that the transferrin attached itself onto the reticulocyte surface before releasing its iron. Moreover, iron-rich transferrin was bound to the reticulocyte surface more readily than was iron-free transferrin. The release of iron by the transferrin into the reticulocyte was abolished if the latter was treated with trypsin. These results, as well as the inability of albumin and γ-globulin to be bound to the reticulocyte membrane, gave rise to the concept that the reticulocyte membrane contains receptor sites that specifically combine with iron-rich transferrin.

Kinetic studies have shown that a rabbit reticulocyte has approximately 300,000 transferrin receptor sites when rabbit transferrin is used as the iron donor (Baker and Morgan, 1969). If human transferrin is used, the number of functional sites is apparently only one-half that number (Lane, 1972). The free energy of interaction was only -7.3 kcal/mole, which is indicative of a rather weak interaction. The specificity of the active sites toward iron-rich transferrin apparently depends on differences in the conformations of the two proteins (Kornfeld, 1969; Lane, 1972).

Some investigators believe that iron-saturated transferrin enters the intracellular compartment to donate its iron to the reticulocyte and is then refluxed back into the extracellular medium. This was first proposed by Morgan and Appleton (1969) on the basis of radioautographic data following the incorporation of iron-saturated ^{125}I-transferrin into reticulocytes. This theory is also consistent with the findings of Martinez-Medellin and Schulman (1972) that all iron found in reticulocyte stroma could be accounted for as transferrin-bound iron, and the fact that a significant fraction of the ^{125}I-transferrin was recovered from the intracellular compartment following its incubation with reticulocytes (Martinez-Medellin, 1972; Sly *et al.*, 1975). Since compounds disrupting cell membrane microtubules are known to interfere with the binding of transferrin by reticulocytes (Hemmaplardh *et al.*, 1974), it may be possible that the iron-saturated transferrin finds its way into the intracellular compartment of the reticulocyte via the microtubules.

Though the above peculiarities of iron incorporation into the reticulocyte are not necessarily inconsistent with the existence of specific iron-rich transferrin receptors in the reticulocyte membrane, it is certainly much easier to fit such a receptor concept into a simpler scheme of iron incorporation first proposed by Jandl and Katz (1963) and later elaborated upon by Morgan (1964): (1) the formation of a weak bond between iron-rich transferrin and reticulocyte membrane, (2) the development of a stronger bond between the two entities presumably involving the specific binding site, (3) the transfer of iron to the reticulocyte, which depends on cellular metabolism, and (4) the release of transferrin from the receptor site.

The pathway of iron incorporation into hemoglobin in the human reticulocyte has been followed by Speyer and Fielding (1974; Fielding and Speyer, 1974). Several iron-containing macromolecules were identified by these workers, both in the membrane and the cytosol of human reticulocytes following the incubation of iron-labeled transferrin with the cells. It was postulated that iron-rich transferrin first combines with the specific binding site on the membrane surface to form the transferrin-receptor complex (also called component B2). The iron is then transferred to a membrane component B1, from which it may be channeled to another membrane component called A (apparently a dead end), or to a cytosol component C, from which the iron is used for hemoglobin synthesis. Little transferrin was found in the cytosol, contrary to the experience of Martinez-Medellin and Schulman (1972) and of Sly *et al.* (1975) with the rabbit reticulocyte system. It is noteworthy that it was possible to extract B2 from reticulocyte membranes using buffer systems with the Triton X-100 detergent. The work of these authors thus does not appear to support the idea that transferrin must of necessity penetrate the reticulocyte membrane before it can lose its iron.

The rest of this paper deals with the extraction and some properties of the rabbit reticulocyte receptor-transferrin complex using human transferrin as the marker.

MATERIALS & METHODS

RETICULOCYTE MEMBRANES

The production of reticulocytes in rabbits, their incubation with ^{125}I-transferrin, the isolation of labeled membranes, and the method of labeling of iron-saturated transferrin with ^{125}I were previously described (Sly *et al.*, 1975b). Transferrin was labeled to the extent of approximately one iodine atom per protein molecule. The labeled membrane fraction was extracted with a 1% Triton X-100 solution buffered with 2 m*M* phosphate at pH 7.4 (Speyer and Fielding, 1974). The extract was then directly placed on a DEAE column as described in Results.

MATERIALS

Sepharose-type gel filtration media were purchased from Pharmacia Corporation (Piscataway, New York). Polyacrylamide gel chemicals with the electrophoresis apparatus, as well as DEAE Bio-Gel A and Bio-Beads SM2 were purchased from Bio-Rad Laboratories (Richmond, California). The standards used in gel electrophoresis experiments and gel filtration runs were β-galactosidase, β-glucuronidase (both from Worthington Corporation, Freehold, New Jersey), bovine colostrum IgG (prepared by the method of Smith, 1946), transferrin (Behringwerke, A. G., Marburg, West Germany), thyroglobulin, cytochrome C, and ovalbumin (the latter three from Sigma Chemical Corp., St. Louis, Missouri). Triton X-100 detergent was also a product of Sigma Chemical Corporation. All common laboratory reagents were obtained from Fisher Scientific Co. (Chicago, Illinois) and were of reagent grade.

ANALYTICAL PROCEDURES

Amino acid analyses were done on reduced and carboxymethylated proteins using the Beckman 120C apparatus equipped with the "long" and "medium" columns. Prior to analysis, proteins were hydrolyzed with 5.8 *M* HCl at 110°C for 24 hr in a nitrogen atmosphere. Ultracentrifugation was done in a Spinco Model E apparatus at 52,000 rev/min at 20°C. The conditions are described in Results.

Polyacrylamide gel electrophoresis was done either in the presence of dissociating and reducing reagents (sodium dodecyl sulfate, urea, and mercaptoethanol) or in the presence of sodium dodecyl sulfate only. The analysis in the presence of reducing and dissociating agents was done essentially by the method of Booyse *et al.* (1972). Samples were prepared by concentrating the required amount of material in an Amicon MMC apparatus (Lexington, Massachusetts) fitted with XM 50 membranes, to a volume of approximately 40 μl, then adding the reducing-dissociating solvent to a final protein concentration of 5 mg/ml. Staining was done with Amido black, and after mechanical destaining, the gels were scanned in a densitometer attached to the Zeiss PMQ-II spectrophotometer. Occasionally, staining for lipids was performed with Fat Red 7B (obtained from Pfizer Diagnostics Division, New York). The Pfizer apparatus, plates, and buffer were used for agarose gel electrophoresis.

All spectrophotometric determinations were carried out in a Zeiss PMQ-II spectrophotometer. Radioactive counting was done in a Beckman Biogamma Counter (II) (Palo Alto, California). The time of counting was adjusted so as to give an error not greater than 3 to 5%. Triton X-100 concentrations were determined by the method of Garewal (1973) and protein concentrations were determined without interference by Triton X-100, by the method of Lowry *et al.* (1951). Carbohydrate analyses were done as follows: hexose and fucose by the method of Dische (1955), hexosamine by the method of Winzler (1955), and sialic acid by the method of Warren (1959).

RESULTS

Reticulocyte membrane extracts labeled with ^{125}I-transferrin were placed on a DEAE-Biogel A column, and its constituents were eluted in the stepwise manner (Fig. 1). The bulk of the protein was not retained by the column (component A). The material with most ^{125}I counts was in component B, which was identified as free transferrin. Component D was deemed to be the transferrin-receptor complex. It was rechromatographed on a Sepharose 4B column (Fig. 2), which served to remove the bulk of extraneous protein material. The next purification step was carried out on a Sepharose 6B column (Fig. 3A), where the radioactive material eluted showed a substantial shoulder. The shoulder was identified as being due to free transferrin, which had apparently dissociated from the receptor during the purification process.

The last step in the purification scheme is shown in Fig. 3B, where it is observed that the protein material eluted corresponded exactly to radioactivity eluted. This was deemed to represent the transferrin-receptor complex and was subjected to further physical-chemical studies. The overall yield of

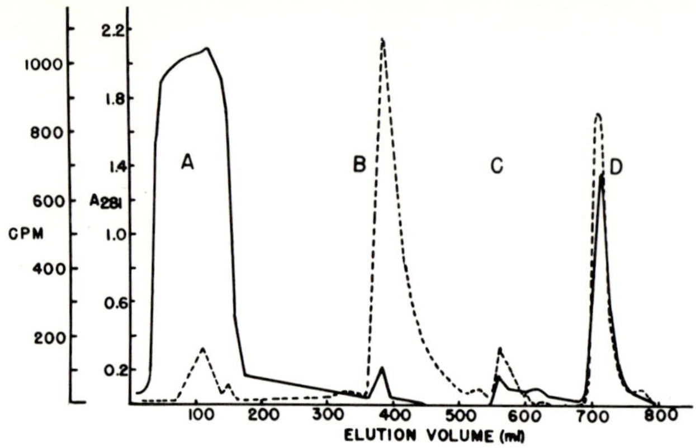

FIGURE 1 Chromatography of the ^{125}I-transferrin labeled reticulocyte membrane extract on DEAE-Biogel A column. Membranes (from 20 ml of packed red cells, 70% reticulocytes) were extracted three times with a 1% Triton X-100-2 mM phosphate, pH 7.4 buffer. 95 ml of the extract were adsorbed onto a 2 × 28 cm column equilibrated with 0.1% Triton X-100-10 mM phosphate buffer at pH 7.4. The column was then washed with the above buffer to elute component A, each fraction measuring 10 ml in volume. Component B was eluted with 0.1% Triton X-100-22.5 mM phosphate, pH 6.8 (introduced after the elution volume reached 325 ml). Component C was eluted with 0.1% Triton X-100-40 mM phosphate, pH 6.5 (introduced at 510 ml). Finally, component D was eluted with 0.1% Triton X-100-50 mM phosphate - 100 mM NaCl at pH 6.0. Solid line, absorption at 281 nm; broken line, radioactivity (cpm).

the purified receptor-transferrin complex was 0.7 mg per 100 ml packed red cells with a 70% reticulocyte count.

Agarose gel electrophoresis of the complex (Fig. 4) showed that at pH 8.6 it migrated as a broad band slightly ahead of transferrin, and it can therefore be termed a fast-beta protein with respect to its electrophoretic mobility. When the receptor-transferrin complex was chromatographed on a Sepharose 6B column standardized with proteins with known molecular weights, it was eluted as a component with a molecular weight of approximately 188,000 (Fig. 5).

Polyacrylamide gel electrophoresis results are shown in Fig. 6. In the absence of reducing agents and urea, but in the presence of sodium dodecyl sulfate, the receptor-transferrin complex showed the pattern seen in Fig. 6A. The sample resolved itself into a heavy component indicated as REC-TF with a molecular weight of near 200,000, a component labeled

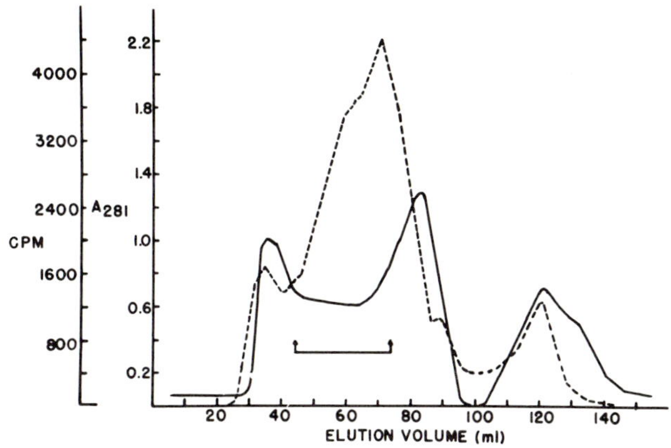

FIGURE 2 Chromatography of component D (Fig. 1) on a Sepharose 4B column. Concentrated (Amicon apparatus) component D in a volume of 3 ml was applied to a 1.5 × 85 cm Sepharose 4B column equilibrated with 0.25% Triton X-100-2 mM phosphate buffer at pH 7.4. Void volume was 32 ml, elution was effected with the above buffer, and 2.5 ml fractions were collected every 25 min. Solid line, absorption at 281 nm; broken line, radioactivity (cpm). Fraction indicated by arrows was retained for further purification.

as TF and having a molecular weight of near 80,000, and lighter components. Determination of radioactivity in these fractions showed that ^{125}I-transferrin was present only in the first two components. The 200,000 dalton component was therefore identified as the receptor-transferrin complex, whereas the 80,000 component was identified as free transferrin. Apparently, the receptor-transferrin complex had partially dissociated into free transferrin and lighter components under these conditions.

The results obtained with polyacrylamide gel electrophoresis in a reducing-dissociating solvent are shown in Fig. 6B. It is seen that the 200,000 dalton component was no longer present. Instead, the electropherogram showed a component with a molecular weight of 80,000, designated as TF and containing all the radioactivity; component A with a molecular weight of 48,000; component B with a molecular weight of approximately 17,000; and component C, which did not stain with amido black, yet absorbed light during the tracing procedure because of its opaqueness. It was, however, stainable with the fat positive dye. It was concluded that the 80,000 dalton component was free transferrin and that reticulocyte components A and B were subunits of the receptor molecule, whereas

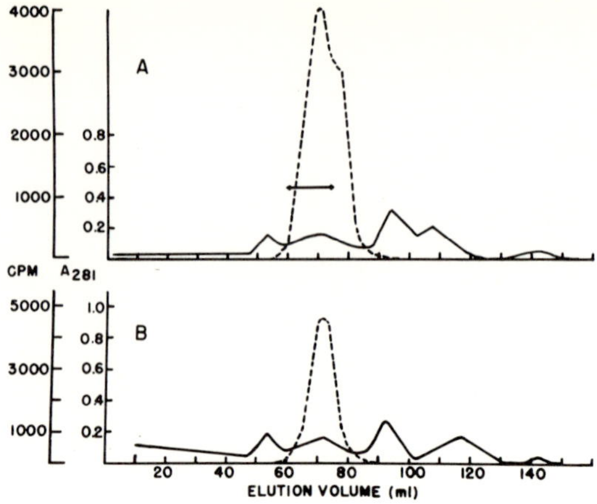

FIGURE 3 (a) Chromatography of the designated fraction from Fig. 2 on a Sepharose 6B column. The concentrated fraction in a volume of 2 ml was applied to a 1.5 × 100 cm column of Sepharose 6B equilibrated with 0.25% Triton X-100-2 mM phosphate buffer at pH 7.4. Elution was effected with the same buffer, whereby 2.5 ml fractions were collected every 30 min. Void volume was 51 ml. Solid line, absorption at 281 nm; broken line, radioactivity (cpm). Fraction selected for final purification is designated by the double arrow. (b) Final purification of the receptor-transferrin complex. A pool of several fractions obtained as in Fig. 3a was rechromatographed on a Sepharose 6B column as above.

component C was lipid and apparently also a component of the reticulocyte receptor molecule.

Ultracentrifugal analysis was also performed on the receptor-transferrin complex at a concentration of 2 mg/ml using the 0.26% Triton X-100-0.1 *M* NaCl - 0.05 *M* phosphate buffer at pH 7.4. The receptor-transferrin complex moved as a single broad boundary, but at a rate slower than that for free transferrin, even though its molecular weight was much higher than that of transferrin. This difference may be accounted for by the presence of lipid in the receptor-transferrin complex, or by its possible large degree of asymmetry, or both.

Amino acid and carbohydrate analyses were performed on the receptor-transferrin complex. Assuming a lipid contribution of about 5% (see Discussion), it is possible to estimate the contribution of transferrin to the total amino acid and carbohydrate content of the receptor-transferrin complex, and to calculate the amino acid and protein-bound carbohydrate con-

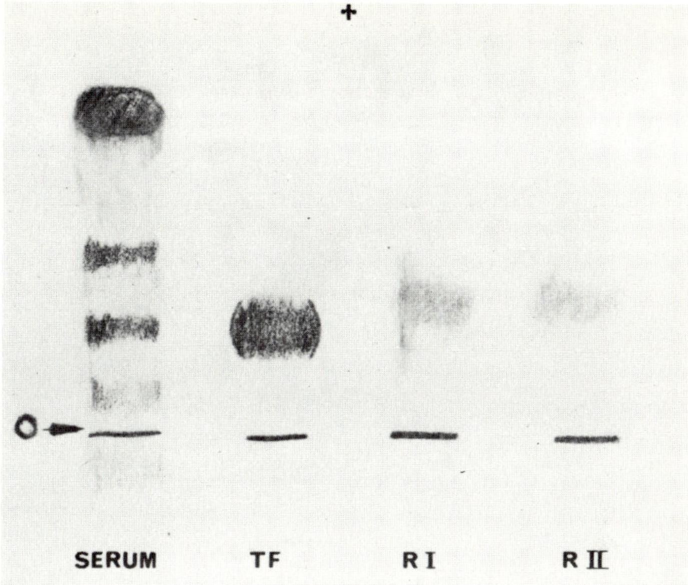

FIGURE 4 Agarose gel electrophoresis of the receptor-transferrin complex. Buffer was 0.25% Triton X-100-0.05 M barbital-0.035% EDTA at pH 8.6. The run proceeded at 90 V for 45 min in the Pfizer apparatus. Staining was done with amido black. Approximately 40 μg of receptor-transferrin were used in each experiment. TF indicates different receptor-transferrin preparations.

tent of the receptor molecule. This result is shown in Table I, which indicates that the receptor had an amino acid composition quite different from that of transferrin.

DISCUSSION

We have purified the rabbit reticulocyte transferrin receptor molecule in the form of the receptor-transferrin complex. It was judged by the criteria of Sepharose 6B chromatography and agarose gel electrophoresis to be of sufficient purity to justify further physical-chemical characterization.

There is no doubt that the receptor molecule extracted from the rabbit reticulocyte membranes in this work is analogous to the B-2 molecule of Speyer and Fielding (1974) extracted from human reticulocyte membranes. However, the rab-

TABLE I Amino Acid and Carbohydrate Content of the Glycoprotein Portion of the Rabbit Reticulocyte Receptor Molecule[a]

Component	Receptor	Transferrin[b]
Lys	2.4	10.4
His	1.8	3.6
Arg	4.4	5.6
Asp	11.9	12.3
Thr	3.0	4.0
Ser	8.1	4.7
Glu	13.2	11.0
Pro	5.1	4.1
Gly	5.2	3.7
Ala	6.2	5.4
Val	3.6	5.9
Met	0.6	1.4
Ile	1.8	2.3
Leu	11.8	8.9
Tyr	3.8	5.0
Phe	6.1	5.5
Hexose	4.6	2.6
N-Acetyl hexosamine	4.4	2.2
Sialic acid	1.0	1.5
Fucose	1.0	0.1

[a]Cystine and tryptophan were not estimated. In grams/100 grams.

[b]From Schultze and Heremans (1966), ignoring one-half cys, trp, and NH_3.

bit material appears to be substantially smaller than its human counterpart: the human material was estimated to have a molecular weight of near 150,000, whereas we estimate the size of the rabbit reticulocyte receptor to be near 108,000 (188,000 less 80,000, which is the molecular weight of transferrin, assuming a one to one interaction).

Polyacrylamide gel electrophoresis experiments yielded certain information in regard to the molecular architecture of the receptor molecule. It appears to have a subunit with a molecular weight of nearly 48,000 (component A) and subunit(s) with molecular weights of nearly 17,000 (component B). Semiquantitative estimates of the various components shown in Fig.

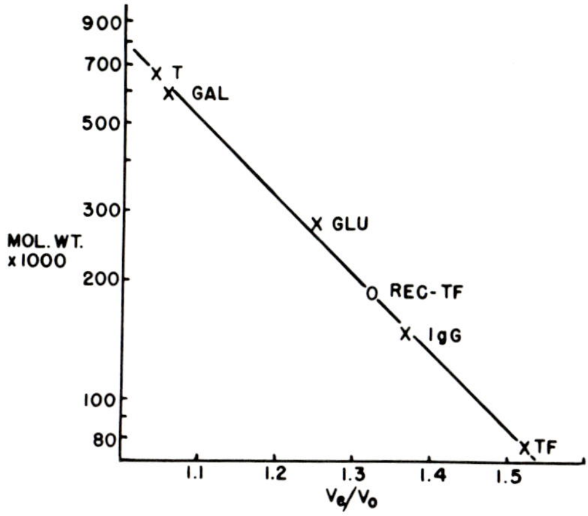

FIGURE 5 Standardization of a Sepharose 6B column with proteins of known molecular weight and the place of elution of the receptor-transferrin complex. Column size was 1.5 × 100 cm, void volume was 51 ml, and V_e/V_0 indicates the ratio of elution volume (at the peak) to void volume. T, GAL, GLU, REC-TF, IgG, and TF indicate thyroglobulin, β-galactosidase, β-glucuronidase, receptor-transferrin complex, bovine colostrum IgG, and transferrin, respectively.

6B indicated that the sum of components A and B would account for about 55% of the total stainable protein present. The amount of A is approximately equal to that of B. We therefore may come to the conclusion that each receptor molecule consists of one subunit A (48,000 daltons), and three subunits B (total 51,000 daltons). These subunits may be bonded, in part, by physical interactions, since sodium dodecyl sulfate was able to effect a partial dissociation of the receptor-transferrin complex (Fig. 6A). Total dissociation, however, required reducing conditions, thus indicating also the presence of disulfide linkages.

The amount of lipid present in the receptor molecule has not been rigorously determined. However, a reasonable estimate can be made by adding the molecular weight contributions of transferrin and the receptor molecule protein subunits, then subtracting these from 188,000. Lipid contribution thus comes to approximately 9000 daltons, or about 5% of the receptor-transferrin complex.

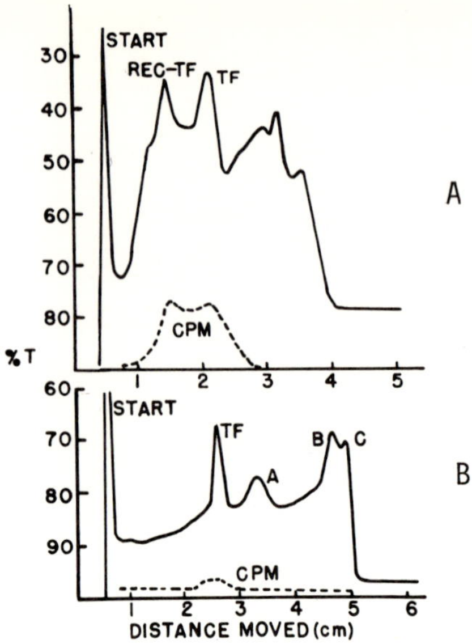

FIGURE 6 Tracings of polyacrylamide gel electropherograms of the transferrin-receptor complex using 5% gels. Protein used: approximately 125 μg. (a) Electrophoresis in 0.05 M barbital-0.035% EDTA-0.1 M sodium dodecyl sulfate at pH 8.6. The experiment was performed at a constant current of 6.2 mA per gel (approximately 110 V) for 80 min. After staining and tracing, gel was segmented and each segment counted for ^{125}I. Solid line, amido black stain; broken line, radioactivity (cpm). REC-TF indicates receptor-transferrin complex; TF indicates free transferrin. (b) Electrophoresis in reducing-dissociating buffer (urea, mercaptoethanol, and sodium dodecyl sulfate). The experiment was performed at constant current at 1.8 mA/gel (approximately 7 V) for 18 hr. After staining and tracing, the gels were segmented and radioactivity determined therein in the Biogamma II counter. Solid line, protein stain; broken line, radioactivity. TF designates free transferrin.

SUMMARY

A transferrin-receptor complex was isolated from rabbit reticulocyte membranes by extraction with Triton X-100- containing buffer. It was reasonably homogeneous by the methods of agarose gel electrophoresis and chromatography on a Sepharose 6B column. The complex had a molecular weight of near 188,000. The receptor molecule itself was an aggregate of one subunit with a molecular weight of 48,000, three subunits with molecular weights of 17,000 each, and an as yet unidentified lipid material. It also contained protein-bound carbohydrate consisting of hexose, hexosamine, sialic acid, and fucose.

ACKNOWLEDGMENTS

This work was supported by American and Chicago Heart Associations Grant No. 73-698.

REFERENCES

Aisen, P., and Brown, E. B. Structure and Function of transferrin. *Progr. Hematol. 9*: 25-56, 1975.

Aisen, P., and Leibman, A. Citrate-mediated exchange of Fe^{3+} among transferrin molecules. *Biochem. Biophys. Res. Comm. 32*: 220-226, 1968.

Baker, E., and Morgan, E. H. The kinetics of the interaction between rabbit transferrin and reticulocytes. *Biochem. 8*: 1133-1141, 1969.

Bezkorovainy, A., and Zschocke, R. Structure and function of transferrin. I. Physical, chemical, and iron-binding properties. *Arzneimittelforsch. 24*: 476-485, 1974.

Booyse, F. M., Hoveke, T. P., Kisieleski, D., and Rafelson, M. E. Mechanism and control of platelet-platelet interactions. *Microvas. Res. 4*: 199-206, 1972.

Dische, Z. New color reactions for determination of sugars in polysaccharides. *Meth. Biochem. Anal. 2*: 313-358, 1955.

Eldor, A., Manny, N., and Izak, G. The effect of transferrin-free serum on the utilization of iron by rabbit reticulocytes. *Blood 36*: 233-238, 1970.

Fielding, J., and Speyer, B. E. Iron transport intermediates in human reticulocyte and the membrane binding site of iron-transferrin. *Biochim. Biophys. Acta 363*: 387-396, 1974.

Garewal, H. S. A procedure for the estimation of microgram quantities of Triton X-100. *Anal. Biochem. 54*: 319-324, 1973.

Hemmaplardh, D., Kailis, S. G., and Morgan, E. H. The effects of inhibitors of microtubule and microfilament function on transferrin and iron uptake by rabbit reticulocytes and bone marrow. *Br. J. Haematol. 28*: 53-65, 1974.

Jandl, J. H., Inman, J. K., Simmons, R. L., and Allen, D. W. Transfer of iron from serum iron-binding protein to human reticulocytes. *J. Clin. Invest. 38*: 161-185, 1959.

Kornfeld, S. The effect of metal attachment to human apotransferrin on its binding to reticulocytes. *Biochim. Biophys. Acta 194*: 25-53, 1969.

Lane, R. S. Transferrin-reticulocyte binding: evidence for the functional importance of transferrin conformation. *Br. J. Haematol. 22*: 309-317, 1972.

Lowry, O. H., Rosenbrough, N. J., Farr, A., and Randall, R. J. Protein measurement with the Folin phenol reagent. *J. Biol. Chem. 193*: 265-275, 1951.

Martinez-Medellin, J., and Schulman, H. M. The kinetics of iron and transferrin incorporation into rabbit erythroid cells and the nature of stromal-bound iron. *Biochim. Biophys. Acta 264*: 272-284, 1972.

Morgan, E. H. The interaction between rabbit, human, and rat transferrin and reticulocytes. *Br. J. Haematol. 10*: 442-452, 1964.

Morgan, E. H., and Appleton, T. C. Autoradiographic localization of ^{125}I-labeled transferrin in rabbit reticulocytes. *Nature 223*: 1371-1372, 1969.

Schultze, H. E., and Heremans, J. F. *In* "Molecular Biology of Human Proteins." Vol. I, Elsevier, Amsterdam, 1966, p. 212.

Sly, D. A., Grohlich, D., and Bezkorovainy, A. Transferrin in the reticulocyte cystosol. *Biochim. Biophys. Acta 385*: 36-40, 1975(a).

Sly, D. A., Grohlich, D., and Bezkorovainy, A. Transferrin receptors from reticulocyte membranes and cytosol. *In* "Proteins of Iron Storage and Transport in Biochemistry and Medicine," R. R. Crichton, North-Holland Publ. Co., Amsterdam, 1975, p. 141(b).

Smith, E. L. The immune proteins of bovine colostrum and plasma. *J. Biol. Chem. 164*: 345-357, 1946.

Speyer, B. E., and Fielding, J. Chromatographic fractionation of human reticulocytes after uptake of doubly labeled [^{59}Fe, ^{125}I] transferrin. *Biochim. Biophys. Acta 332*: 192-200, 1974.

Walsh, R. J., Thomas, E. D., Chow, S. K., Flunarty, R. G., and Finch, C. A. Iron metabolism. Heme synthesis *in vitro* by immature erythocytes. *Science 110*: 396-398, 1949.

Warren, L. The thiobarbituric acid assay of sialic acids. *J. Biol. Chem. 234*: 1971-1975, 1959.

Winzler, R. J. Determination of serum glycoproteins. *Meth. Biochem. Anal. 2*: 279-311, 1955.

MOLECULAR PROBES FOR THE MECHANISM OF D-GLUCOSE TRANSPORT ACROSS CELLULAR MEMBRANES

N. F. TAYLOR
G. L. GAGNEJA

Department of Chemistry
University of Windsor
Windsor, Ontario
Canada

INTRODUCTION

Since the discovery of the natural occurrence of fluoroacetic acid (1) and the elucidation of the biochemical mode of toxic action of this compound by Peters and his co-workers (2), considerable progress in the biochemistry of synthetic and naturally occurring carbon-fluorine compounds has occurred. Some idea of the extent of these studies is reflected in two recent symposia (3, 4) and a comprehensive review of the biological aspects of inorganic and organic fluorides (5). The rationale for the replacement of the hydroxyl group by fluorine has been discussed in detail elsewhere (6, 7a). In short, the unique physical properties of the carbon-fluorine bond (8) together with the comparative similarity of bond length (9, 10) size, (11) and electronegativity (12) (Table I) of covalently bonded fluorine (C-F) and oxygen (C-O-H) atoms suggested that deoxyfluorosugars would be excellent carbohydrate analogs in biological systems. Results in the area of

TABLE I Comparison of Physical Parameters of Some Elements Covalently Bonded to Carbon

Element	Bond length[a] (CH_3-X) Å	Van der Waals[b] radius (Å)	Electroneg-[c] ativity
H	1.09	1.20	2.1
F	1.39	1.35	4.0
O (in OH)	1.43	1.40	3.5
Cl	1.77	1.80	3.0
S (in SH)	1.82	1.85	2.5

[a](9) (10). [b](11). [c](12).

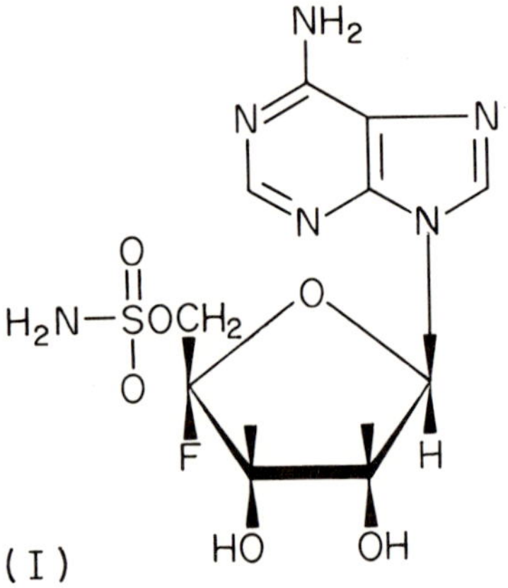

(I)

fluorocarbohydrate metabolism and enzyme specificity support this contention (13).

Apart from the discovery of the antibiotic nucleocidin (14), which possesses a ribose moiety (I) with fluorine substituted at C_4', no other naturally occurring fluorocarbohydrates have been reported. A wide range of synthetic deoxyfluorosugars, however, is now available (15) and their use as metabolic and transport probes will undoubtedly continue to be exploited.

Other chemical probes such as 2,4,6-trinitrobenzene sulphonic acid (TNBS) and 1-fluoro-2,4-dinitrobenzene (FDNB) have now been used extensively to provide information about the asymmetric distribution of phospholipids in membranes (16).

A similar approach has been made in order to characterize the receptor groups of the carrier protein associated with glucose transport in the human erythrocyte (17) and a partial purification of this carrier protein, which is considered to be transmembrane and attached to the hydrophobic core of the membrane, has been reported (18). Further studies may permit the structural elucidation of this protein. An alternative method of obtaining information about the nature of the carrier-protein interaction with glucose relies upon the comparative determination of kinetic parameters of transport for glucose and glucose analogs. The half-saturation constants (K_x) or the inhibitor constants (K_i) so obtained may be considered a measure of the location of the important binding sites of the nolecule to the carrier protein. Although the exact nature of the carrier protein is unknown and (based on kinetic data) the translocational mechanism for the transport of glucose across the human erythrocyte membrane is still controversial (19, 20), the saturation kinetics obtained for glucose and glucose analogs are in accordance with a facilitated mechanism that allows the binding of the sugar molecule to one or more sites of a receptor protein in the membrane. A comprehensive study of this binding has been undertaken by a number of workers (21-25) and it has become apparent that the important binding sites for efficient glucose transport reside at the ring oxygen (at C_5) and the hydroxyl groups at C_1 (β-orientation) and C_3.

The idea of multipoint contact between glucose and protein through hydrogen bonding was introduced some time ago (26), and some of the first studies used to explore the specificity of this binding to an enzyme using halogenated sugar derivatives as probes were carried out by Helferich *et al.* (27). Since a hydrogen bond between an -OH group and a protein may use either the hydrogen of the -OH group or that of a receptor group (X) in the protein to form the hydrogen bridge (Fig. 1), then it follows that since fluorine can only hydrogen bond in one direction, deoxyfluorosugars may be used to probe the location and direction of hydrogen bonding between the sugar and the protein. Thus the presence of a hydrogen bond at each position in the sugar may be investigated by comparing the kinetic transport or inhibitor constants for sugars in which the hydroxyl group has been replaced by fluorine or hydrogen. If the presence of the oxygen of the hydroxyl group is necessary for hydrogen bonding, then stereospecific substitution by fluorine should reveal only minor changes in the half-saturation (K_x) or inhibitor constants (K_i). Similarly, substitution of this hydroxyl group by hydrogen should prevent this type of bonding and be reflected by an increase in K_x or K_i value (i.e., a decrease in binding affinity).

Specific inhibitors of glucose transport may also yield information about the stereospecific requirements of the sugar-carrier protein binding sites. This is especially the case with cytochalasin B (28), which has been shown to bind specifically at the site(s) of glucose transport in the human erythrocyte with $K_D \sim 10^{-7}$ *M* (29) and $K_i = 1.2 \times 10^{-7}$ *M* (30).

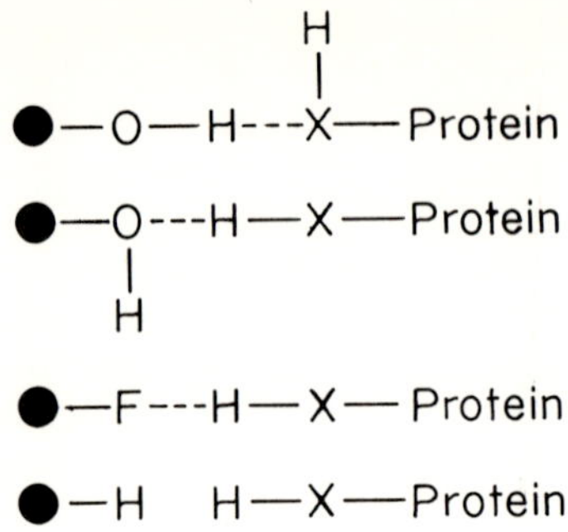

FIGURE 1 Possible hydrogen bonds between hydroxyl, fluorine substituted sugars (o), and protein (X = O, S, and N).

The purpose of this paper is (a) to discuss the use of fluorodeoxymonosaccharides as probes for the stereospecific bonding of mediated glucose transport in hamster intestine and the human erythrocyte, and (b) to propose a model for the mode of inhibition of glucose transport by cytochalasin B in the human erythrocyte that is consistent with the binding requirements of glucose for the carrier protein in the membrane.

BINDING REQUIREMENTS OF ACTIVE SUGAR TRANSPORT IN HAMSTER INTESTINE

As a result of previous studies by Crane (31) and others (32, 33), Barnett *et al.* (7bd) have systematically examined the effectiveness of a number of fluorinated hexoses as inhibitors of sugar transport in everted hamster intestine. As can be seen (Table II) the tissue/medium ratios for D-galactose derivatives modified at C_6, D-galactose, and 6-deoxy-6-fluorogalactose were relatively well transported whereas 6-deoxygalactose was not. This result implicates the oxygen of the hydroxyl group at C_6 for hydrogen bonding. The K_i values, which were shown to be competitive for methyl α-D-glucopyranoside by Lineweaver-Burk plots, also reflect the relative binding affinity of the analogs, the lower K_i value indicating the more effective binding of the sugar to the carrier protein. A similar analysis of K_i values for glucose, 3-deoxy-3-fluoroglucose, and 3-deoxylgucose with galactose as a substrate indicates the importance of the hydroxyl group at C_3 (Table II) for binding. Other tissue/medium ratio studies

TABLE II Transport Parameters for Fluorinated Galactose and Glucose Analogs in Everted Hamster Intestine[a]

Sugar	Tissue/medium ratio	K_i (mM)
Galactose	8.3 ± 0.25	3[b]
6-Deoxy-6-fluoro-galactose	4.3 ± 0.8	7[b]
6-Deoxygalactose	1.5 ± 0.15	16[b]
Glucose	16.4 ± 0.4	2.3[c]
3-Deoxy-3-fluoro-glucose	26.6 ± 3.6	2.5[c]
3-Deoxyglucose	2.1 ± 0.02	24.0[c]
2-Deoxy-2-fluoro-glucose	0.4 ± 0.2	∞,[d] 65.0[e]
2-Deoxyglucose	0.2 ± 0.02	∞,[d] 65.0[e]

[a]Experimental details and data from Barnett et al. (7b-d).
[b]Using methyl α-D-glucopyranoside as substrate.
[c]Using galactose as substrate.
[d]Using D-galactose as substrate.
[e]Using L-glucose as substrate.

with modification at C_1 and C_5 of galactose derivatives using methyl β-D-galactopyranoside, methyl β-thio-D-galactopyranoside, 1-deoxy-D-galactose, and 5-thio-D-glucose also indicated that the sugar ring oxygen at C_5 was not involved but the one at C_1 (β-orientation) probably was implicated. Unfortunately, β-D-galactosyl fluoride was not used in this study. In accordance with the rather special assignment of D-gluco configuration given to the oxygen function at C_2 by Crane (31), neither the 2-deoxy-D-glucose nor 2-deoxy-2-fluoro-D-glucose binds effectively using D-galactose or L-glucose as substrates (Table II). This loss of transport was much more dramatic than that observed with any other modified position of the glucose molecule and it was concluded that a special type of bond was necessary to effect binding at C_2. The mechanism of this transport system is considered to be linked to a sodium pump (32). This hypothesis for active transport requires the sugar to bind reversibly to the membrane and without net utilization of energy. In order to interpret the results of their binding studies, Barnett has proposed that a covalent bond at C_2 may be formed by cleavage of an intramolecular

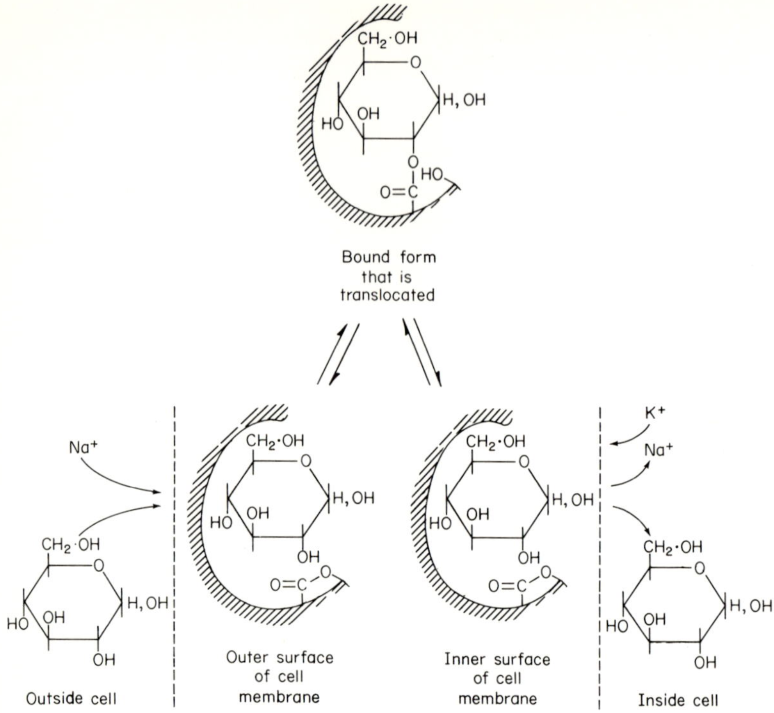

FIGURE 2 A possible mechanism for the active transport of D-glucose into the hamster intestine [from Barnett (7a)].

ester on the membrane (Fig. 2) as well as the hydrogen-bonded sites at C_1, C_3 of D-glucose although, if this model is correct, the covalent bond is probably of paramount importance for efficient transport.

BINDING REQUIREMENTS FOR THE TRANSPORT OF D-GLUCOSE ACROSS THE HUMAN ERYTHROCYTE MEMBRANE

Using the optical method of Sen and Widdas (22), which allows the rate of exit of sugar from preloaded erythrocytes to be followed, the modified rate equation of Miller (34) allows a linear plot of $1/V$ against the external concentration of the sugar to give $1/V_{max}$ as the intercept on the y axis and $-K_x$ as intercept on the x axis, where K_x is a half-saturation constant for the sugar-carrier complex. This constant may be

TABLE III Exit Transport Parameters of Monosaccharides across Human Erythrocyte Membrane at 37°C[a]

Sugar	K_x (mM)	V_{max} (mmol $liter^{-1} min^{-1}$)
Glucose	3.9	640
3-Deoxy-3-fluoro-glucose	2.3	600
3-Deoxyglucose	15.3	795
4-Deoxy-4-fluoro-glucose	5.0	600
4-Deoxyglucose	?	?
5-Thioglucose	15.0	500

[a]For experimental details see reference 35.

considered as a measure of the binding affinity of sugar for the carrier system analogous to the K_m value in Michaelis-Menton kinetics. Thus, comparatively low and high K_x values indicate high- and low-affinity binding, respectively. The results obtained for the exit of D-glucose, 3-deoxy-3-fluoro-D-glucose, and 3-deoxy-D-glucose (Table III) indicate that replacement of the oxygen function at C_3 of D-glucose by fluorine does not significantly change the K_x value or the V_{max} for the carrier protein. In contrast, 3-deoxy-D-glucose has lost this ability to hydrogen bond at C_3 and consequently has a lower affinity for the carrier protein (higher K_x value). The importance of the oxygen function at C_4 for hydrogen bonding remains debatable. Although 4-deoxy-4-fluoro-D-glucose is well transported, the 4-deoxy-D-glucose was not available for comparison. The result with 5-thio-D-glucose (Table III), in which the ring oxygen at C_5 in D-glucose has been replaced by sulfur, is included to illustrate the importance of the ring oxygen as a possible hydrogen bonding site. To provide further evidence that fluorine is hydrogen bonded with the carrier at C_3 in glucose, the apparent dissociation energies of the glucose-carrier and the 3-deoxy-3-fluoro-D-glucose-carrier complexes were compared by plotting log K_x values for the sugars at four different temperatures against the reciprocal of these temperatures. A ΔE value can be obtained from the slope (Fig. 3) and hence an apparent dissociation energy $\Delta H^*_{app.}$ at 37°C can be calculated (36). The values obtained for D-glucose and 3-deoxy-3-fluoro-D-glucose

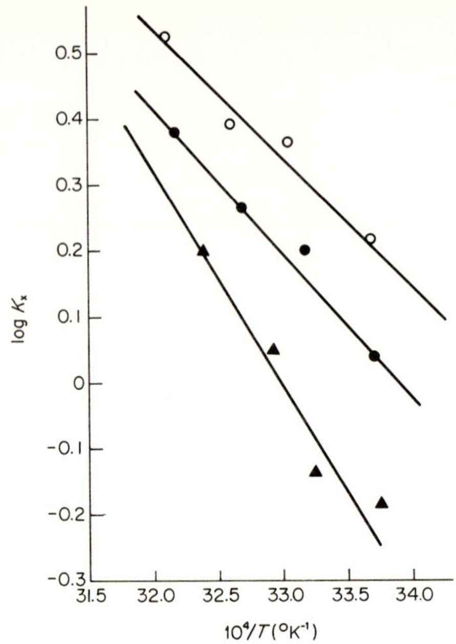

FIGURE 3 Estimations of the 3-deoxy-3-fluoro-D-glucose-carrier dissociation energy [from Riley and Taylor (35)]. ▲H (kJ/mol) at 37°C: ●, 34.3;●36.0;▲55.0. ΔE is obtained from the slope of the plot of log K_x versus 1/T, and ΔH* from the relationship $\Delta E = -\Delta H^* + (RT/2.303R)$, where R is the gas constant and T is the absolute temperature.*

are 47.8 ± 4.1 and 4.1 ± 7.1 kJ/mol (±S.E.M.), respectively, and did not differ significantly ($P > 0.05$). These results are similar to the value 42 kJ calculated for glucose previously (22) and may reflect the sum of the hydrogen bond forces between the sugar and the saturated transport carrier sites. Using fluorinated glucoses as probes but by an entirely different method, extensive studies by Barnett *et al.* (37) have demonstrated the importance of the oxygen functions at C_1 (β-orientation) and C_3 in glucose for binding to the carrier. Thus, a comparison of the inhibition of the poorly transported L-sorbose entry in the human erythrocyte by D-glucose, deoxyfluoro- and deoxy-D-glucose analogs may be considered to reflect the affinity of the sugars for the carrier system. For example, 1-deoxy-D-glucose and α-D-glucosyl fluoride were poor competitive inhibitors (relatively high K_i values) whereas β-D-glucosyl fluoride had a K_i value closer to

TABLE IV Inhibition of L-Sorbose Entry into Human Erythrocytes by Glucose Analogs[a]

Sugar	K_i *(mM, 25°C)*
Glucose	*7.6*
1-Deoxyglucose	*76.0*
α-D-glucosyl fluoride	*77.0*
β-D-glucosyl fluoride	*15.4*
3-Deoxyglucose	*71.5*
3-Deoxy-3-fluoro-glucose	*6.9*
6-Deoxyglucose	*6.7*
6-Deoxy-6-fluoro-glucose	*1.2*

[a]*For experimental details see reference 37.*

D-glucose (Table IV), which suggests the importance of a β-oriented hydroxyl group at C_1 of glucose. Likewise, the K_i values for 3-deoxy-D-glucose and 3-deoxy-3-fluoro-D-glucose implicate the oxygen function at C_3 in hydrogen bonding. Unlike the case of glucose transport in hamster intestine, the oxygen function at C_6 does not appear to be involved in hydrogen bonding since there is no major change in the K_i values for 6-deoxy-D-glucose and 6-deoxy-6-fluoro-D-glucose.

Taken together, these studies with fluorinated sugars constitute further evidence for the existence of at least three stereospecific hydrogen bonding oxygen atoms located at C_1, C_3, and the ring oxygen at C_5 of β-D-glucopyranose and probably in the C1-conformation (38) (Fig. 4).

THE CYTOCHALASINS AS TRANSPORT PROBES

Cytochalasin A(CA) and cytochalasin B(CB) are two fungal metabolites found in *Helminthosporium dematioideum* (28) known to exert a diverse action on cellular movement (39). They are also known to inhibit phagocytosis (40), platelet aggregation (41), clot refraction (42), thyroid secretion (43), cytoplasmic streaming (44), release of growth hormone (45), and glucose transport in chick fibroblasts (46, 47) and Hela cells (48). Although only minor differences in structure be-

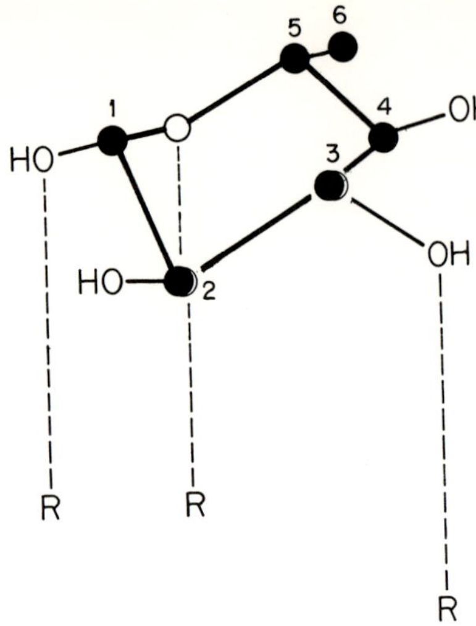

FIGURE 4 β-D-glucopyranose C1-conformation. R, protein receptor sites; -, hydrogen bonds.

tween CA (II, R=O) and CB (II, R=H, OH) exist, major differences in their mode of inhibitory action on glucose transport are becoming apparent. Thus, CA inhibits the energy-dependent uptake of metabolizable sugars by intact yeast cells, while the transport of nonmetabolizable hexoses by a carrier-mediated process was unaffected (49). Recently, it has been shown that CA inhibits the growth and sugar uptake in yeast whereas CB does not. A partial explanation for this difference in action is considered to be due to the fact that unlike CB, CA is shown to be an active sulfhydryl reagent that inhibits isolated plasma membrane bound ATP-ases (50). In this connection, it is of interest to note that CA at C_1, C_2, C_3, and C_4 (II, R = O) has the same atomic grouping and reactive centers as *N*-ethylmaleimide (III), which is a well-known sulfhydryl-reactive agent.

Evidence for the specific binding of CB to the human erythrocyte membrane was first demonstrated by Lin and Spudich (29). They showed that CB binds rapidly and reversibly to some high-affinity binding sites ($K_D \cong 10^{-7}$ *M*) located in the erythrocyte membrane. This binding was competitively inhibited (K_i, 50 m*M*) up to 80-90% by glucose and the effectiveness of the sugar to compete is directly related to its affinity for the glucose transport system. Recently a model for the mode of binding of CB to the glucose-carrier protein has been proposed (30). A molecular model of CB reveals a stable

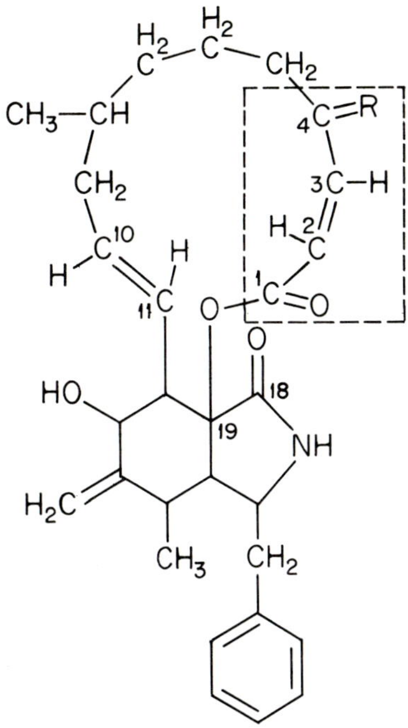

(II)

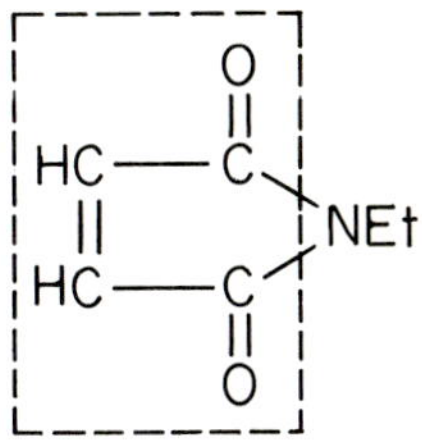

(III)

conformation in which four oxygen atoms, located at C_1, C_{19}, C_{18}, and C_4 (Fig. 5a) are superimposable on those located at C_5, C_1, C_2, and C_3 of the C1-conformation of β-D-glucopyranose (Fig. 6a). At least three of these oxygen atoms in glucose are considered to be critical for stereospecific hydrogen bonding to corresponding receptor sites A, D, and C of the carrier protein (Fig. 6a). In this respect, therefore, CB may be considered a glucose analog. Although CB has eight asymmetric carbon atoms only those located at C_4 and C_{19} are implicated in binding. Application of the sequential rule (51) indicates that the R-configuration of the oxygen functions at C_4 and C_{19} (Fig. 5a) should be adopted for superimposition at sites C and A, respectively, of β-D-glucopyranose (Fig. 6a). Alternatively, the S-configuration of oxygen functions at C_4 and C_{19} of CB (Fig. 5b) would also still permit the superimposition at sites C and A, respectively, of the inverted form of β-D-glucopyranose (Fig. 6b). The actual stereochemical assignment of naturally occurring CB at C_4 and C_{19} has not yet been determined. Experimental evidence for this model is based on the fact that other lipophilic compounds with a similar geometrical distribution of oxygen functions to those found in CB might also competitively in-

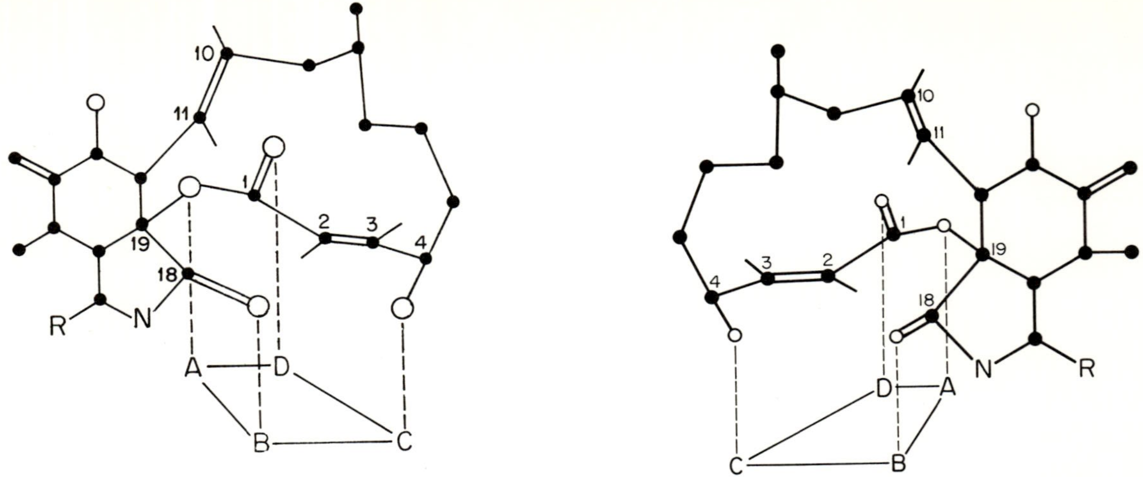

FIGURE 5 (a) Model of cytochalasin B with C_4 and C_{19} in R configuration. O, oxygen functions; ●, carbon atoms; . . . , hydrogen bonds; R, CH_2Ph. A, B, C, and D represent protein receptor sites. (b) The mirror image of the model of cytochalasin B shown in Fig. 5a. C_4 and C_{19} are in the S configuration.

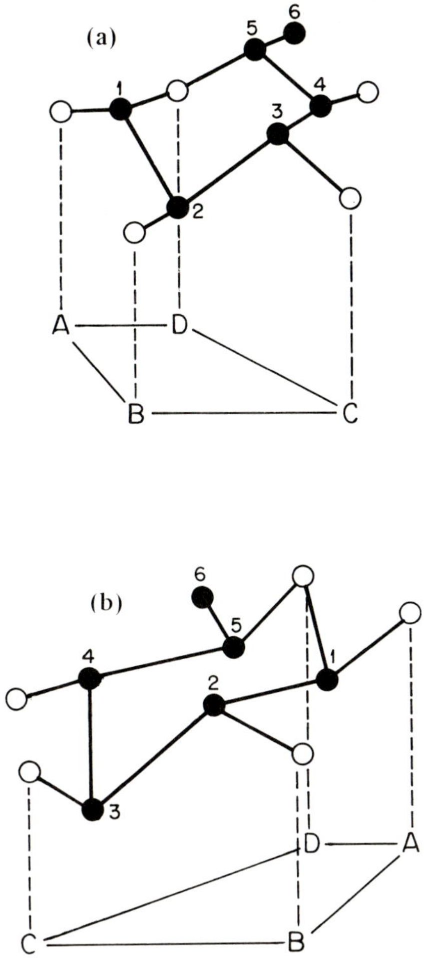

FIGURE 6 (a) Model of C1-conformation of β-D-glucopyranose. (b) The inverted model. ●, carbon atoms. O, oxygen functions, . . ., hydrogen bonds. A, B, C, and D represent protein receptor sites.

hibit glucose transport. This prediction is substantiated by the fact that hydrocortisone (IV) (Fig. 7), corticosterone (V), prednisolone (VI), and 11β-hydroxyprogesterone (VII) are all competitive inhibitors of glucose transport in human erythrocytes (Table V). Moreover, as the number of relevant oxygen functions decreases in this series of steroids a corresponding decrease in the affinity of the inhibitor for the glucose transport system is reflected by the increase in K_i values. Recently we have shown (Taylor, N. F. and Gagneja, G. L., 1976

(IV) R' = OH, R" = OH
(V) R' = OH, R" = H
(VI) R' = H, R" = OH, double bond at C_1-C_2
(VII) R' = H, R" = H

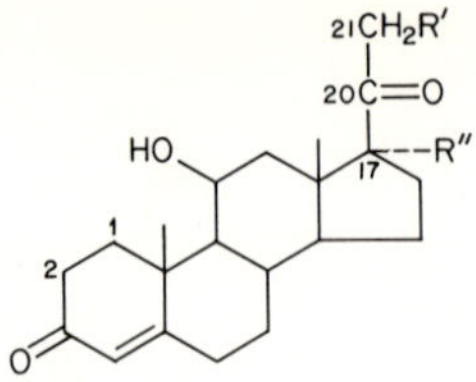

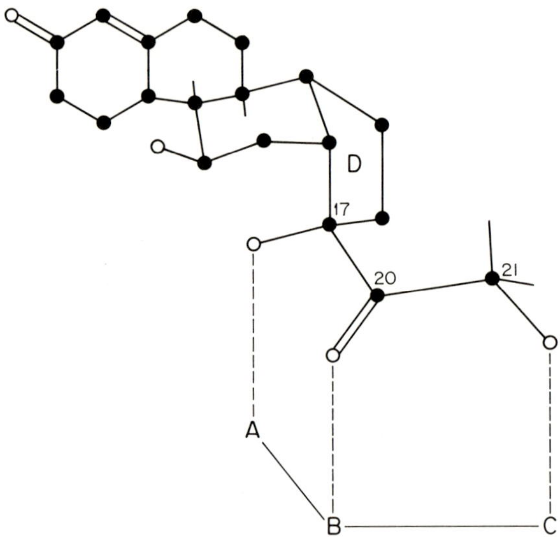

FIGURE 7 Model of hydrocortisone (IV). ●, carbon atoms; O, oxygen functions; . . ., hydrogen bonds. A, B, and C represent protein receptor sites.

unpublished) that alloxan, which also has a distribution of hydrogen bonding sites (Fig. 8) similar to those found in CB, can act as a competitive inhibitor of glucose transport with $K_i = 1.85 \times 10^{-6}$ *M* (Fig. 9). Further support for this model is based on estimations of the apparent ΔE and Δ* for glucose exit in the presence of CB, alloxan, hydrocortisone, corticosterone, prednisolone, and 11β-hydroxyprogesterone. For example, glucose exit at four different temperatures in the presence of CB is shown (Fig. 10) and the plot of the K_i values obtained against the reciprocal of these temperatures allows ΔE to be calculated from the slope and hence ΔH* (Fig. 11). It is of interest to note that as the effectiveness of

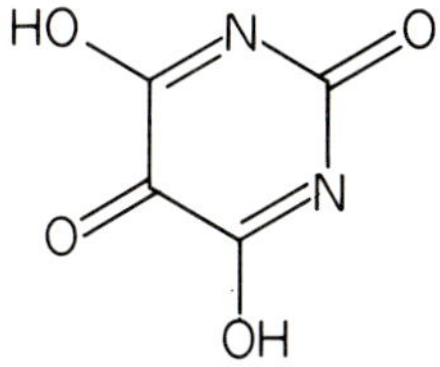

Alloxan. $K_i = 1.85 \times 10^{-6}$ M

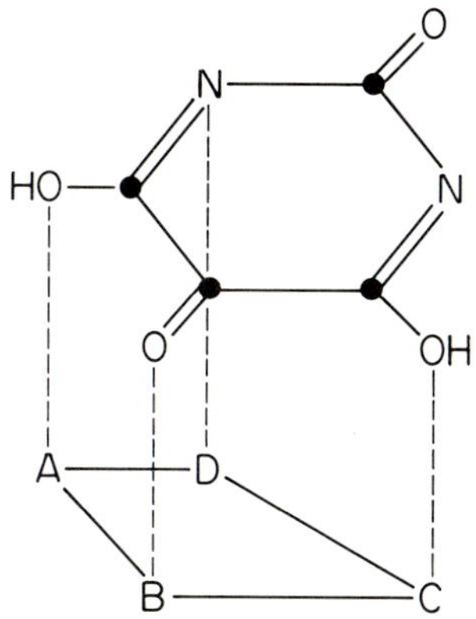

FIGURE 8 Model of alloxan, showing hydrogen bonding sites possible at the oxygen and nitrogen atoms. A, B, C, and D represent protein receptor sites.

the inhibitor decreases, the ΔH^*_{gi} values for the glucose-carrier complex increases from 15.02 kJ/mol in the presence of CB to 42.0 kJ/mol in the absence of inhibitor (ΔH^*_g) (Table V). The number of glucose-carrier sites has been estimated as 3×10^5/cell (29) and our thermodynamic data may be considered, therefore, to reflect in part the number of carrier sites occupied by glucose in the presence of inhibitors. Thus, the stronger the affinity of the inhibitor (smaller K_i value) for the carrier protein, the smaller the number of glucose occupied sites and hence the lower ΔH^*_{gi} values observed (Table V). The thermodynamic data also correlate in the same way with the number of hydrogen bonding sites present in the inhibitor and the extent of glucose-occupied carrier sites.

The above kinetic and thermodynamic results are consistent with the proposed model of the structure-action relationship between CB and β-D-glucopyranose and partially explain why CB is such an effective competitive inhibitor of facilitated glucose transport. Moreover, this model for the mode of ac-

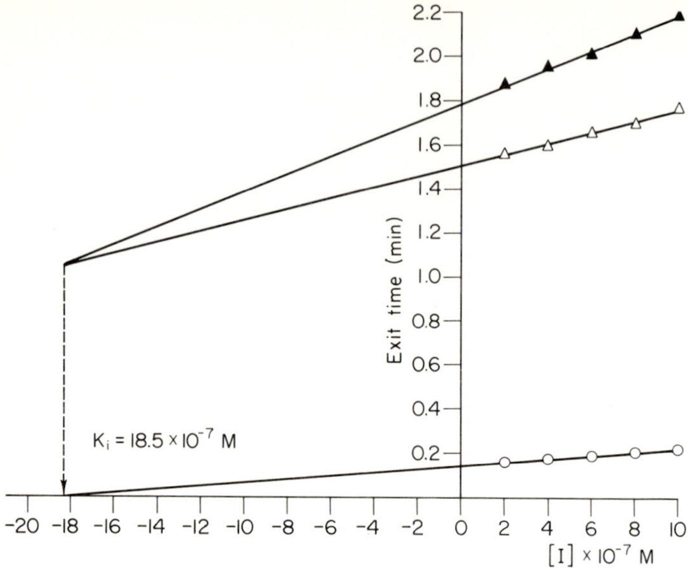

Figure 9 The inhibition of glucose transport in human erythrocytes by alloxan at 37°C. Dixon plot; ▲, external 30 mM glucose; Δ, external 50 mM glucose; alloxan concentrations in 2, 4, 6, 8 and 10 × 10^{-7} M; O, alloxan inhibition of glucose exit into saline, pH 7.4. Experimental procedures as previously reported (30).

tion of CB on glucose transport suggests that a larger number of apparently unrelated molecules may act as analogs with respect to protein receptor sites. This competitive binding, especially to energy-dependent glucose transport carriers found in a number of membranes, may significantly affect the regulation of glucose metabolism in the cell (52). In this connection it has been reported that both Δ^1-tetrahydrocannabinol (VIII) and cannabidiol (IX), the active components of marihuana, are inhibitors of glucose transport ($K_i \sim 10^{-6}$ M) in the human erythrocyte (53). These results may account for the hyperglycemic changes that have been attributed to this drug (54). An examination of the structures (IV) and (V) reveals the presence of two oxygen functions (O^*) that are superimposible on the oxygen atoms at C_1 and C_3 of β-D-glucopyranose (Fig. 6a) and associated with hydrogen bonding to the carrier protein. The inhibition of glucose transport by (VIII) and (IX) may be explained, therefore, by this structural similarity. In the kinetic studies, however, it should be pointed out that the question of whether competitive in-

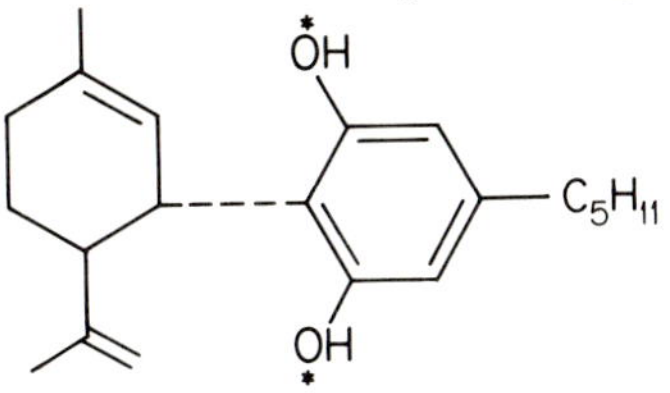

(VIII)

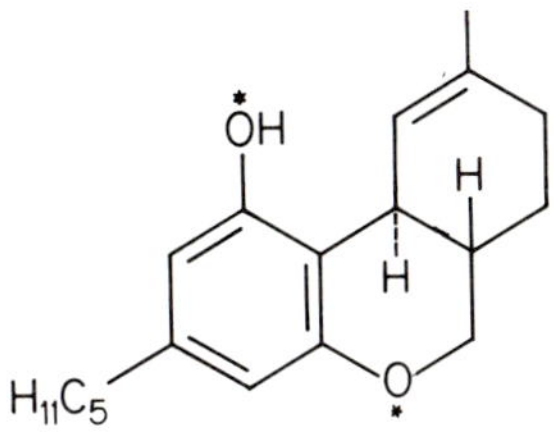

(IX)

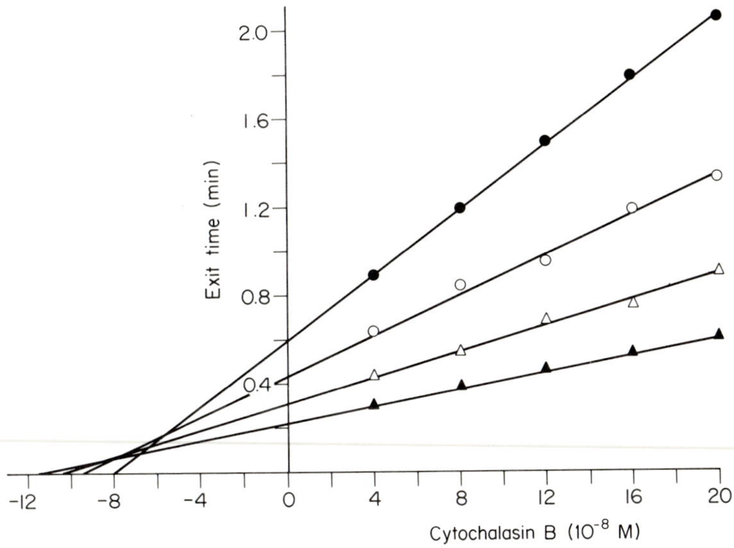

FIGURE 10 Determination of the K_i value for cytochalasin B at four different temperatures. ●, *20.5°C, K_i = 0.82 × 10^{-7} M;* ○, *25.2°C, K_i = 0.96 × 10^{-7} M;* △, *30.5°C, K_i = 1.1 × 10^{-7} M;* ▲, *37°C, K_i = 1.2 × 10^{-7} M.*

TABLE V Apparent Dissociation Energies of the Glucose-Carrier of the Human Erythrocyte in the Presence of Various Inhibitors at 37°C[a]

Compound	Hydrogen bonding sites	K_i (M)	ΔH_g^* (kJ)	ΔH_{gi}^* (kJ)
Glucose	4		42.0	
Glucose plus				
Cytochalasin B	4	1.2×10^{-7}		15.02
Alloxan	4	1.85×10^{-6}		19.16
Hydrocortisone	3	2.0×10^{-4}		27.72
Corticosterone	2	4.0×10^{-4}		39.48
Prednisolone	2	4.0×10^{-4}		40.32
11β-Hydroxy-progesterone	1	6.0×10^{-4}		41.58

[a]ΔH_g^* and ΔH_{gi}^* are the apparent dissociation energies of glucose in the absence and presence of inhibitors and obtained from the slope of the plot of log K_x or log K_i values vs $1/T$ at four different temperatures where slope $\times$ 2.303$R = \Delta H^* + RT$ (R = 8.314 J mol^{-1} K^{-1} and T is absolute temperature). K_i values were determined as previously reported (30).

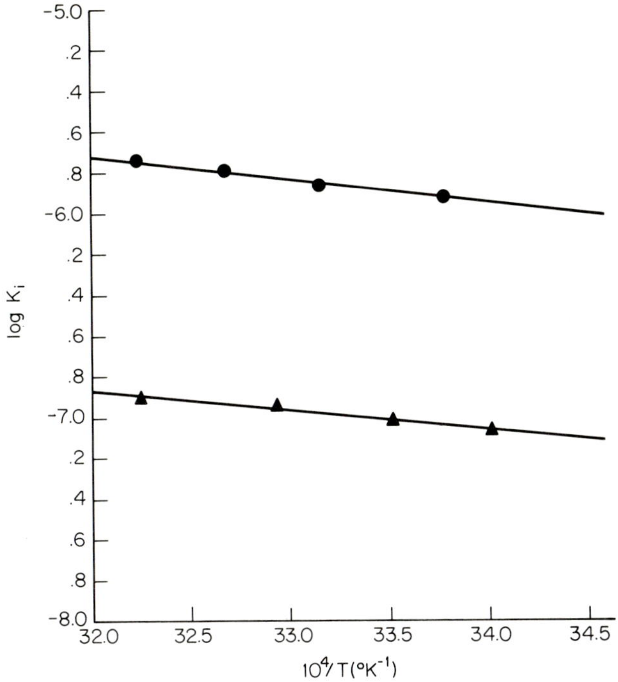

FIGURE 11 Estimations of the apparent ΔH^* of glucose in the presence of cytochalasin B and alloxan. ΔH_{gI} (KJ/mol): at 37°C: ▲, cytochalasin B 15.66; ●, alloxan 19.16 (see Table V).

hibition or otherwise of glucose by (VIII) or (IX) occurs was not rigorously established (53). Further tests should be made to experimentally ascertain the kinetic nature of this inhibition before any assignment of the binding sites on the cannabiloids can be made.

ACKNOWLEDGMENTS

This work was supported in part by the National Research Council of Canada. One of us (NFT) thanks the Division of Carbohydrate Chemistry of the American Chemical Society and Dr. R. E. Harmon, Symposium Chairman, for the invitation to present this work.

REFERENCES

1. Marais, J. S. C., *Onderstepoort. J. Vet. Sci. Anim. Ind.* (1944) *20*, 67-73.
2. Peters, R. A., Wakelin, R. W., Buffa, P., and Thomas, L. C., *Proc. Roy. Soc.* (1953) *B140*, 497-506.
3. *Ciba Fdn. Symposium* (1972), "Carbon-Fluorine Compounds: Chemistry, Biochemistry and Biological Activities," pp. 1-417. Associated Scientific Publishers, New York.
4. *ACS Symposium Series No. 28* (1976), "Biochemistry Involving Carbon-Fluorine Bonds" (R. Filler, ed.), pp. 1-214, American Chemical Society, Washington, D.C.
5. *Handbook Exp. Pharmakol.* (1970) *20*, Parts 1 and 2.
6. Taylor, N. F., in *Ciba Fdn. Symp.* (1972), pp. 215-238, Associated Scientific Publishers, New York.

7a. Barnett, J. E. G., in *Ciba Fdn. Symp.* (1972), pp. 95-115, Associated Scientific Publishers, New York.

7b. Barnett, J. E. G., Jarvis, W. T. S., and Munday, K. A., *Biochem. J.* (1968) *109*, 61.

7c. Barnett, J. E. G., Ralph, A., and Munday, K. A., *Biochem. J.* (1969) *114*, 569.

7d. Barnett, J. E. G., Ralph, A., and Munday, K. A., *Biochem. J.* (1970) *118*, 843.

8. Sharpe, A. G., in *Ciba Fdn. Symp.* (1972), pp. 33-54, Associated Scientific Publishers, New York.
9. *Chemical Scoiety* (1958), Tabl-s of Interatomic Distances and Configuration in Molecules and Ions. Special Publication No. 11, p. S12. London: Chemical Society.
10. *Chemical Society* (1965), Tables of Interatomic Distances and Configuration in Molecules and Ions. Suppl. 1956-1959, Special Publication No. 18, p. S145. London: Chemical Society.
11. Pauling, L., *The Nature of the Chemical Bond* (1940), 2nd ed. p. 189. London: Oxford Univ. Press.
12. Pauling, L., *The Nature of the Chemical Bond* (1940), 2nd ed. p. 64. London: Oxford Univ. Press.
13. Taylor, N. F., Romaschin, A. D., and Smith, D. A., in *ACS Symposium Series No. 28* (1976), "Biochemistry Involving Carbon-Fluorine Bonds" (R. Filler, ed.), pp. 90-116. American Chemical Society, Washington, D. C.
14. Morton, G. O., Lancaster, J. E., Van Lear, G. E., Fulmor, F., and Meyer, W. E., *J. Amer. Chem. Soc.* (1969) *91*, 1535.
15. Kent, P. W., in *Ciba Fdn. Symposium* (1972), "Carbon-Fluorine Compounds," pp. 169-208. Associated Scientific Publishers: New York.
16. Gordesky, S. E., *TIBS* (1976) *1* (No. 9), 208.
17. Bloch, R., *J. Biol. Chem.* (1974) *249*, 1814.

18. Kallenberg, A., and Walker, C., *J. Biol. Chem.* (1976) *251*, 1582.
19. Lieb, W. R., and Stein, W. D., *Biochim. Biophys. Acta* (1972) *265*, 187.
20. LeFevre, P. G., *Ann. N.Y. Acad. Sci.* (1975) *264*, 398.
21. LeFevre, P. G., *Pharmacol. Rev.* (1961) *13*, 39.
22. Sen, A. K., and Widdas, W. F., *J. Physiol. (London)* (1962) *160*, 392.
23. Lacko, L., and Burger, M., *Biochem. J.* (1962) *83*, 622.
24. Baker, G. F., and Widdas, W. F., *J. Physiol. (London)* (1972) *271*, 10p.
25. Kahlenberg, A., and Dolansky, D., *Can. J. Biochem.* (1972) *50*, 638.
26. Gottschalk, A., *Adv. Carbohydr. Chem.* (1950) *5*, 49.
27. Helferich, B., Grimler, S., and Gnüchtel, A., (1937) *Hoppe-Seyler's Z. Physiol. Chem. 248*, 85.
28. Aldridge, D. C., Armstrong, J. J., and Speake, R. N., *J. Chem. Soc.* (1967), 1667.
29. Lin, S., and Spudich, J. A., *J. Biol. Chem.* (1974) *249*, 5778.
30. Taylor, N. F., and Gagneja, G. L., *Can. J. Biochem.* (1975) *53*, 1078.
31. Crane, R. K., *Physiol. Rev.* (1960) *40*, 789.
32. Bihler, I., Hawkins, K. A., and Crane, R. K., *Biochim. Biophys. Acta* (1962) *59*, 94.
33. Wilson, T. H., and Landan, B. R., *Am. J. Physiol.* (1960) *198*, 99.
34. Miller, D. M., in "Red Cell Structure and Function" (Jamieson, G. A., and Greinwalt, T. A., eds.), pp. 240-292. Lippincott, Philadelphia and Toronto, 1969.
35. Riley, G. J., and Taylor, N. F., *Biochem. J.* (1973) *135*, 773.
36. Dixon, M., and Webb, E. G., *Enzymes* (1964), 2nd ed., pp. 145-166, Longmans Green, London.
37. Barnett, J. E. G., Holman, G. D., and Munday, K. A., *Biochem. J.* (1973) *131*, 211.
38. Reeves, R. E., *Adv. Carbohydr. Chem.* (1951) *6*, 107.
39. Wessells, N. K., Spooner, B. J., Ash, J. F., Bradley, M. D., Ludens, M. A., Taylor, E. L., Wren, J. T., and Yamada, K. M., *Science* (1971) *171*, 135.
40. Allison, A. C., Davis, P., and DePetris, S., *Nature New Biol.* (1971) *232*, 153.
41. Shepro, D., Belamarich, F. A., Robblee, L., and Chao, F. C., *J. Cell Biol.* (1970) *47*, 544.
42. Majno, G., Bouvier, C. A., Gabbiani, G., Ryan, C. B., and Statkov, P., *Thromb. Diath. Haemorrh.* (1972) *28*, 49.
43. Williams, J. A., and Wolff, J., *Biochem. Biophys. Res. Commun.* (1971) *44*, 422.

44. Thomas, D. des. S., Lutzac, M., and Manavatha, E., *Nature*, (1974) *249*, 140.
45. Macpherson, M. A., Scholfield, J. G., *FEBS Lett*. (1972) *24*, 45.
46. Kletzien, R. F., Purdue, J. F., and Springer, A., *J. Biol. Chem*. (1972) *247*, 2964.
47. Kletzien, R. F., and Purdue, J. F., *American Society for Cell Biology Abstract* (1974), 171a.
48. Mizel, S. B., and Wilson, L., *J. Biol. Chem*. (1972) *247*, 4102.
49. Kuo, S. C., and Lampen, J. O. *Ann. N.Y. Acad. Sci*. (1974) *235*, 137.
50. Kuo, S. C., and Lampen, J. O., *Biochimica Biophys. Acta*. (1975) *389*, 145.
51. Cahn, R. S., and Ingold, C. K., *J. Chem. Soc*. (1951), *612*.
52. Elbrink, J., and Bihler, I., *Science* (1975) *188*, 1177.
53. Schurr, A., Sheffer, N., Graziani, Y., and Livne, A., *Biochem. Pharmacol*. (1974) *23*, 2005.
54. Podolsky, S., Pattavina, C. G., and Amaral, M. A., *Ann. N.Y. Acad. Sci*. (1971) *191*, 54.

CONTROL OF CELL GROWTH BY NUCLEOSIDE EFFLUX THROUGH THE MEMBRANE*, †

PAUL W. WIGLER

Memorial Research Center
University of Tennessee Center for the Health Sciences
Knoxville, Tennessee

Studies with the membrane transport inhibitor, 5-bromodeoxyuridine 5'-methylphosphonate (BrdUrd-OPO$_2$Me), are used as the basis for a hypothesis on the control of cell growth by nucleoside efflux through the membrane. The hypothesis suggests that BrdUrd-OPO$_2$Me binds to the active site of the membrane-carrier protein for thymidine (dThd) and deoxyuridine (dUrd) and "traps" the carriers at the outer surface of the cell membrane. The reduced availability of carriers at the inner membrane surface inhibits the efflux of deoxyribonucleosides. The blockage of efflux of dThd and dUrd produces an increased internal concentration of thymine and thymidine triphosphate (dTTP). Since dTTP is an inhibitor of several en-

*This work was supported in part by grant CA14322 from the NIH, USPHS.

†The following abbreviations are used: BrdUrd-OPO$_2$Me, ammonium 5-bromodeoxyuridine 5'-methylphosphonate; BrudUrd-OPO$_2$[^{3}H]Me, the ^{3}H-labeled compound; 5F[6-^{3}H]dUrd, 5-fluoro [6-^{3}H]deoxyuridine; [6-^{3}H]dUrd, 2'-deoxy[6-^{3}H]uridine; [*methyl*-^{3}H]dThd, 2'-deoxy[*methyl*-^{3}H]thymidine; OMP, orotidylate; UMP, uridylate; Urd, uridine; dUMP, dUDP, the 5'-mono- and diphosphates of 2'-deoxyuridine; dCMP, dCDP, dCTP, the mono-, di-, and triphosphates of 2'-deoxycytidine; dTMP, dTTP, the mono- and triphosphates of 2'-deoxythymidine; dUrd, 2'-deoxyuridine; dThd, 2'-deoxythymidine; FdUrd, 5'fluoro 2'-deoxyuridine; FdUMP, FdUTP, the mono- and triphosphates of FdUrd; BrdUrd, 5-bromo 2'-deoxyuridine; Ura, uracil; and Thy, thymine.

zymes in the pathway for the de novo synthesis of DNA, elevated levels of dTTP can produce a feedback inhibition of cell division. The hypothesis also suggests that rapidly growing cells excrete dThd and dUrd into the growth medium, but resting cells retain nucleosides. Transport experiments show that BrdUrd-OPO$_2$Me inhibits the uptake of tritium labeled fluorodeoxyuridine, bromodeoxyuridine, and dUrd into L5178Y cells. By contrast, the BrdUrd-OPO$_2$Me does not inhibit the uptake of dThd. This finding suggests that there is some structural specificity to the interaction between the transport inhibitor and the active site of a membrane carrier protein for deoxyribonucleosides.

INTRODUCTION

The aim of this paper is to present a hypothetical mechanism for the control of cell growth by nucleoside efflux through the membrane. Although this control mechanism is in no sense proved, it provides an intellectual framework for the design of new experiments. The hypothetical control mechanism can be stated as follows:

(1) Since dThd$^+$ can be readily synthesized *de novo*, the uptake of dThd into growing culture cells is not a requirement of growth.

(2) Growing cells may produce an oversupply of dThd or dUrd that is excreted into the growth medium.

(3) An increase in the rate of efflux of dUrd or dThd by growing cells decreases the internal pools of the deoxyribonucleosides and dTTP.

(4) Inhibition of dUrd or dThd efflux by high cell density or a membrane transport inhibitor increases the internal pools of Ura, Thy, and the deoxyribonucleosides.

(5) Elevated levels of dUrd or dThd in the cytoplasm lead to an increase in the concentration of dTTP or other nucleotides.

(6) Elevated levels of dTTP or other nucleotides inhibit the growth of culture cells.

(7) There will be a delay in time between the inhibition of transport and the inhibition of cell division.

If the foregoing hypothesis is to be taken seriously, the control mechanism must lead to certain expectations. It is well known that most culture cells do not require the addition of nucleosides to the culture medium for optimal growth. Nevertheless, carrier-mediated transport of deoxyribonucleosides is faster in growing cells than in resting cells (1). Clearly, the facilitated transport in growing cells is not used for the salvage of dThd or dUrd from the medium.

The release of dUrd into growth medium by growing neoplastic cells has been demonstrated (2). The effect of changes in the rate of deoxyribonucleoside efflux on the internal concentrations of dUrd, dThd, and dTTP has not been demonstrated. The inhibition of cell growth by dThd and thymidine nucleotides, however, is a well-documented phenomenon (3). The membrane transport inhibitor BrdUrd-OPO_2Me is an appropriate agent to test the nucleoside efflux hypothesis. The BrdUrd-OPO_2Me may bind the active site of the membrane-carrier protein for dThd or dUrd and "trap" the carriers at the outer surface of the cell membrane. The reduced availability of carriers at the inner membrane surface inhibits the efflux of deoxyribonucleosides. Perhaps a blockage of efflux of dThd or dUrd produces an increased internal concentration of thymine and dTTP. Since dTTP is an inhibitor of several enzymes in the pathway for the *de novo* synthesis of DNA (4), elevated levels of dTTP can produce a feedback inhibition of cell division.

Transport experiments show that BrdUrd-OPO_2Me inhibits the uptake of tritium labeled FdUrd, BrdUrd, and dUrd into L5178Y cells (5). By contrast, the BrdUrd-OPO_2Me does not inhibit the uptake of dThd. This finding suggests that there is some structural specificity to the interaction between the transport inhibitor and the active site of a membrane-carrier protein.

THE PATHWAYS FOR INCORPORATION OF CYTOSINE AND THYMINE INTO DNA

A consideration of the metabolic pathways inside a living cell as a closed system is an over simplification of the available information; the influence of the cellular environment should not be neglected. Metabolites in the growth medium that are readily interchanged with internal metabolites can produce significant effects on the physiology of cell growth. The interference with this exchange of metabolites between medium and cell, by the binding of membrane transport inhibitors at the outer cell surface, could also affect cell division. The internal-external balance of metabolites is somehow linked in a feedback control system to the biosynthesis of DNA.

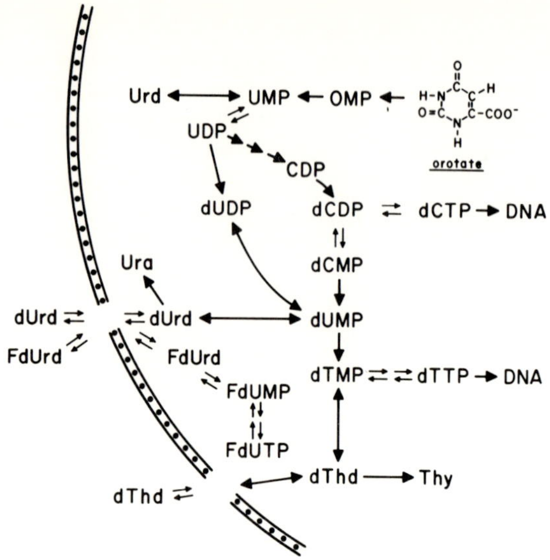

FIGURE 1

Pathways for the utilization of orotate, dUrd, and dThd for the biosynthesis of DNA are summarized in Fig. 1. All of the reactions of Fig. 1 are catalyzed by known enzymes (6); the membrane transport system for dUrd and dThd has been described (7).

All of the pathways in Fig. 1 can be utilized to label the nucleotides in the DNA of culture cells. For example, [^{14}C]orotate is converted to labeled OMP and the nucleotides in the *de novo* pathway for DNA-thymine and DNA-cytosine are labeled (8). The uptake of labeled Urd produces a labeling pattern similar to labeled orotate. On the other hand, the administration of labeled dUrd and dThd produce labeling in the dTTP pathway, but virtually no labeling in the dCTP pathway (9, 10). The uptake of labeled BrdUrd will produce a labeling pattern the same as dThd (11).

The metabolism of FdUrd is different from that of dThd and BrdUrd because FdUMP is not a potent substrate for thymidylate kinase and FdUTP is not incorporated into DNA (12).

Several of the reactions of Fig. 1 are essentially nonreversible in living cells. In this group are included the CDP → dCDP reaction (cytidine diphosphate reductase), dUMP → dTMP (thymidylate synthetase), and dTTP → DNA-thymine (DNA polymerase). The reactions dUrd → dUMP and dThd → dTMP are reversible; these are catalyzed by thymidine kinase in the forward direction and a phosphomonoesterase or 5'-nucleotidase in the reverse direction. The conversion of dTMP to dTTP is also reversible; this reaction is catalyzed by thymidylate kinase and dTDP kinase in the forward direction and by a nucleotide pyrophosphatase in the reverse direction.

The reactions in Fig. 1 can provide an explanation of the delayed growth inhibition observed with a transport inhibitor, such as BrdUrd-OPO_2Me (13). There is an extensive literature on membrane transport inhibitors (14); one of the best examples is colchicine (15), an inhibitor of the transport of Urd and dThd. Ilan and Quastel (16) showed that when cells are exposed to colchicine for 24 hr DNA synthesis is inhibited. If the colchicine treatment is less than one hour, the synthesis of DNA and cell division are unaffected (15). The interaction of colchicine with microtubules (15) could account for the inhibition of cell growth observed with this compound. Another explanation of the growth inhibition with colchicine is that the compound blocks the membrane efflux of a natural inhibitor of DNA synthesis.

Acronycine is an alkaloid derivative that blocks the uptake of Urd and dThd; this is accompanied by a delayed inhibition of cell growth (17). There is one similar aspect of the literature on acronycine as well as colchicine. All the transport inhibition experiments are based on the influx of Urd and dThd and not on efflux data. The efflux experiments are relatively difficult because the culture cells must be preloaded with labeled nucleosides. Excretion rates are then monitored. Since Urd and dThd are degraded to Ura and Thy or incorporated into RNA and DNA, the efflux rate is difficult to measure. The excreted nucleosides must be identified in the medium and quantitated separately from the other excreted metabolites.

Another interesting study on the influx of dThd showed that when hamster cells move from G_1 to the S phase of division, the capacity of cells to transport dThd is enhanced (18). The addition of cytochalasin B to cells in G_1 produces an inhibition of dThd transport. Cytochalasin B inhibits the uptake of Urd as well as dThd (19). Estensen (20) has reported the inhibition of cytoplasmic division by cytochalasin B, but the experiments were run for 15 hr, which is not long enough to show a delayed inhibition of DNA synthesis.

An increase in the concentration of dThd (or dUrd) in the cytoplasm could lead to an increase in dTTP (see Fig. 1). If the internal level of dTTP is too high, the growth of cells that are undergoing division is inhibited (21). Apparently, dTTP is a feedback inhibitor of several enzymes in the DNA synthetic pathways, cytidine diphosphate reductase (3), thymidine kinase (22), and thymidylate kinase (4).

Rao has shown that growing cells excrete dUrd (2); thus, a blockage of nucleoside efflux could produce a feedback control mechanism for cell growth. One question that is raised in Fig. 1 is whether there are two different transport systems for dUrd and dThd with overlapping specificities. The transport of dUrd and dThd is mediated by a single carrier in Novikoff rat hepatoma cells (7). In this discussion it is assumed that dUrd and dThd are transported by one transport system in L5178Y cells.

THE CARRIER-MEDIATED TRANSPORT SYSTEM FOR dThd and dUrd

There is an important difference between the kinetics of an enzyme reaction and the kinetics of a membrane transport system. The addition of a reaction product to an enzyme reaction produces an inhibition of the rate of conversion of substrate to product (23). Inhibition of an enzyme reaction by the product is due to the exposure of the active site to a single solvent phase that contains both substrate and product. For the carrier protein of a transport system, however, the active site may be exposed to two different solvent phases: the cell cytoplasm and the growth medium. The cytoplasm dUrd (or dThd) may be considered the substrate of the transport reaction and the dUrd in the medium is the product. When dUrd (or another compatible substrate) is added to the medium, the rate of efflux of internal dUrd is accelerated. This kinetic phenomenon has been called "accelerated exchange diffusion" (24). Product acceleration of membrane transport is due to the exposure of the carrier active site to one solvent phase that contains substrate and a different solvent phase that contains product.

Figure 2 may explain the product acceleration of transport and the effect of inhibitors, such as BrdUrd-OPO$_2$Me. This scheme for the transport of dUrd is similar to those described by Goldman (24). The influx reaction, conversion of product to substrate, will be considered first. The dUrd enters the membrane system and forms a complex with a carrier protein (reaction 1). The dUrd-carrier complex penetrates the cell membrane (reaction 2), and the dUrd is unloaded by the carrier into the interior of the cell (reaction 3).

In the direction for efflux, the dUrd substrate is loaded by carrier in reaction 3. The complex is transported to the outer surface in reaction 2 and the dUrd is unloaded into the medium in reaction 1. If cells are preloaded with labeled nucleoside and incubated with no nucleoside in the growth medium, there is a "lag" in the efflux of nucleoside. The

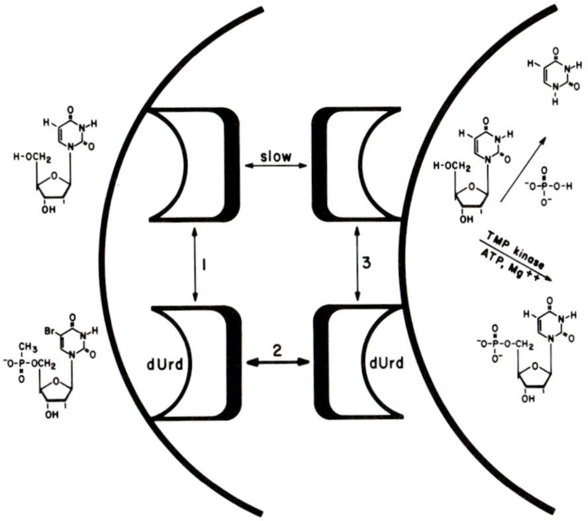

FIGURE 2

addition of a compatible transport substrate to the medium produces "accelerated exchange diffusion" in the direction of efflux.

The model for transport of Fig. 2 is based on the concept that the dUrd-carrier complex is transported much more rapidly from one side of the membrane to the other side than is the unloaded carrier ("slow" reaction). If there is no appropriate transport substrate available outside the cell, the unloaded carrier protein is retarded in movement from the outside membrane surface to the inside (24). A finite time delay is required for the unloaded carrier to move to the inside surface (slow reaction). The reduced number of carriers at the inner membrane surface may lead to a lag in the efflux of dUrd or other nucleosides.

The model also shows how an inhibitor that is not a transport substrate, such as BrdUrd-OPO_2Me, can produce a "decelerated exchange diffusion" in the direction of efflux. The inhibitor can trap the transport carrier protein on the outer surface of the membrane to inhibit the efflux of a substrate. In this mechanism it is assumed that BrdUrd-OPO_2Me from the medium forms a complex at the active site of the carrier protein; the complex traps the carrier at the outer surface of the cell membrane.

A carrier-BrdUrd-OPO_2Me complex at the outer surface decreases the number of carriers available at the inner membrane surface to catalyze the efflux of nucleoside substrates.

Thus, BrdUrd-OPO_2Me in the medium could block the efflux of dUrd and dThd through the membrane.

Although the kinetic phenomenon of decelerated exchange diffusion for efflux has not been studied in much detail, the observation of Pickard and Paterson (25) supports the validity of this concept. The efflux of Urd from human erythrocytes is inhibited by 0.6 μ*M* 4-nitrobenzylthioinosine in the medium (25). An inhibition of dThd efflux by an S-nitrobenzyl derivative of guanosine has also been demonstrated (26).

The chemical structure of BrdUrd-OPO_2Me is given in Fig. 2. The report of Wigler and Schluter (5) shows that BrdUrd-OPO_2Me can form a complex with the active site of a membrane-carrier protein for the transport of dUrd by L5178Y cells. It should be noted that decelerated exchange diffusion for efflux will not depend on the relative affinities to the active site of the cytoplasm substrate in comparison with the inhibitor. This follows from the fact that the substrate and inhibitor are not competing at the same membrane surface. The inhibitor is in one solvent phase, the medium; and the efflux substrate is in another solvent phase, the cell cytoplasm.

If it is assumed that the cell does not compensate for the inhibition of dUrd efflux, the effect of BrdUrd-OPO_2Me will be to increase the cytoplasm concentration of dUrd, Ura, dUMP, dTMP, and dTTP (see Fig. 1). Inhibition of dThd efflux will increase the internal concentration of dThd, Thy, dTMP, and dTTP.

THE EFFLUX OF NUCLEOSIDES AND CELL GROWTH

Many culture cell lines are capable of the *de novo* synthesis of dUrd, dThd, and other nucleosides; for these lines the addition of nucleosides to the growth medium is not required for cell viability. The importance of nucleoside transport systems to cells that have no metabolic requirement for the influx of nucleosides is difficult to comprehend. Perhaps the physiological function of certain transport systems is to enhance the efflux of nucleosides rather than to salvage exogenous nucleosides. One example of this may be the systems for the excretion of the minor ribonucleosides from tRNA, such as pseudouridine (27), ribosylthymine, and 4-thiouridine. The minor ribonucleosides of tRNA are synthesized at the polynucleotide level, and are only observed in the cytoplasm as catabolic products of tRNA. These breakdown products are not ordinarily reutilized for nucleic acid synthesis and may even be inhibitors of cell growth. For example, 4-thiouridine and 2-thiouridine are inhibitors of cell growth (28). An efflux mechanism for the excretion of the minor ribonucleosides from tRNA may therefore be quite important for the maintenance of optimal cell growth.

There have been several reports of a marked elevation in the excretion of certain nucleosides into the urine of patients with advanced cancer. For example, Waalkes *et al.* (27) have demonstrated a high level of pseudouridine and two methylated nucleosides in the urine of patients with solid tumor malignancies. Thus, the cellular efflux of nucleosides from tRNA may be very efficient in tumor cells. Since pseudouridine is a nonmetabolized substrate for animal cells (29), this compound would be a good substrate for cellular efflux studies.

Although the correlation of membrane transport systems with cell growth is important, the study of culture cells is complicated in comparison with erythrocytes. Human erythocytes have been used for studies of the membrane transport of dUrd and dThd because the nucleosides are not metabolized by these cells. One difficulty in comparing the results on erythrocytes with data on culture cells is that there is a single transport system for Urd, dThd, and dUrd in red cells (26). In culture cells, however, the transport system for Urd may be separate from the transport system for dThd and dUrd (14). In the report of Cass and Paterson (26), erythrocytes were preloaded with labeled dThd, the cells were resuspended in a medium that contained a transport inhibitor, and the rate of efflux was determined. These authors showed that the transport inhibitor, 6-[(2-hydroxy-5-nitrobenzyl) thio]-guanosine can bind to the transport-carrier protein at the outer surface of the erythrocyte (see Fig. 2). This complex traps the carrier at the outer surface of the membrane and inhibits the efflux of dThd.

A subline of L1210 mouse leukemia cells unable to metabolize deoxycytidine was used for nucleoside efflux studies (30). Cells were preloaded with labeled deoxycytidine and resuspended in fresh medium that contained the transport inhibitor persantin. It was observed that the binding of the persantin at the outer cell surface produces an inhibition of the efflux of labeled deoxycytidine into the medium.

There does not appear to be a report available on the effect of density inhibition or a transport inhibitor on the intracellular levels of dTTP. Under certain metabolic circumstances, a block in the efflux of dThd could produce an increase in the dThd and dTTP concentration. The report by Lowe and Grindey (21) of an elevation of dTTP by dThd in the medium supports this concept. On the other hand, elevated levels of dThd and dTTP may be prevented in some cells with a high dThd catabolism rate by thymidine phosphorylase (31); another factor will be the rate of utilization of dTTP for DNA synthesis (8). If the overall rate of synthesis of dTMP from orotate (Fig. 1) is greater than the *sum* of the catabolism to thymine and the incorporation into DNA, the internal levels of dTTP will increase.

Lowe and Grindey (21) correlated the dThd inhibition of L1210 cell growth with changes in the internal concentrations of deoxynucleoside triphosphates. An increased concentration of dThd in the medium produced an increase in the dTTP pool. The increase in dThd (and dTTP) to a growth inhibition of 10% produced a reduction (compared with controls) in the concentration of dCTP. Up to a growth inhibition of 40% the dATP pool was similar to controls, but at higher dTTP concentrations the dATP pool decreased. The authors concluded that the dThd inhibition of the growth rate is approximately proportional to the sum of the concentrations of dCTP and dATP (21).

The information available on cellular transport and growth changes for dThd is fairly extensive in comparison with dUrd. Treatment of L1210 cells with dUrd inhibits the uptake and incorporation of labeled dThd into DNA (9). Thus, an increase in dUrd could produce an increase in dTTP (Fig. 1) and feedback inhibition of growth. Rao (2) demonstrated that growing papilloma and Vx-2 carcinoma cells excrete dUrd into the medium. If elevated levels of dUrd (and dTTP) provide a means of regulating growth, the efflux of dUrd may be related to the unrestricted growth of the tumor cells. In further experiments, Rao (2) demonstrated that embryonic, hyperplastic, and normal skin cells do not release dUrd into the growth medium. Goldman (10) has shown that the treatment of L-cell mouse fibroblasts with methotrexate leads to an enhanced efflux of labeled dUrd into the medium.

The primary site of biological action of BrdUrd-OPO$_2$Me appears to be at a cell membrane transport carrier system for dUrd (and dThd) (5). BrdUrd-OPO$_2$Me is not a substrate for a carrier mediated system, but slowly diffuses into the cell to achieve a concentration level approximately equal to that of the BrdUrd-OPO$_2$Me in the medium. Experiments with tritium-labeled BrdUrd-OPO$_2$Me have shown that, within the first two hours of exposure, the uptake of inhibitor by the cells is not detectable. After a 96 hr exposure to 25 μM BrdUrd-OPO$_2$[^{3}H]Me (52 mCi/mmole), a significant uptake is observed. The uptake observed was 8600 cpm/10^6 cells (5); this corresponds to a final internal concentration of inhibitor that is approximately equal to the inhibitor concentration in the medium.

Several experiments were performed to detect an inhibitory effect of BrdUrd-OPO$_2$Me at metabolic sites in addition to the transport system. The inhibitor had no effect on the phosphorylation of dThd by the thymidine kinase in cell-free systems (5). When culture cells were treated with labeled dThd, BrdUrd-OPO$_2$Me had no effect on labeling of the pools of dTMP, dTDP, and dTTP. Culture cells were treated with [6-^{14}C]orotate in combination with BrdUrd-OPO_2Me, but the inhibitor had no effect on the labeling ratio for ^{14}C incorporation into DNA-cytosine and DNA-thymine (32). In other experiments, it was shown that there is no detectable incorporation of the tritium label from BrdUrd-OPO_2[^{3}H]Me into cellular DNA.

The transport inhibitor BrdUrd-OPO_2Me produces a delayed inhibition of the growth of L5178Y cells (5). This observation confirms the results of a study on the inhibition of Chinese hamster V-79 cells (13). The growth of 1.5×10^5 L5178Y cells/ml for 96 hr at 37° in Fischer's medium will yield 5.0×10^6 cells/ml. If the medium contains 25 μ*M* BrdUrd-OPO_2Me, the final count is 2.5×10^5 cells/ml. Thus, the inhibitor produces a 20-fold reduction in cell yield after 96 hr, in comparison with the controls.

RATIONALE FOR THE USE OF BrdUrd-OPO_2Me

A handicap in the use of many nucleoside analogs and derivatives for mechanism studies with culture cells is the catabolism of the analogs to inactive products (31). The *N*-glycosyl linkage between the pyrimidine group and the deoxyribose moiety is particularly susceptible to cleavage by a pyrimidine deoxynucleoside phosphorylase. One method of maintaining the integrity of the glycosyl linkage is the use of a C-ribosyl nucleoside analog (29, 33) and another approach is to add a polar group to the 5' position of the deoxyribosyl moiety.

The 5'-methylphosphonate derivate of BrdUrd was chosen for metabolic studies because of the structural similarity to a nucleoside phosphate (see Fig. 3). The compound, 5-bromodeoxyuridine 5'-phosphate is shown as an example; this compound is susceptible to hydrolysis by a phosphomonoesterase enzyme. (Once the 5'-phosphate group has been removed, the N-glycosyl linkage is readily cleaved by thymidine phosphorylase.) Studies of phosphomonoesterases with ^{18}O-labeled substrates have shown that when the phosphate ester is hydrolyzed, bond cleavage occurs between the phosphorus atom and the oxygen atom but not at the C-O bond (34). This suggests an enzymic attack of a negatively charged hydroxyl group at the positively charged phosphorus atom (Fig. 3).

To prevent the cleavage of the P-O bond by phosphomonoesterase, an obvious approach is to reduce the positive charge at the phosphorus atom with an electron-releasing substituent. The most convenient substituent for this purpose is a methyl group. This suggests that the methylphosphonate ester of a

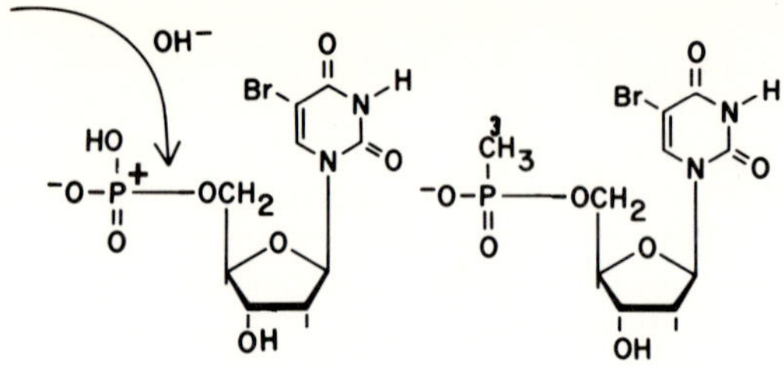

FIGURE 3

nucleoside should be highly resistant to a phosphomonoesterase enzyme (35); BrdUrd-OPO_2Me is also resistant to phosphorolysis by thymidine phosphorylase. Another advantage to the use of methylphosphonate derivatives is the ease of addition of isotope labels to the compounds. The transport inhibitor BrdUrd-OPO_2[^{3}H]Me is shown with tritium labeling in Fig. 3.

To verify the stability of BrdUrd-OPO_2Me to enzymic attack, the compound was prepared with tritium labels in the methyl group at the 5'-position. First, methylphosphonic acid was labeled by a tritium exchange reaction. The purified [^{3}H]methylphosphonate was attached to the 5'-hydroxyl group of a suitably blocked BrdUrd derivative. After deblocking, the BrdUrd-OPO_2[^{3}H]Me was isolated and purified. Phosphomonoesterase action on BrdUrd-OPO_2[^{3}H]Me would release the nucleoside BrdUrd which is a highly active compound for membrane transport and for the enzymes in the dThd to DNA-thymine pathway.

A thin layer chromatographic system was developed that permits the separation of BrdUrd-OPO_2[^{3}H]Me and [^{3}H]methylphosphonic acid (5). The marker compounds mixed with growth medium were applied in a streak to silica gel sheets; the chromatograms were developed with methanol:chloroform (70:30, v/v). The experimental procedures have been reported (5); the results are given in Fig. 4.

The markers for BrdUrd-OPO_2[^{3}H]Me and [^{3}H]methylphosphonate were mixed with growth medium, the solvent was removed under vacuum, methanol was added, and the mixture was applied to channel B of the chromatogram (Fig. 4). The location of the tritium-labeled zones shows that the two compounds are separated by this chromatographic procedure. The tritium in the labeled compounds was detected by placing the chromatograms in contact with X-ray film. The divalent anion [^{3}H]methylphosphonate was located in a zone (R_F 0.12) close to the origin of the chromatogram. The marker for the monovalent anion of BrdUrd-OPO_2[^{3}H]Me was located at R_F 0.53. This is the only method that produced a separation in tests of several

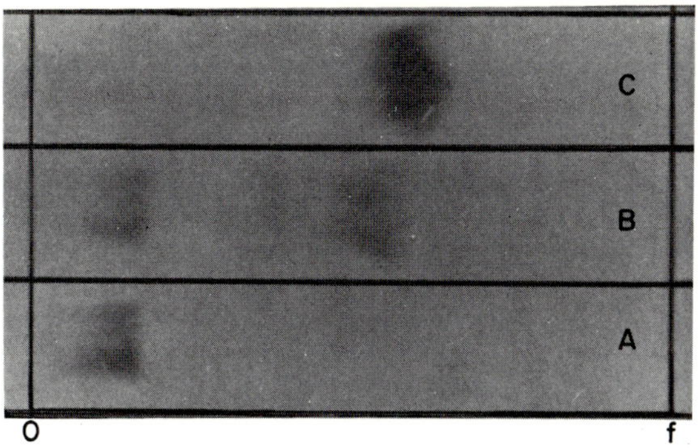

FIGURE 4 Effect of a 96 hr treatment with L5178Y cells on BrdUrd-OPO$_2$[^{3}H]Me in growth medium. The suspension was grown to a final density of 2.5 × 10^5 cells/ml, the cells were removed, and the growth medium was concentrated to a syrup. The syrup was suspended in methanol and applied to Eastman 13181 Chromagram sheets for development with methanol:chloroform (70:30, v/v). The tritium-labeled compounds were detected by placing the chromatograms in contact with Kodak SB54 X-ray film. The origin is indicated by an O and the front is indicated by an f. A marker of [^{3}H]methylphosphonic acid was detected in channel A, and the two markers in a mixture of the labeled acid plus BrdUrd-OPO$_2$[^{3}H]Me were detected in channel B. The experimental sample of BrdUrd-OPO$_2$Me, treated with L5178Y cells for 96 hr, was detected in channel C. The total absence of detectable labeled breakdown product, [^{3}H]methylphosphonic acid, in channel C should be noted.

different standard chromatographic systems for nucleosides and nucleotides. A marker of [^{3}H]methylphosphonate in growth medium was applied to channel A.

An inoculum of L5178Y cells was incubated for 96 hr at 37° in growth medium that contained BrdUrd-OPO$_2$[^{3}H]Me. The cells were removed by centrifugation and the growth medium was frozen. The medium was thawed, the solvent was removed under vacuum, and the residue was resuspended in a minimal amount of methanol. The medium from cultured cells was applied to the chromatogram in channel C. It may be noted that only a single zone of tritium-labeled compound was detected in a zone that corresponds to BrdUrd-OPO$_2$[^{3}H]Me. No [^{3}H]methylphosphonate could be detected in channel C. Thus, the transport inhibitor was not cleaved at the methylphosphonate ester bond during a 96 hr time period.

INHIBITION OF DEOXYRIBONUCLEOSIDE INFLUX WITH BrdUrd-OPO_2Me

Cultures of L5178Y mouse leukemia cells were grown at 37° for three days in a mixture of 50 ml 2.0 m*M* L-glutamine in Fischer's medium for leukemic cells of mice (Modified, 10×), 7.5 ml 0.89 *M* $NaHCO_3$, 50 ml heat inactivated horse serum, 5 ml 10,000 units/ml penicillin and 10,000 μg/ml streptomycin solution in saline, and 388 ml sterile water. The cell suspension was diluted threefold into fresh growth medium that contained BrdUrd-OPO_2Me; this mixture was incubated in a culture bottle at 37° for 1.0 hr in the dark. The medium for the control cells did not contain BrdUrd-OPO_2Me. The cell density of the suspension was approximately 5×10^5 cells/ml. After the prior incubation in a sterile solution of BrdUrd-OPO_2Me, a stock solution of labeled deoxyribonucleoside was added to prepare the final concentrations of [6-^{3}H]dUrd (100 mCi/mmole), 5F[6-^{3}H]dUrd (50 mCi/mmole), and [*methyl*-^{3}H]dThd (360 mCi/mmole) shown in Table I. The cell suspensions were pulsed for 0.5 hr at 37° with one of the labeled nucleosides and the suspension was diluted with cold Ringer's solution. The cells were isolated by centrifugation and the supernatant was discarded.

In the 5F[6-^{3}H]dUrd experiment of Table I, transport was terminated after 0.5 hr by mixing each cell suspension with a 50 μ*M* acronycine solution; this transport inhibitor prevented the cellular efflux of 5F[6-^{3}H]dUrd during the cell-rinsing procedure. It was necessary to use a transport inhibitor during the rinsing procedure for the separation (by centrifugation and resuspension) of the cells from the 5F[6-^{3}H]dUrd in the medium. (The acronycine was not used in certain control experiments; under these conditions the results for the uptake of 5F[6-^{3}H]dUrd were erratic and depended on the speed of manipulation during the centrifugation step.) The cells were isolated again and each cell sample was mixed with 5% $HClO_4$. The mixture was heated at 80° for 1 hr; the digest was cooled and mixed with LSC Cocktail. (Radioactivity measurements were performed with a liquid scintillation spectrometer.) The tritium counts taken up by the inhibited cells (U_i) and the counts taken up by control cells (U_c) were determined. The U_i/U_c ratios for three different substrates are given in Table I.

The results of Table I show that with [6-^{3}H]dUrd as the transport substrate, BrdUrd-OPO_2Me produces an inhibition of 57% at an (I)/(S) ratio of 6.1. By contrast, the inhibitor produces a very small (8.0%) inhibition of the uptake of [*methyl*-^{3}H]dThd at an (I)/(S) ratio of 18.3. The comparison between dThd uptake and FdUrd uptake is also interesting; with

TABLE I The Effect of a 1-hr Prior Incubation with BrdUrd-OPO_2Me on the 30 min Uptake of [6-^{3}H]dUrd, 5F[6-^{3}H]dUrd, and [methyl-^{3}H]dThd into L5178Y cells[a]

Substrate	*(I), μM*	*(S), μM*	*(I)/(S)*	U_i/U_c
[6-^{3}H]dUrd[b]	9.8	1.6	6.1	0.43
5F[6-^{3}H]dUrd[c]	18.0	1.0	18.0	0.23
[methyl-^{3}H]dThd[b]	110.0	6.0	18.3	0.92

[a]*The following symbols are used: (I) represents the concentration of BrdUrd-OPO_2Me; (S) represents the concentration of the transport substrate [6-^{3}H]dUrd, 5F[6-^{3}H]dUrd, or [methyl-^{3}H]dThd.*

[b]*U_i represents the 30 min uptake and incorporation into the DNA of inhibited cells; U_c represents DNA incorporation into control cells.*

[c]*U_i represents the 30 min uptake into inhibited cells and U_c represents the uptake into control cells. The uptake was terminated with an acronycine solution.*

the latter substrate, BrdUrd-OPO_2Me produces a 77% inhibition at an (I)/(S) ratio of 18.0.

When the influx substrate is labeled dThd, an 18-fold excess of BrdUrd-OPO_2Me in the medium does not inhibit the uptake of dThd into L5178Y cells. This finding suggests that the Michaelis constant (K_M) for the transport of dThd is much lower than the inhibition constant for the transport inhibitor. Thus, dThd in the medium can displace BrdUrd-OPO_2Me from the active site of the carrier protein. On the other hand, BrdUrd-OPO_2Me inhibits the uptake of dUrd (Table I) and may displace dUrd from the carrier site. These observations are consistent with the report on Novikoff cells that the K_M for transport of dThd is lower than the K_M for transport of dUrd (7).

Apparently, when dThd and BrdUrd-OPO_2Me compete for the carrier site at the outer cell surface, the transport inhibitor has almost no inhibitory effect. The results of Table I on membrane transport provide an explanation for the protective effect of dThd in the medium on the growth inhibition with BrdUrd-OPO_2Me. Wigler and Lozzio (13) reported that 10 μ*M* BrdUrd-OPO_2Me produces a marked inhibition of cell growth after four days. When culture cells are incubated in a mixture of 10 μ*M* dThd and 10 μ*M* BrdUrd-OPO_2Me, however, the

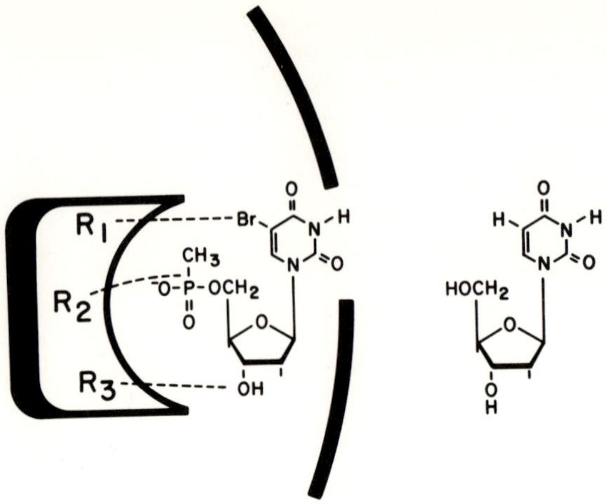

FIGURE 5

colony survival is approximately 90% compared with the controls. The displacement of BrdUrd-OPO_2Me by dThd from the carrier site at the outer membrane surface could account for the protective action of dThd. The efflux system for dThd and dUrd at the membrane of BrdUrd-OPO_2Me inhibited cells would be reactivated by the presence of dThd in the medium.

A hypothetical model for the complex formed between BrdUrd-OPO_2Me and the membrane carrier protein is given in Fig. 5. The bromine atom of the inhibitor may interact with a group (R_1) in the active site of the carrier protein. This is a reversible reaction in the dark; all of the experiments reported by Wigler and Schluter (5) were performed in the dark. If the complex of Fig. 5 is exposed to light, however, a debromination reaction may occur that would lead to an irreversible complex (13). The model suggests that BrdUrd-OPO_2Me may sensitize the carrier to radiation inactivation. This is similar to the enzymic phenomenon reported by Cysyk and Prusoff (36) and may explain the strong inhibitory effect of light *plus* BrdUrd-OPO_2Me on the growth of V-79 Chinese hamster cells (13).

In Fig. 5 the structure of BrdUrd-OPO_2Me is compared with the structure of a natural substrate, dUrd. It is suggested that the negative charge on the methylphosphonate group, interacting with group R_2 blocks the transport of the inhibitor complex across the cell membrane.

Cysyk and Prusoff (36) have shown that the treatment of thymidine kinase with 5-iodo 2'-deoxyuridine (IdUrd) and ultraviolet light produces an irreversible loss of enzyme activity. The substrate dThd protects the enzyme from inactivation by IdUrd plus uv light. Since IdUrd is a competitive substrate, these authors have suggested that the uv light reaction occurs when IdUrd is in a complex with the kinase active site. The photolysis of the IdUrd-kinase complex leads to a deiodination reaction and the formation of a free radical of dUrd. The free radical reacts with an amino acid in the active site to form a covalent bond between the dUrd moiety and thymidine kinase. Experiments with [^{14}C]-labeled IdUrd showed that the pyrimidine moiety is irreversibly attached to the enzyme (36). Thus, IdUrd can be used as a photoaffinity label of the active site of thymidine kinase.

The mechanism of the photosensitization of thymidine kinase (36) may be similar to a mechanism of inactivation of a dUrd carrier protein. Consider the model of the complex between BrdUrd-OPO$_2$Me and the active site of a carrier protein shown in Fig. 5. The treatment of this complex with light (13) may produce a debromination reaction and covalent bond formation between the pyrimidine moiety and group R_1 of the carrier site. BrdUrd-OPO$_2$Me may be useful, therefore, as a specific photoaffinity label for a membrane carrier protein. Apparently the carrier protein is protected from the inhibitory effects of BrdUrd-OPO$_2$Me by the competitive substrate dThd (13).

SUMMARY

The phenomenon of a delayed inhibition of cell growth with BrdUrd-OPO$_2$Me has been known for over five years (13). This phenomenon could only be explained by the formation of a complex between BrdUrd-OPO$_2$Me and some bioreceptor site on the cell surface or inside the cell. Presumably, the BrdUrd-OPO$_2$Me receptor complex must interfere with the binding of the bioreceptor to some natural substrate that influences cell division. In this way, a complex of bioreceptor site and inhibitor modifies the metabolic pathways. This metabolic change produces a new physiological status in the growing cell.

Some of the early experiments showed that the exposure of BrdUrd-OPO$_2$Me inhibited cells to light produces a marked enhancement of the growth inhibitory effect (13). The incorporation of BrdUrd into DNA is known to produce a light sensitization. Thus, the marked effect of BrdUrd-OPO$_2$Me plus light was interpreted to indicate that BrdUrd-OPO$_2$Me penetrates the cell membrane and is incorporated into the DNA. This conclusion is totally incorrect. When BrdUrd-OPO$_2$[^{3}H]Me was tested, it was observed that no detectable amounts of labeled compound entered the L5178Y cells within the first few hours and there was no incorporation into DNA even after several days of treatment with BrdUrd-$OPO_2[^3H]Me$. The growth inhibitory effect persisted nonetheless, and could readily be demonstrated with Chinese hamster V-79 cells, the L5178Y cells, human chronic myeloid leukemia cells, and WI-38 human embryo cells (32).

The protective effect of 10 μM thymidine, from the cell growth inhibition by 10 μM BrdUrd-OPO_2Me, was also a puzzling phenomenon (13). This finding suggested that BrdUrd-OPO_2Me interferes with a reaction in the following pathway: medium dThd $\rightarrow$ cytoplasm dThd $\rightarrow$ dTMP $\rightarrow$ dTTP $\rightarrow$ DNA-thymine. The results of Wigler and Schluter (5) showed that for L5178Y cells, this conclusion was false. The answer to the problem of BrdUrd-OPO_2Me action mechanism lies in some deeper understanding of the membrane transport systems.

It was important to show that BrdUrd-$OPO_2[^3H]Me$ remains intact after prolonged exposure to living cells (5). Then it was observed that the bioreceptor site for BrdUrd-OPO_2Me could be localized at a membrane transport carrier for dUrd and related analogs. Although BrdUrd-OPO_2Me does not interfere with the influx of dThd, the compound may inhibit the efflux of dUrd and dThd.

Apparently, dThd can displace BrdUrd-OPO_2Me from the active site of the transport carrier. Thus, BrdUrd-OPO_2Me in the medium does not inhibit dThd influx and BrdUrd-OPO_2Me in the cytoplasm does not inhibit dThd efflux.

In a situation where the two compounds are on opposite sides of the cell membrane, however, BrdUrd-OPO_2Me may inhibit dThd transport. BrdUrd-OPO_2Me in a nucleoside-free growth medium may trap the transport carrier at the outer membrane surface of the L5178Y cell. The reduced availability of carriers at the inner membrane surface inhibits the efflux of the growth inhibitor dThd. The protective effect of dThd in the medium may be explained by a reactivation of the transport system for dThd efflux.

Although the conceptualization of this problem has advanced significantly in the last two years, much remains to be done. It will be important to demonstrate a BrdUrd-OPO_2Me induced decelerated exchange diffusion for efflux. The identification of labeled nucleotides that collect inside the BrdUrd-OPO_2Me inhibited cells is also important. Finally, it is necessary to find out which of the elevated nucleotide pools can be correlated with an inhibition of cell division.

ACKNOWLEDGMENTS

The help of Albert W. Biggs, M. D., who facilitated my attendance at the symposium, is gratefully acknowledged. It would not have been possible to conceptualize the action mechanism of BrdUrd-OPO_2Me without several recent publications. The most important of these are the correlation of membrane transport and cell growth by Pardee (1); a clear description of membrane transport by Goldman (24); studies of dThd transport by Plagemann and others (7, 14); the discovery of decelerated exchange diffusion for efflux with transport inhibitors, by Paterson and co-workers (25, 26); the effect of elevated dThd on intracellular nucleotides and cell growth by Lowe and Grindey (21); and finally, the description of a photoaffinity label by Prusoff and his collaborators (36). Clearly, the nucleoside efflux hypothesis depends on the interconnection of many new ideas.

REFERENCES

1. Pardee, A. B., *In Vitro* (1971) *7*, 95; Sander, G., Pardee, A. B., *J. Cell. Physiol.* (1972) *80*, 267.
2. Rao, P. R., *Experientia* (1971) *27*, 440.
3. Morris, N. R., Fischer, G. A., *Biochim. Biophys. Acta* (1963) *68*, 84.
4. Hosono, R., Kuno, S., *Eur. J. Biochem.* (1975) *57*, 177.
5. Wigler, P. W., Schluter, D. N., *Cancer Res.* (1976) *36*, 3207.
6. Cleaver, J. E., "Thymidine Metabolism and Cell Kinetics," p. 44, North-Holland, Amsterdam, 1967.
7. Plagemann, P. G. W., Erbe, J., *J. Cell. Physiol.* (1974) *83*, 337.
8. Goodman, J. I., Potter, V. R. *Mol. Pathology* (1973) *9*, 297.
9. Bruckner, H. W., Schreiber, C., Waxman, S., *Cancer Res.* (1975) *35*, 801.
10. Goldman I. D., *Mol. Pharmacol.* (1974) *10*, 257.
11. Bresnick, E., Thompson, U. B., *J. Biol. Chem.* (1965) *240*, 3967.
12. Heidelberger, C., *Progr. Nucleic Acid Res. Mol. Biol.* (1965) *4*, 1.
13. Wigler, P. W., Lozzio, C. B., *J. Med. Chem.* (1972) *15*, 1020.
14. Plagemann, P. G. W., Richey, D. P., *Biochim. Biophys. Acta* (1974) *344*, 263.

15. Mizel, S. B., Wilson, L., *Biochemistry* (1972) *11*, 2573.
16. Ilan, J., Quastel, J. H., *Biochem. J.* (1966) *100*, 448.
17. Dunn, B. P., Gout, P. W., Beer, C. T., *Cancer Res.* (1973) *33*, 2310.
18. Everhart, L. P., Rubin, R. W., *J. Cell Biol.* (1974) *60*, 434.
19. Plagemann, P. G. W., Estensen, R. D., *J. Cell Biol.* (1972) *55*, 179.
20. Estensen, R. D., *Proc. Soc. Exp. Biol. Med.* (1971) *136*, 1256.
21. Lowe, J. K., Grindey, G. B., *Mol. Pharmacol.* (1976) *12*, 177.
22. Adams, R. L. P., *Exp. Cell Res.* (1969) *56*, 49.
23. Wigler, P. W., *J. Biol. Chem.* (1968) *243*, 3466.
24. Goldman, I. D., "Drug Resistance and Selectivity," E. Mihich, Ed., p. 299, Academic Press, New York, 1973.
25. Pickard, M. A., Paterson, A. R. P., *Can. J. Biochem.* (1972) *50*, 839.
26. Cass, C. E., Paterson, A. R. P., *J. Biol. Chem.* (1972) *247*, 3314.
27. Waalkes, T. P., Gehrke, C. W., Zumwalt, R. W., Chang, S. Y., Lakings, D. B., Tormey, D. C., Ahmann, D. L., Moertel, C. G., *Cancer* (1975) *36*, 390.
28. Lozzio, C. B., Wigler, P. W., *J. Cell. Physiol.* (1971) *78*, 25.
29. Wigler, P. W., Bindslev, B., Breitman, T. R., *J. Carb. Nucleosides-Nucleotides* (1974) *1*, 307.
30. Kessel, D., Hall, T. C., *Biochim. Biophys. Acta* (1970) *211*, 88.
31. Cooper, G. M., Dunning, W. F., Greer, S., *Cancer Res.* (1972) *32*, 390.
32. Wigler, P. W., Schluter, D. N., *IRCS Medical Sci. Supplement* (1975) *3*, 8.
33. Fox, J. J., Abstracts 172nd National Meeting, American Chemical Society, (1976) San Francisco, Aug. 29-Sept. 3, paper no. 37.
34. Nordlie, R., Lardy, H., *The Enzymes* (1962) *6*, 3.
35. Holy, A., *Collect. Czech. Chem. Commun.* (1967) *32*, 3713.
36. Cysyk, R., Prusoff, W. H., *J. Biol. Chem.* (1972) *247*, 2522.

CELL SURFACE AS A TARGET FOR CHEMOTHERAPY POTENTIAL INHIBITORS OF BIOSYNTHESIS OF THE PROTEIN-CARBOHYDRATE LINKAGE IN GLYCOPROTEINS

B. PAUL
W. KORYTNYK

Department of Experimental Therapeutics
Roswell Park Memorial Institute
Buffalo, New York

Numerous studies have shown significant differences between the cell surface properties and composition of normal cells on one hand and of neoplastic or transformed cells on the other. The nature of these changes has been the subject of extensive investigations in different laboratories (1-3), some of which have been reviewed in this volume (4-6). These differences may well be the basis for the selectivity of some cancer chemotherapeutic agents; they also play a role in host immune responses against cancer cells. There is considerable evidence that most of the cell surface changes that have been observed are due to the cell membrane glycoconjugates, the carbohydrate portion of which is believed to be responsible for many properties associated with the cell surface, such as the antigenicity, cell-to-cell contact, differentiation, and the "homing" of cells. This view is supported by experiments in which the cell surface carbohydrates were removed by specific glycosidases. Thus a dramatic change in the homing of lymphocytes on L-fucosidase treatment has been demonstrated by Gessner and Ginsburg (7). Along similar lines, it has been shown that lymphocytes that have been treated with neuraminadase undergo transformation on treatment with lectins (8).

Another approach to the modification of cell surface glycoconjugates has been by carbohydrate analogs that either could be incorporated as fraudulent components into glycoconjugate or could interfere with the biosynthesis of the latter. Such an effect has been observed with 2-deoxy-D-glucose, which was found to alter the agglutinability of cells (9, 10).

Another carbohydrate analog, 2-deoxy-2(2-fluoroacetamido)-α-D-glucopyranose, was found to modify glycoprotein biosynthesis (11, 12). The antitumor effects of some carbohydrates, such as D-glucosamine and L-fucose is well documented, (13, 14) but their mode of activity is largely unknown. More recently, certain fluorinated carbohydrate analogs were found to have antitumor activity (15, 16).

Based on these findings we have initiated a program in the area of synthesis of potential inhibitors or modifiers of the cell surface glycoconjugates. Different analogs of L-fucose, D-glucosamine, D-galactosamine, and sialic acid have been synthesized with the object of affecting certain biosynthetic pathways in the glycoconjugate biosynthesis (17-19).

In addition, we have attempted to affect the plasma membrane directly, by synthesizing different glycosides of *N*-acetyl-D-glucosamine with hydrophobic aglycons (20).

Still another approach, the object of which is to inhibit the first step in the attachment of the carbohydrate to protein, has also been developed in our laboratory. This will be the subject of this symposium report. Part of this work has been presented earlier (21).

The key compound representing the carbohydrate-protein link of many glycoproteins is 2-acetamido-1-*N*(4-L-aspartyl)-2-deoxy-β-D-glucopyranosylamine (*1*) (Asn-GlcNAc) (22).

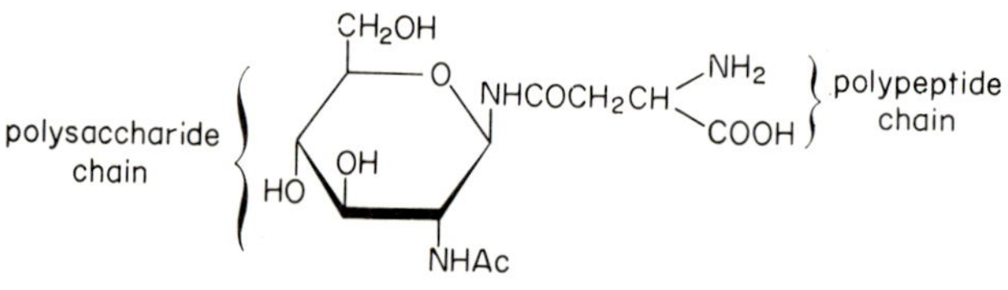

(*1*)

Although details of the first step in the attachment of the GlcNAc moiety to the protein are not known (23, 24), we have undertaken the synthesis of some analogs of (*1*) in the hope of obtaining specific inhibitors of the first step of glycoprotein biosynthesis. As an initial approach we have synthesized several 1-*N*-substituted derivatives of 2-acetamido-2-deoxy-3,4,6-tri-*O*-acetyl-β-D-glucopyranosylamine (*6*). The key intermediate 6 for the synthesis of these analogs was obtained in excellent yield starting from *N*-acetyl-D-glucosamine. Treatment with acetylchloride (*2*) gave acetochloroglucosamine (*3*) (25).

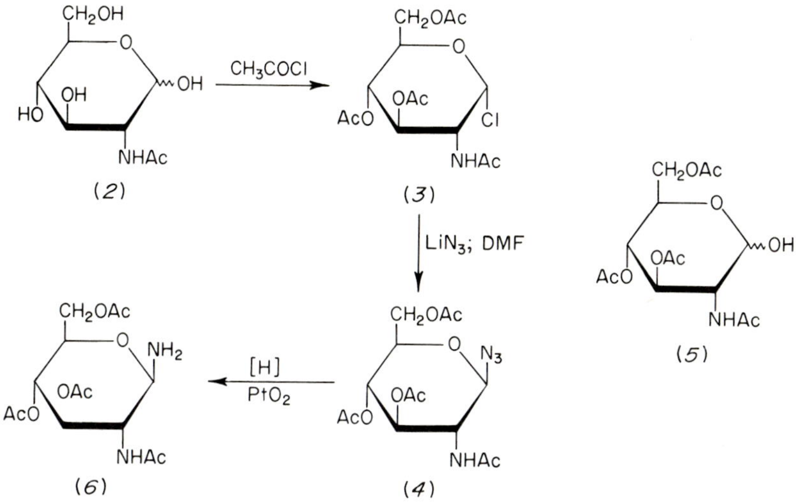

Scheme 1

We found that the best method for the synthesis of the azide (*4*), giving yields of 75 to 80%, was achieved by reacting *3* with dried finely powdered lithium azide in dry DMF at 75 to 80° for 4 hr. The use of silver azide led to explosions, while the use of sodium azide in different solvent gave inferior yield along with 2-acetamido-3,4,6-tri-*O*-acetyl-2-deoxy-D-glucopyranose (*5*) as a byproduct. The ^{1}H NMR spectrum of (*4*) indicated a diaxial arrangements of H(1) and H(2) hydrogens, as shown by the high coupling constant of 9 Hz. Hydrogenation of (*4*) over Adams catalyst at room temperature and atmospheric pressure gave (*6*) in good yield (80%). Since (*6*) is unstable, readily forming isomeric dimers (*7*) and (*8*), different melting points and optical rotations have been reported. We find that the compound shrinks and turns brown at 145 to 146° and decomposes at 235 to 240° with $[\alpha]^{24}$D - 22.9 (C 1.15 in $CHCl_3$).

(*6*)

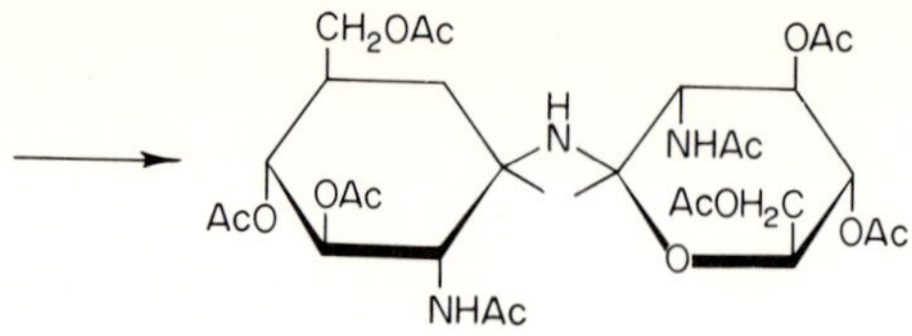

(*7*)

+

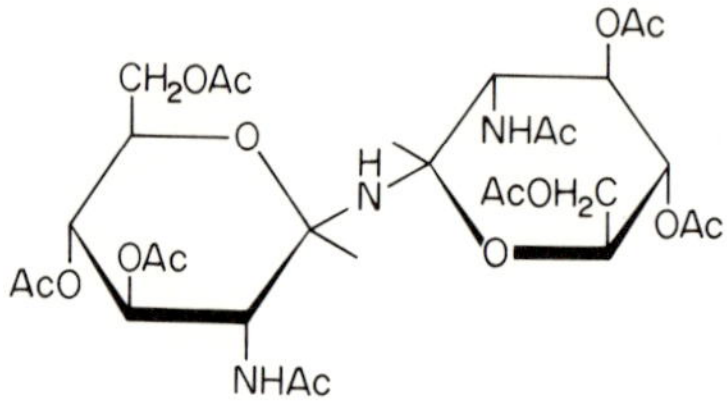

(*8*)

The formation of dimers (*7*) and (*8*) during catalytic reduction of (*4*) was reported by Yamamoto *et al.* (26) and extensively studied by Bolton *et al.* (27). We have reexamined this reaction in order to be able to postulate a mechanism for it and to explain some biological activities of the compound. Thus we found that the formation of dimers occurs on storage of (*6*) at room temperature, or on heating a solution of (*6*) in methanol. It is also formed during silica gel chromatography using methanol-chloroform as an eluant. When a methanol solution of (*6*) is kept at room temperature for eight weeks, a de-O-acetylated dimer (*9*) was obtained. The identity of the product was shown by the acetylation of (*9*) giving products (*7*) and (*8*).

(*6*)

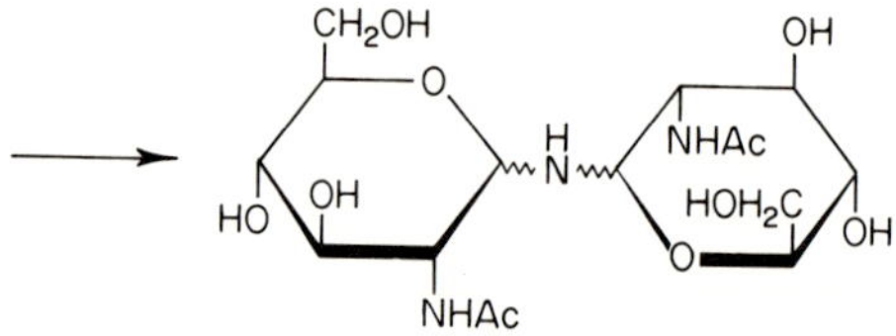

(*9*)

A probable mechanism of the dimer formation is shown in Scheme 2. Initially, *6* is converted into an acyclic immonium ion intermediate (*10*), which then reacts with second molecule of (*6*) to give the intermediate (*11*). The latter undergoes ring closure with the elimination of ammonia to give the acetylated dimers (*7*) and (*8*). The two isomers can be

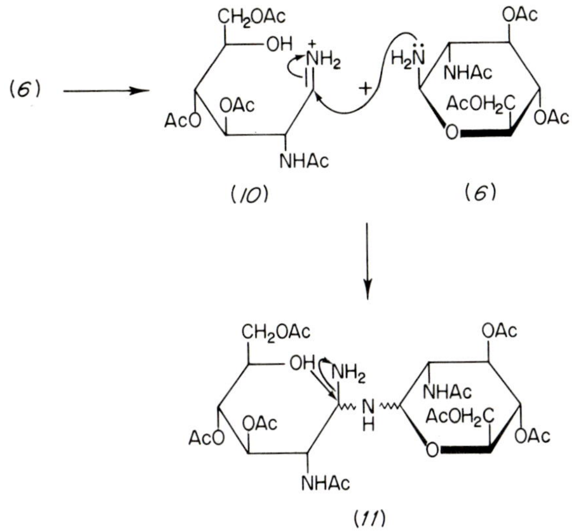

Scheme 2

separated by TLC on silica gel (1:9 methanol:chloroform R_f 0.20 for ββ and R_f 0.33 for αβ isomer) and separated on a silica gel column. The conversion of the ββ isomer to α, β isomer on standing in chloroform was studied initially by ^1H NMR spectroscopy at 100 MHz. Figure 1A shows the NMR spectrum of freshly dissolved (7) and Fig. 1B after 24 hr. It is evident that a substantial change has occurred in this period. This change could be much more clearly defined by ^{13}C NMR spectroscopy at 25.4 MHz. A fresh solution of (7) in $CDCl_3$ (Fig. 2A) shows a very similar ^{13}C NMR spectrum to that of (*6*) (Fig. 2B), indicating that the dimer is symmetrical. A similar simplicity of ^{13}C NMR spectra due to symmetry has been observed for α,α and β,β trehalose (28). The ^{13}C NMR spectrum of the ββ isomer (7) in $CDCl_3$ after being kept for three weeks at room temperature (Fig. 2C) shows a much more complex pattern. C-1, C-2, and C-6 each show distinctly three separate peaks (one peak due to ββ isomer and two peaks due to α,β isomer (^{13}C NMR spectrum of α,β isomer is shown in Fig. 2D), supporting the partial conversion of (*7*) to (*8*) as has been indicated earlier by TLC. Compound (*8*) on the other hand appears to be quite stable.

In spite of the instability of the key intermediate (*6*), we have been able to convert it to various 1-*N*-substituted acyl derivatives (Scheme 3), designed to mimic the Asn-GlnNAc moiety (*1*). Thus (*6*) on treatment with acetic anhydride or chloroacetic anhydride or dichloroacetic anhydride in pyridine gave acetyl (*12*) (27), chloroacetyl (*13*), and dichloroacetyl (*14*) derivatives, respectively. However, (*6*)

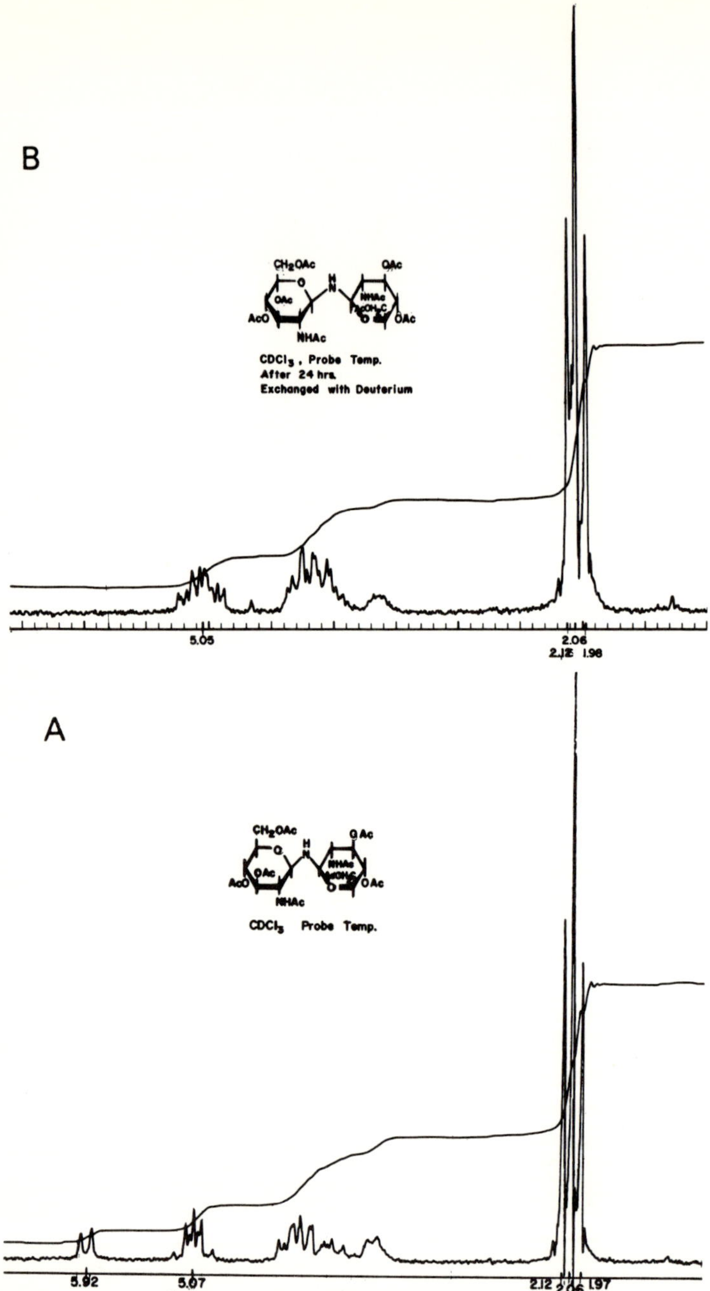

FIGURE 1 (A) ^{1}H NMR spectrum (100 MHz) of the ββ-dimer (7) in $CDCl_3$ showing an NH doublet at δ5.92 and a relatively simple pattern in the anomeric region at δ5.07. (B) Same as above, after 24 hr. The anomeric region around δ5.05 shows a complex pattern even after being exchanged with deuterium.

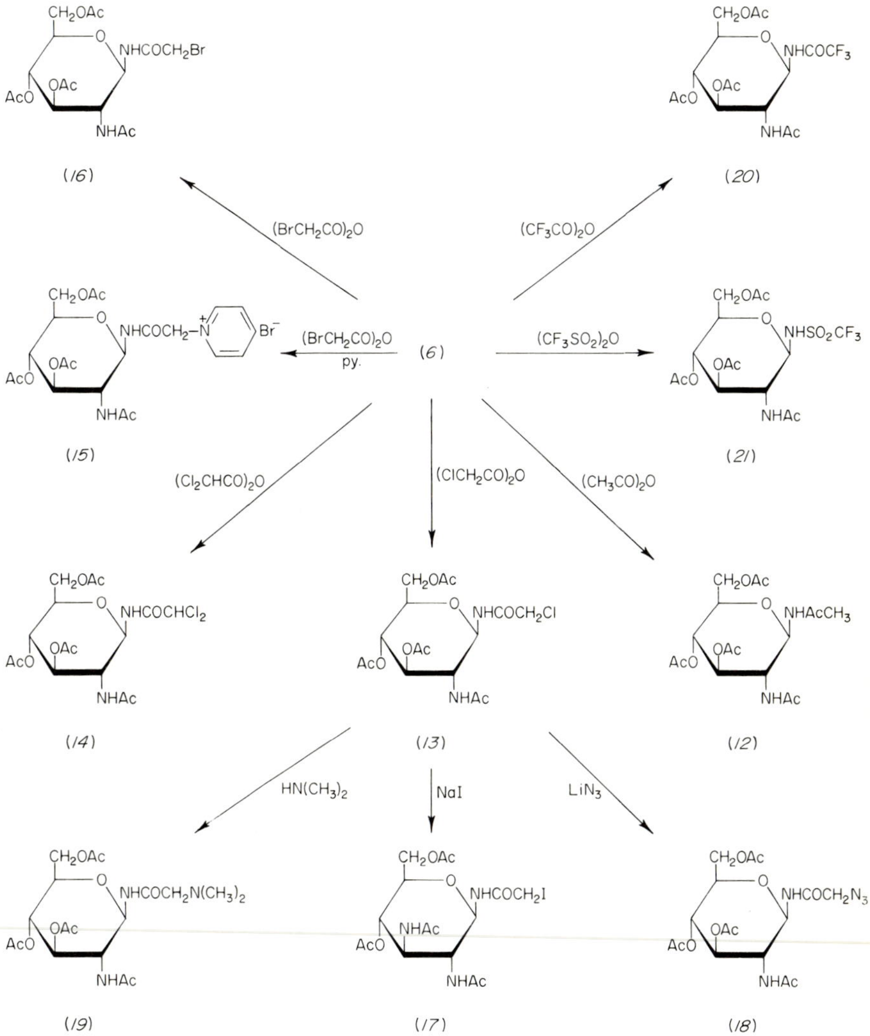
CH2OAc
NHCOCH2Br
AcO
OAc
NHAc
(16)
CH2OAc
NHCOCF3
AcO
OAc
NHAc
(20)
(BrCH2CO)2O
(CF3CO)2O
CH2OAc
NHCOCH2–N+ Br−
AcO
OAc
NHAc
(15)
(BrCH2CO)2O
py.
(6)
(CF3SO2)2O
CH2OAc
NHSO2CF3
AcO
OAc
NHAc
(21)
(Cl2CHCO)2O
(ClCH2CO)2O
(CH3CO)2O
CH2OAc
NHCOCHCl2
AcO
OAc
NHAc
(14)
CH2OAc
NHCOCH2Cl
AcO
OAc
NHAc
(13)
CH2OAc
NHAcCH3
AcO
OAc
NHAc
(12)
HN(CH3)2
NaI
LiN3
CH2OAc
NHCOCH2N(CH3)2
AcO
OAc
NHAc
(19)
CH2OAc
NHCOCH2I
AcO
NHAc
NHAc
(17)
CH2OAc
NHCOCH2N3
AcO
OAc
NHAc
(18)

Scheme 3

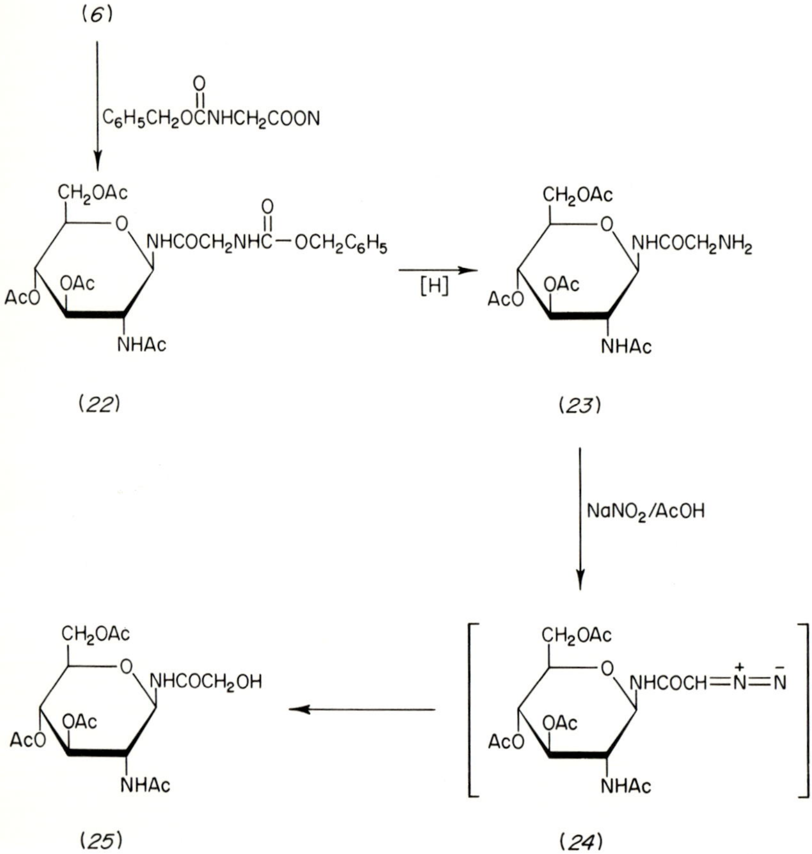
(6)
$C_6H_5CH_2OCNHCH_2COON$
CH_2OAc
$NHCOCH_2NHC-OCH_2C_6H_5$
OAc
AcO
NHAc
(22)
[H]
CH_2OAc
$NHCOCH_2NH_2$
OAc
AcO
NHAc
(23)
$NaNO_2/AcOH$
CH_2OAc
$NHCOCH=\overset{+}{N}=\overset{-}{N}$
OAc
AcO
NHAc
(24)
CH_2OAc
$NHCOCH_2OH$
OAc
AcO
NHAc
(25)

Scheme 4

FIGURE 2 (A) ^{13}C NMR spectrum (25.4 MHz) of the ββ-dimer (7) in $CDCl_3$. Note simplicity of the spectrum indicating symmetry. (B) ^{13}C NMR spectrum of (6) in $CDCl_3$. The spectrum is very similar to that of the dimer (7) in Fig. 2A. (C) Same as in A but after keeping for three weeks. Note complexity of the pattern indicating a mixture of the ββ and αβ isomers. (D) ^{13}C NMR spectrum of the αβ-siomer showing "doubling" of peaks of each monomer due to asymetry.

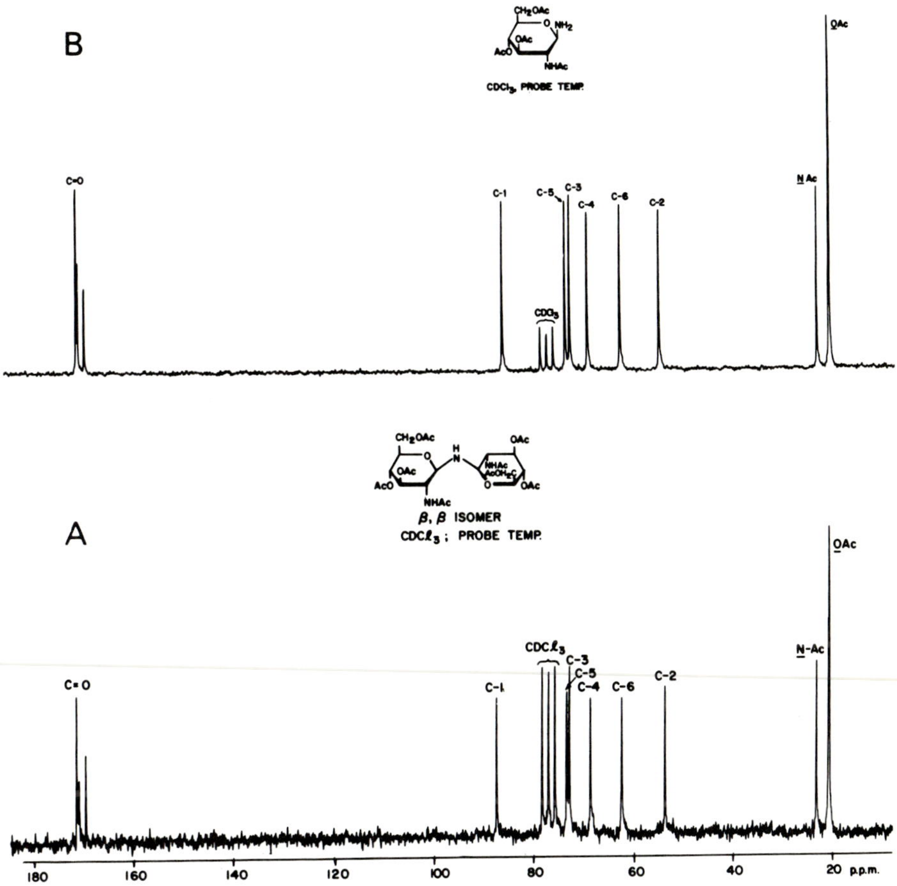

on treatment with bromoacetic anhydride in pyridine gave pyridinium salt of bromoacetyl derivative (*15*). The formation of pyridinium salt could be avoided when the reaction was carried out in methanol and benzene when the desired bromoacetyl derivative (*16*) could be obtained. Iodoacetyl derivative (*17*) was made by reacting (*13*) with dried, finely powdered sodium iodide in acetone. Similarly (*13*) on treatment with dried, finely powdered lithium azide in acetone gave azidoacetyl (*18*) and with dimethylamine in DMF gave dimethylaminoacetyl (*19*) derivatives. Treatment of (*6*) with trifluoroacetic anhydride and trifluoromethane sulfinic anhydride in methylenechloride gave the expected trifluoroacetyl (*20*) and trifluoromethanesulfonyl (*21*) derivatives, respectively.

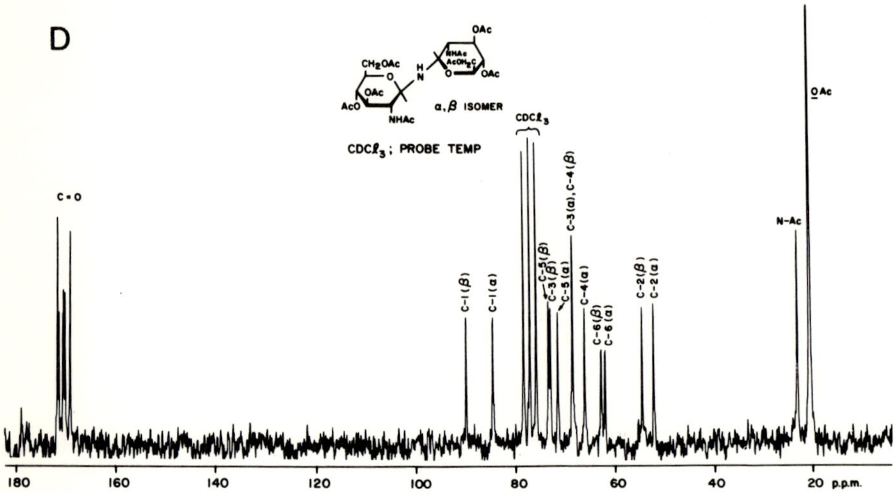

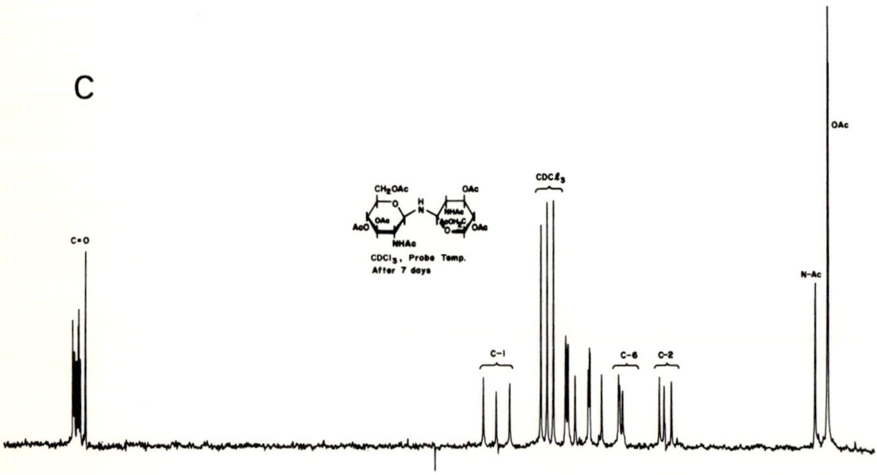

In an attempt to synthesize diazoacetyl derivative of (*6*) (Scheme 4), benzyloxycarbonylglycine was condensed with (*6*) in the presence of dicyclohexylcarbodiimide (DCC) in methylene-chloride to give benzyloxycarbonylglycinamido derivative (*22*) (26). Hydrogenolysis of (*22*) gave glycinamido derivative (*23*) (26), which was then diazotized with $NANO_2$ and acetic acid in the presence of 2 *M* sodium acetate. The crude product showed a diazo peak at 2120 cm^{-1} in the infrared spectrum. However, the diazoacetyl derivative (*24*) was found to be unstable and was converted to hydroxyacetyl (glycolyl) derivative (*25*) during chromatographic separation on a silica gel column.

Most of the compounds synthesized were de-*O*-acetylated by treatment with 10% Et_3N in 50% aqueous methanol at room temperature over a period of 4 to 6 hr.

All these simple analogs of (*1*) were tested for their growth inhibitory activity against mouse mammary adenocarcinoma (TA3) and L1210 cells *in vitro* and inhibitory data are recorded in Table I. The key intermediate (*6*), the chloracetyl (*13*), bromoacetyl (*16*), trifluoracetyl (*20*), and trifluoromethanesulfonyl (*21*) derivatives inhibited growth at 10^{-4} to 10^{-6} *M*. Chloroacetyl (*13*) and bromoacetyl (*16*) derivatives are potential alkylating agents, as in the case of Streptozotocin (29) and *N*-nitroso-*N*-chloroethylurea derivative of glucose (GCNU; chlorozotocin) (30). The finding that (*6*) is inhibitory is intriguing in view of the reactivity of the compound as indicated by its dimer formation. To study its mode of action we have substituted the NH_2 function at position 1 by OH (*5*), SH (*26*), H (*27*), $OCOCH_3$ (α, *28*; β, *29*) and OCH_3 (α, *30* and β, *31*) and also tested 1,3,4,6-tetra-*O*-acetyl-α-D-glucosamine hydrochloride (*32*). All these compounds showed growth inhibitory activity at 10^{-4} *M* except for SH (*26*), H (*27*), and OCH_3 (α, *22* and β, *23*) derivatives, which were inactive at 10^{-3}M. It is interesting to note that none of the de-*O*-acetylated derivatives of the series of analogs listed in Table I showed any inhibitory activity. This is probably due to the insufficient permeability of these analogs.

The similarity of biological activity of (*5*), (*6*), (*28*), (*29*), and (*32*) may indicate the possibility of conversion of these compounds to a common intermediate, possibly to (*5*) in the biological system (Scheme 5). Thus (*5*) can be generated from (*6*), (*28*), and (*29*) by hydrolysis of the amino and acetoxyl function respectively. N→O acetyl migration in (*32*) at neutral or slightly alkaline condition (pH 7.4) could also generate (*5*) in the biophase.

The acylic (aldehydo) form of (*5*) may react with L-cysteine to form a thiazolidine derivative, thus depleting the cells of this amino acid. It has been reported earlier that certain tumor cells require L-cysteine for growth, and

TABLE I Growth Inhibition of Cultured Mouse Mammary Adenocarcinoma (TA3) Cells[a]

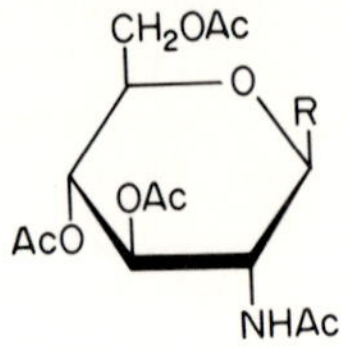

Compound	Compound R=	ID_{50}(M)
(4)	N_3	10^{-3} (∿40%)
(6)	NH_2	3.2×10^{-4}
(12)	$NHCOCH_3$	NI[a]
(13)	NHCO CH_2 Cl	2.6×10^{-4}
(14)	NHCO $CHCl_2$	NI
(15)	NHCO CH_2 +N Br-	NI
(16)	NHCO CH_2 Br	1.5×10^{-5} [b]
(17)	NHCO CH_2 I	NI
(18)	NHCO CH_2 N_3	NI
(19)	NHCO CHN $(CH_3)_2$	NI
(20)	NH $COCF_3$	2×10^{-4}
(21)	NH SO_2 CF_3	7×10^{-6} [b]
(22)	NH CO CH_2 NH CO O-CH_2-	NI
(23)	NH CO CH_2 NH_2	NI
(26)	SH	NI
(27)	H	NI
(29)	O CO CH_3	1.9×10^{-4}
(31)	OCH_3	slight effect at 10^{-3}

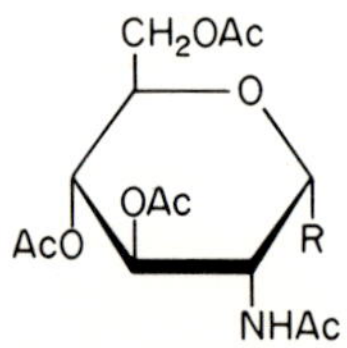

Compound	Compound R=	ID_{50}(M)
(5)	OH	2.4×10^{-4}
(29)	O CO CH_3	3.6×10^{-4}
(31)	OCH_3	NI

[a]NI, no inhibition at 10^{-3} M.
[b]Tested against L1210 cells in culture.

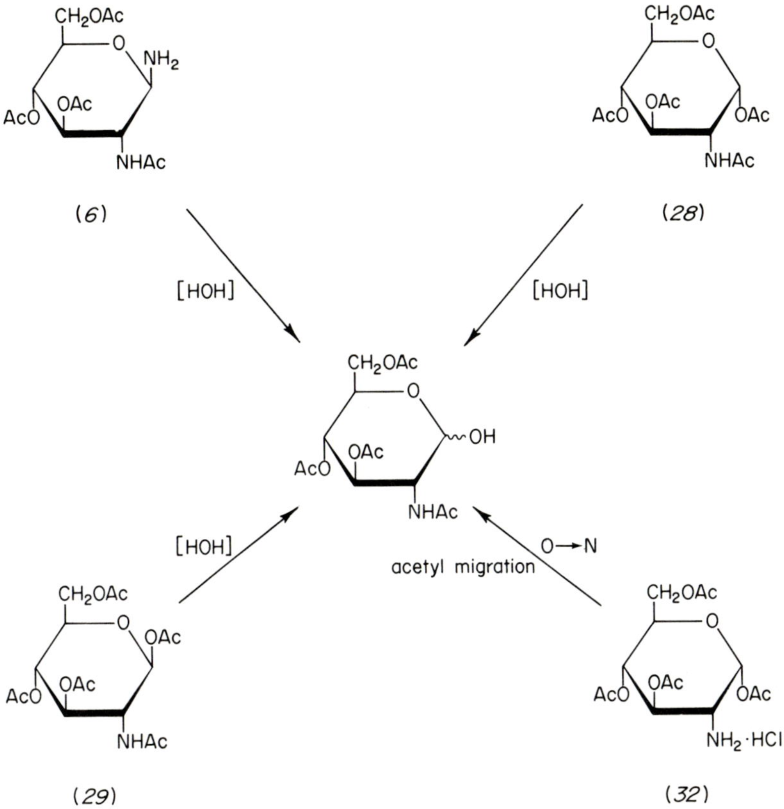

Scheme 5

thus its depletion would result in regression of such tumors (31). A second possibility of the mode of action of (*6*) relates to its conversion to (*5*), as already discussed, followed by de-*O*-acetylation and conversion to UDP-*N*-acetyl-D-glucosamine. An analogous metabolism has been shown to occur with α-pentaacetyl-D-glucosamine (*28*) and results in the depletion of the UTP pool, which may contribute to the observed growth inhibition.

The third possibility considered to explain the inhibitory activity of (*6*) concerns the reactivity of the 1-NH_2 group in forming dimers, under mild conditions, as has been discussed above. Thus a nucleophilic function (such as thio) at the active site of an enzyme or a "receptor" may attack the electron deficient imonium ion (*10*), postulated in the dimerization reaction, forming a covalent bond (Scheme 6, *33*). This is followed by ring closure and elimination of ammonia.

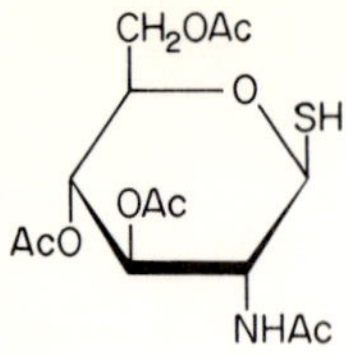

(*26*)

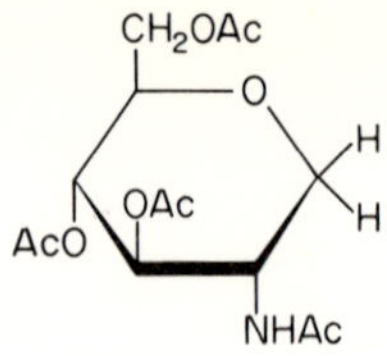

(*27*)

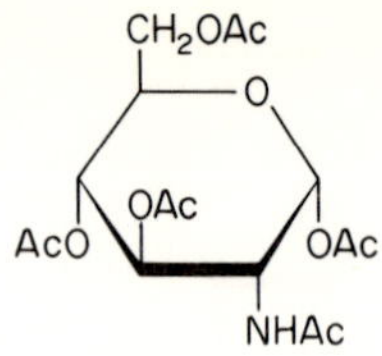

(*28*)

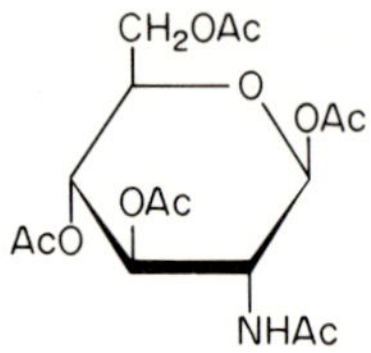

(*29*)

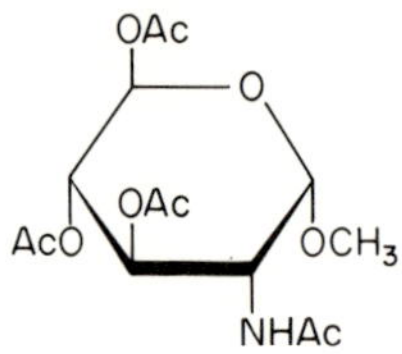

(*30*)

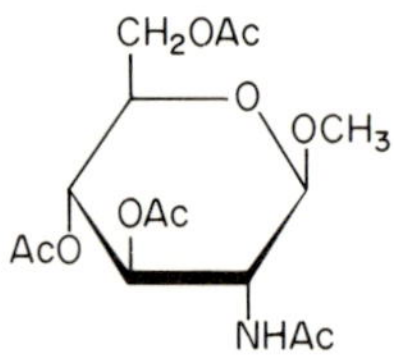

(*31*)

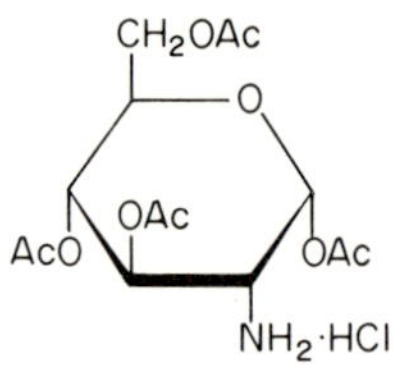

(*32*)

In order to test this hypothesis, (*6*) was reacted with ethanethiol in methanol at room temperature over an extended period of time (two months). The reaction product was found to give five spots on TLC (1:9, MeOH:$CHCl_3$). Chromatography on a silica gel column gave three products, which were identified by elemental analysis, IR, ^1H, and ^{13}C NMR spectroscopy as 1,2-diacetamido-1,1,2-tri-deoxy-1-ethylthio-D-glucosealdehydrol (*35*) and its mono- (*36*) and tri-*O*-acetyl (*37*) derivatives in 35, 25, and 6% yield, respectively (Scheme 7). Similar products were obtained when the reaction was carried out at 60° for 48 hr in a sealed tube.

Initially the reaction proceeds as described in Scheme 6, i.e., by the addition of ethanethiol to the acyclic immonium ion intermediate (*10*). Subsequently, the ring closure which is followed by the elimination of the amino group at position 1 could not take place due to *N*-acetylation. It is probable that this occurs by an intramolecular mechanism as shown in

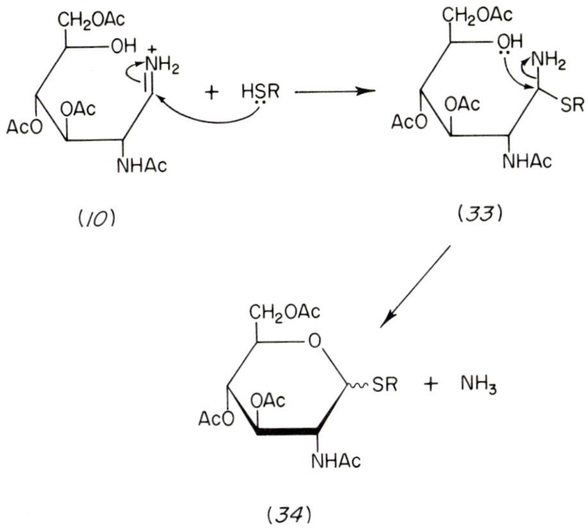

Scheme 6

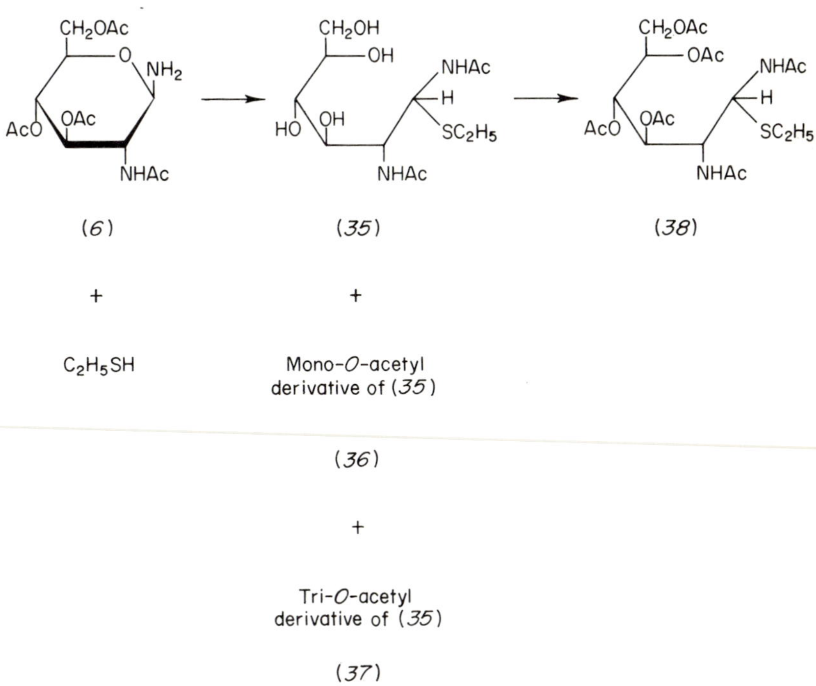

Scheme 7

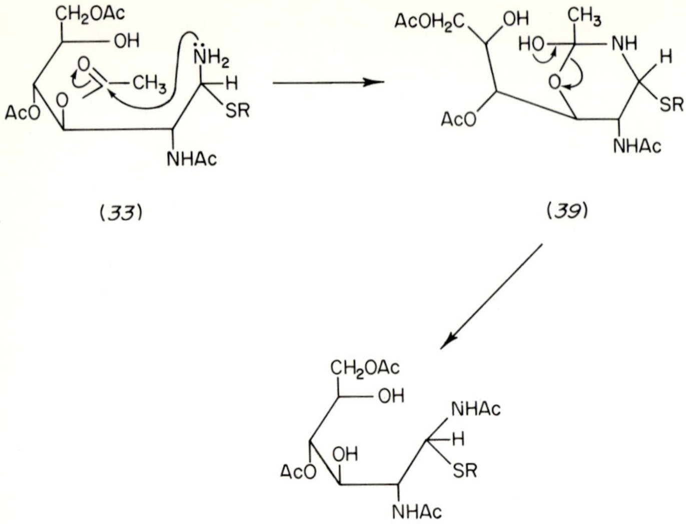

(33) (39) (40)

Scheme 8

Scheme 8. This type of O→N acylmigration is well documented in amino sugar chemistry involving five- or six-membered cyclic intermediates, as shown in (*39*). The thiol compound may also be involved in this reaction, by acting both as an acyl acceptor from the *O*-acetyl group and acyl donor to the nitrogen function. Once this has been accomplished the thiol reagent may be active as a de-*O*-acetylating agent to give the partially and fully de-*O*-acetylated derivatives. Alternatively, de-*O*-acetylation may have occurred as the result of the liberation of ammonia arising during dimer formation in the initial stages of the reaction.

^{1}H and ^{13}C NMR spectra of (*35*) indicate that the product is a mixture of two stereoisomers (Fig. 3). Thus in ^{1}H NMR spectrum (Fig. 3A) there are two superimosed quartets due to the ethyl protons and in the ^{13}C NMR spectrum (Fig. 3B) there are two peaks due to C-4. The formation of the two stereoisomers is rationalized as being due to the backside and frontside attack by the sulfhydryl group on the acyclic imonium ion intermediate. The structures have been further confirmed by converting (*35*) to its hexaacetylderivative (*38*). ^{1}H and ^{13}C NMR spectra (Fig. 4) likewise show the presence of two isomers. Assignments of peaks was facilitated by the preparation of model compounds, such as 2-acetamido-2-deoxy-

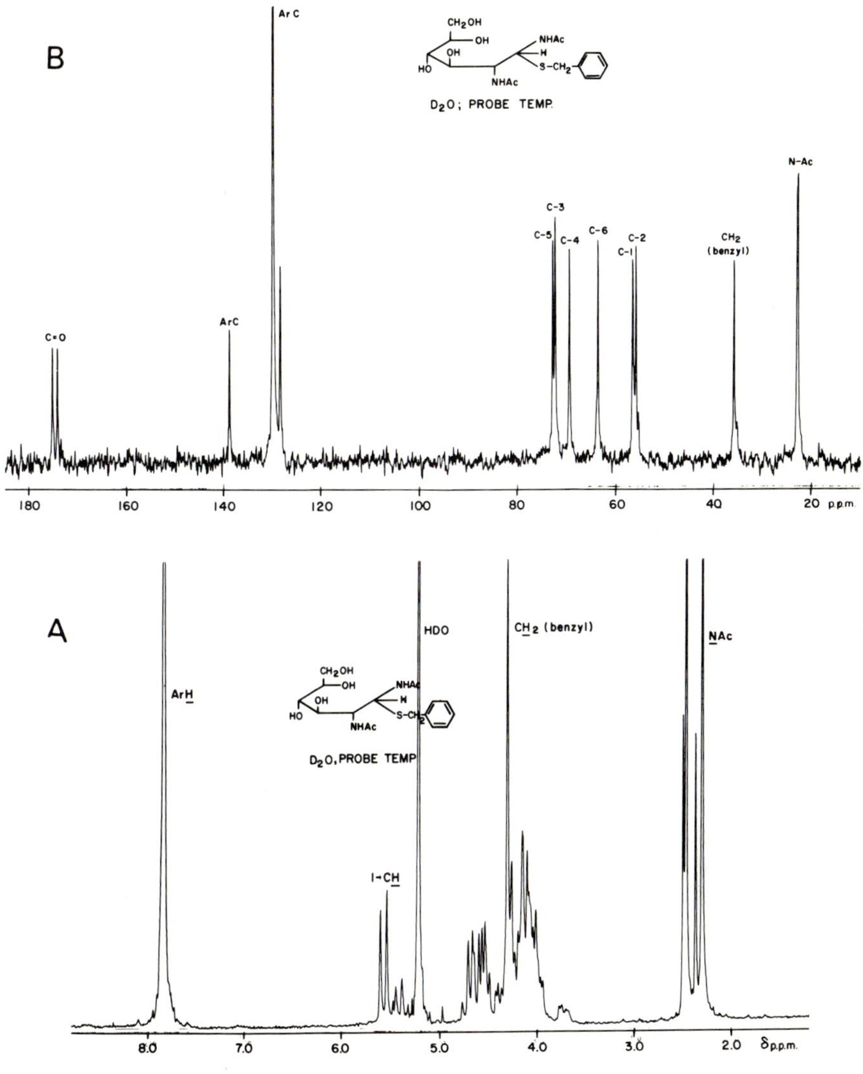

FIGURE 3 (*A*) ^{1}H *NMR spectrum of the ethylthiol adduct (35) in* D_2O *showing the two isomers.* (*B*) ^{13}C *NMR spectrum of (35) in* D_2O.

D-glucose diethylthioacetal and its preacetylated derivative and other acyclic derivatives. Attempts to separate the two stereoisomers by chromatography were not successful.

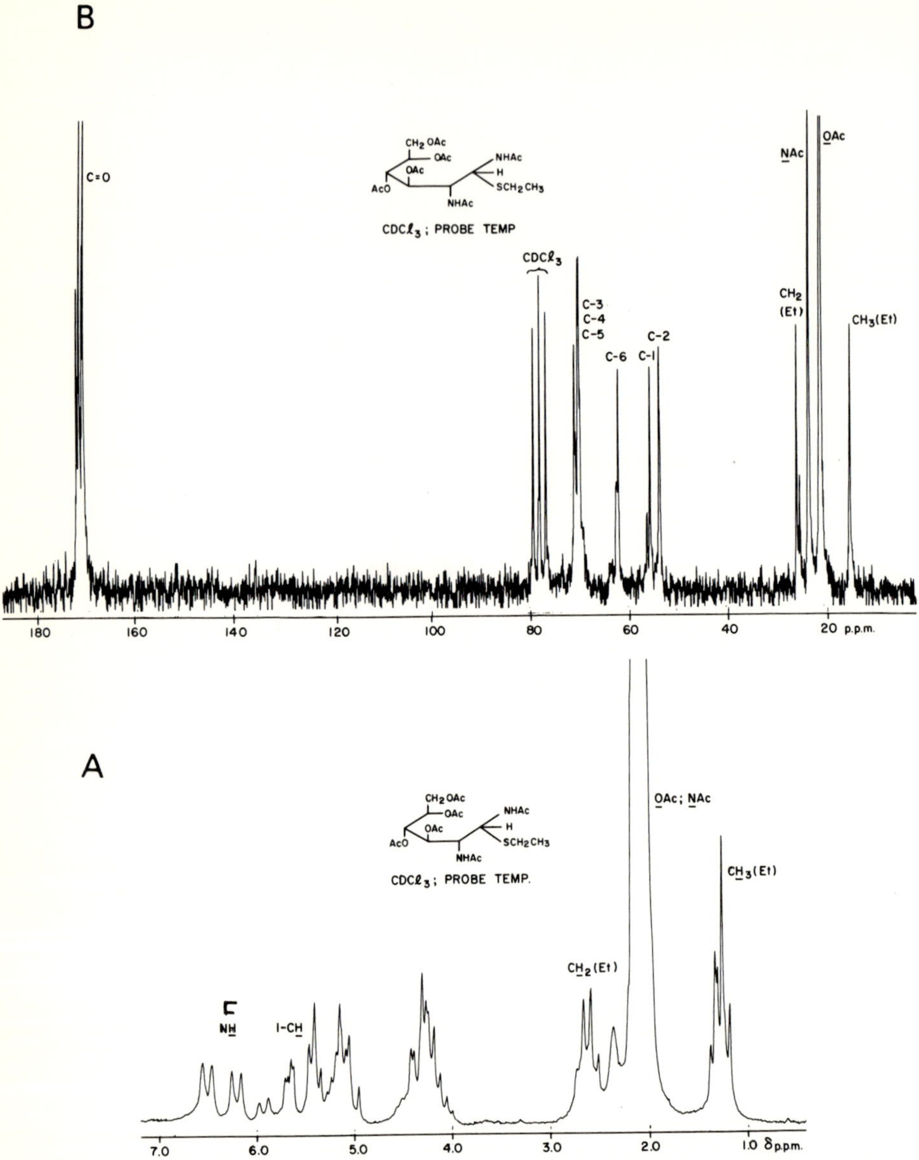

FIGURE 4 (A) ^{1}H and (B) ^{13}C NMR spectra of the acetylated ethylthiol adduct (38) showing the presence of two epimers.

Mass spectral fragmentation pattern (Fig. 5) of the hexa-acetyl derivative shows the expected molecular ion peak m/e 492. The peak at m/e 463 arises from the fission of a carbon-

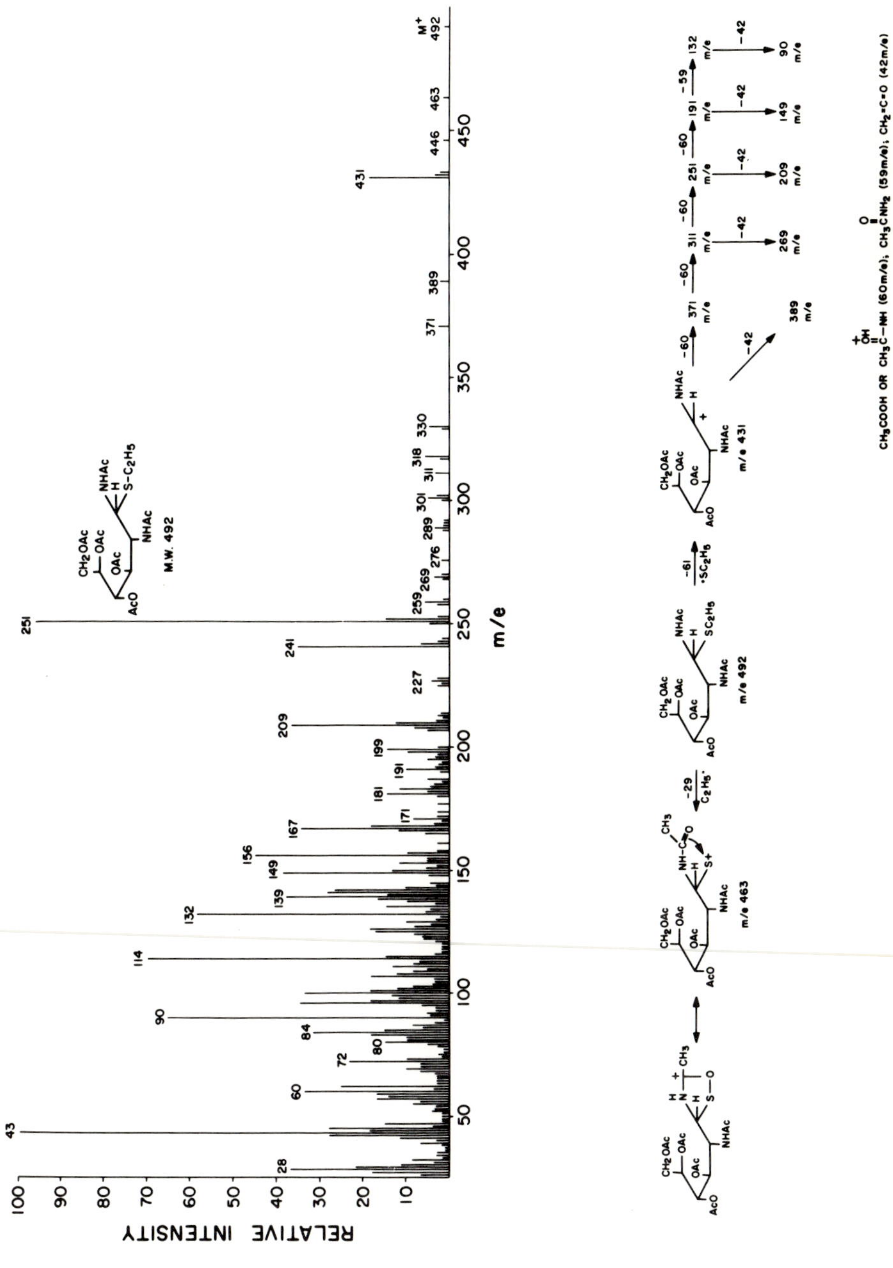
RELATIVE INTENSITY
m/e
M.W. 492
M+ 492
m/e 492
m/e 463
m/e 431
-29 C_2H_5·
-61 ·SC_2H_5
CH_3COOH OR $CH_3\overset{+OH}{C}-NH$ (60m/e); $CH_3\overset{O}{C}NH_2$ (59m/e); $CH_2=C=O$ (42m/e)

FIGURE 5 Mass spectrum of (38) determined with a DuPont 21-491 mass spectrometer, direct inlet, ionizing potential 70 eV, ionizing current 18 μA.

sulfur bond in the molecular ion and elimination of the ethyl group. A stabilized five-membered ring can be drawn for this fragment. The fragment m/e 431 is obtained by the loss of ethanethiol. From this ion species m/e 431, acetic acid or protonated acetamide (m/e 60), acetamide (m/e 59), ketene (m/e 42) is eliminated. The position and the sequence of elimination of acetyl or ketene are not known.

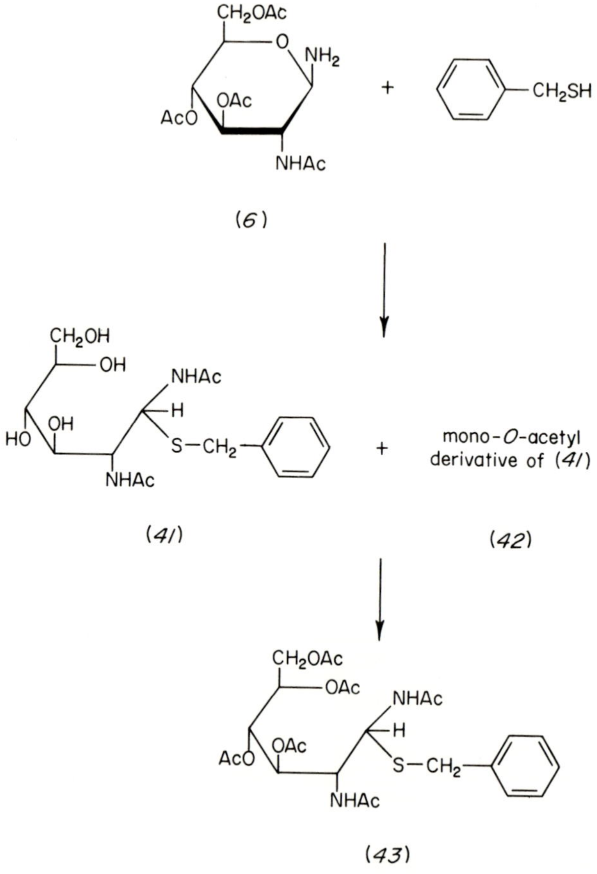

Scheme 9

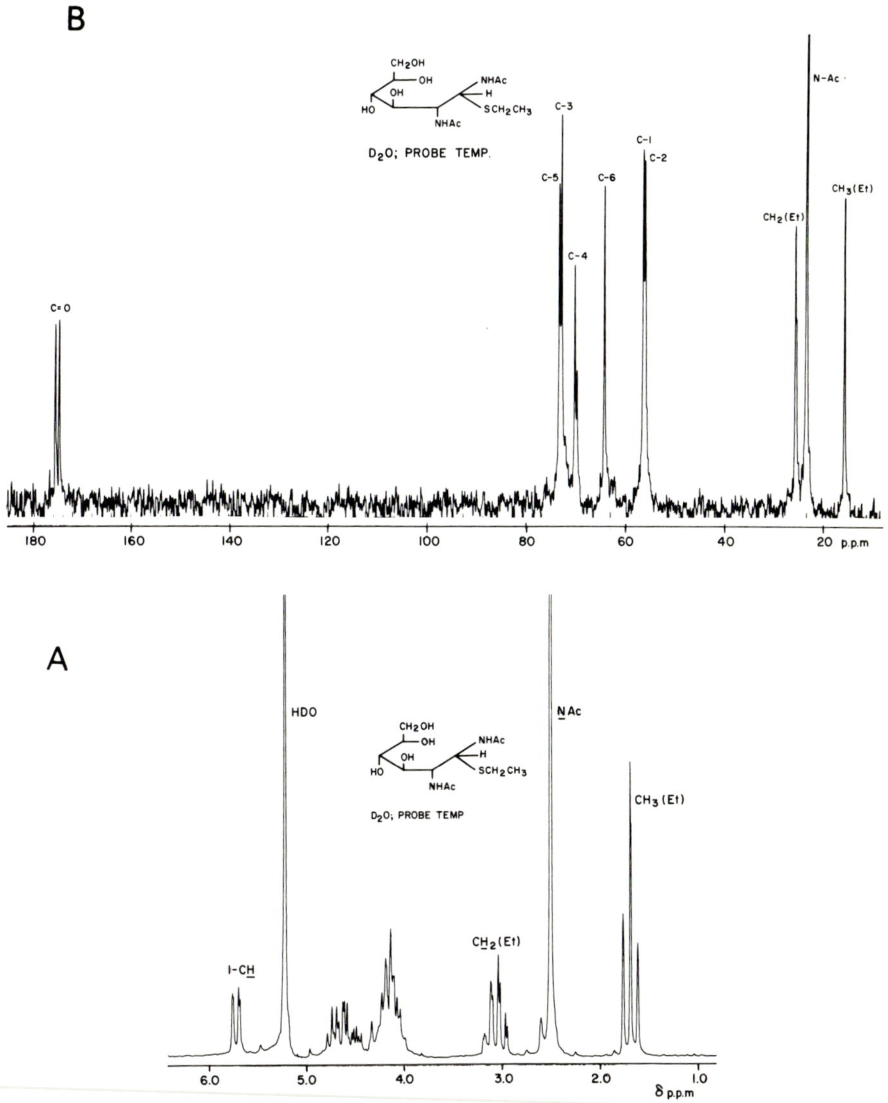

FIGURE 6 (A) ^{1}H and (B) ^{13}C NMR spectra of the benzyl thiol addition product (41) in D_2O.

To study the general applicability of this reaction, (6) was treated with benzylthiol in methanol for a period of two months. The (1:9 MeOH:$CHCl_3$) of the reaction product consisted of four spots. Chromatography on a silica gel column using methanolic chloroform as solvent (10-30%) gave three products identified as 1,2-diacetamido-1,1,2-trideoxy-1-

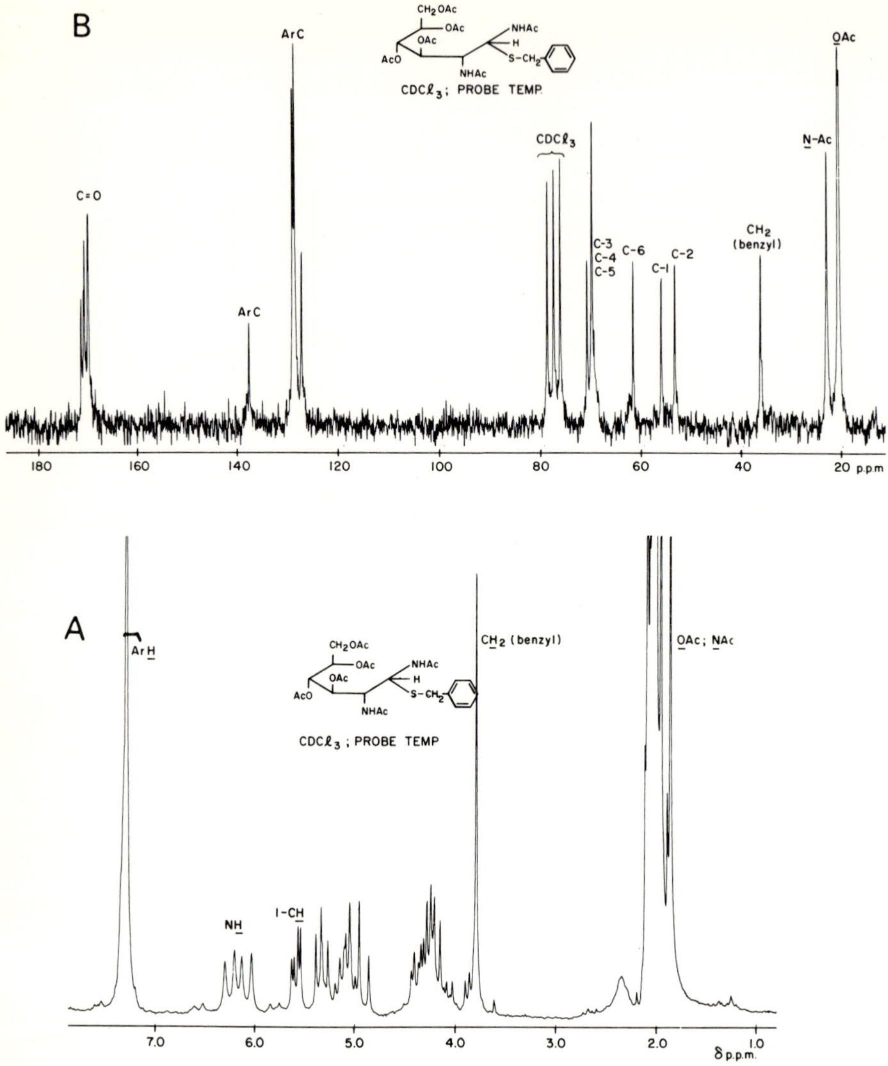

FIGURE 7 (A) ^1H and (B) ^{13}C NMR spectra of the peracetylated thiol addition product (43) in $CDCl_3$.

benzylthio-D-glucose aldehydrol (*41*) and its mono-*O*-acetyl derivative (*42*) and dibenzyldisulfide (Scheme 9). The acyclic structure is confirmed by converting (*41*) to its hexaacetyl derivative (*43*) and also by elemental analysis, IR, ^1H- and ^{13}C-NMR and mass spectral data. ^1H and ^{13}C-NMR spectral data for (*41*) and (*43*) (Figs. 6 and 7) indicate the product to be mixture of two stereoisomers.

Thus the above model experiments show (*6*) to be a potential pseudoalkylating agent that is probably de-*O*-acetylated within the cell, as has been found for many *O*-acetylated derivatives and then reacts with appropriate receptors. Alternative suggestions for its biological activity have also been proposed.

At the beginning of this report we described a number of derivatives of (*6*), some of which were provided with alkylating groups. Some of these derivatives were even more active than the parent compound (*6*) (Table I). The mode of action may be different in each case.

While the target of these studies was to develop specific inhibitors against Asn-GlcNAc linkage formation and hence glycoprotein synthesis, the observed growth-inhibitory effects may be completely unrelated to it. Indeed, the inhibition of the glycoprotein biosynthesis may have little appreciable effect on cell growth and yet results in significant changes in immunogenic properties and social behavior of the cells. In order to answer these questions, more specific test systems are essential. The development of such test systems is being pursued by our colleagues in pharmacology and immunology at Roswell Park Memorial Institute.

ACKNOWLEDGMENTS

We would like to thank Dr. E. Mihich for his active encouragement of the program. This study was supported by USPHS Grants CA-08793, CA-13038 as well as the Institute Core Grant CA-16056. We also thank Mrs. Onda Dodson Simmons for determining the NMR spectra, and Dr. M. T. Hakala, under whose supervision the biological tests were performed.

REFERENCES

1. J. Schultz and R. E. Block (eds.), "Membrane Transformations in Neoplasia," Miami Winter Symposia, Vol. 8, Acad. Press, New York 1974.
2. L. Weiss, The Cell Periphery, Metastasis and Other Contact Phenomena, Vol. 7 of Frontiers of Biology (Neuberger, A. and Tatum, E. L., eds.), North-Holland, Amsterdam 1967.
3. (a) G. L. Nicolson, *Biochim. et. Biophys. Acta. 458*, 1 (1976). (b) G. L. Nicolson and G. Poste, New England J. Med. *295*, 253 (1976).

*4. M. C. Glick, "Surface Membranes of Tumor Cells," (Abstract #114).

*5. J. F. Codington, D. Vanden Eijnden, and R. W. Jeanloz, Structural Studies on the Major Glycoprotein of the TAH3-Ha Ascites Tumor Cell," (Abstract #115).

*6. P. T. Mora, *et. al.*, Pertinence of surface-Membrane Changes in Spontaneous and Virally Transformed Cells to the Balance between Tumorigenicity and Immune Rejection, (Abstract #117).

7. B. M. Gessner and V. Ginsburg, *Proc. Natl. Acad. Sci. USA 52*, 720 (1964).

8. A. Novogradsky and E. Katchalski, *Proc. Natl. Acad. Sci. USA 70*, 2515 (1973).

9. S. Steiner, B. Altenburg, and J. L. Melnick, *J. Natl. Cancer Inst. 52*, 617 (1974).

10. K. M. Hwang and A. C. Sartorelli, *Biochem. Pharmac. 24*, 1149 (1975).

11. A. M. Schulz and P. T. Mora, *Carbohydrate Research 40*, 119 (1975).

12. P. W. Kent and P. T. Mora *in*: "Membrane Mediated Information," Vol. 1, American Elsevier Publishing Co., New York, p. 129.

13. J. H. Quartel and A. Cantero, *Nature 171*, 252 (1953).

14. M. H. Seltzer, J. M. Roseman, D. E. Wolfe, K. C. Tsou, E. E. Miller, and F. E. Rosato, *Growth 33*, 353 (1969).

15. E. M. Bessel, V. C. Courtenay, A. B. Foster, M. Jones, and J. H. Westwood, *Eur. J. Cancer 463* (1973).

16. R. W. Pero, P. Barbiarz, L. Rittman, P. Simon, and T. P. Fondy, Abstracts, 170th Natl. Meeting Am. Chem Soc., Chicago, Illinois, August 1975 FLU015.

17. M. Sharma, R. Bernacki, and W. Korytnyk, *Fed. Proc. 34*, 574 (1975).

18. R. Bernacki, B. Paul, M. Sharma, J. Sufrin, and W. Korytnyk, *Proc. Am. Assn. Cancer Res. 17*, 119 (1976).

19. W. Korytnyk, R. Bernacki, L. Danhauser, M. Hanchak, B. Paul, M. Sharma, and J. Sufrin, *Fed. Proc. 35* (7), 1639 (1976).

20. B. Paul and W. Korytnyk, Abstracts, 172nd ACS National Meeting, San Francisco, California, Aug-Sept. 1976, #MEDI 76.

21. B. Paul, R. J. Bernacki, and W. Korytnyk, Abstracts, 170th Natl. Meeting Am. Chem. Soc., Chicago, Illinois, Aug. 1975, MEDI 44.

22. A. Neuberger and R. D. Marshall, in A. Gottschalk (eds.), Glycoproteins, American Elsevier Publishing Co., New York, 1966, p. 275.

*23. W. Pigman, A. Harp, Abstracts of the 172nd American Chemical Society National Meeting, San Francisco, California, Aug.-Sept. (1976) CARB 112.

*24. B. Fleischer, Abstracts of the 172nd American Chemical Society National Meeting, San Francisco, California, Aug.-Sept. (1976), CARB 113.

25. D. Horton, Organic Synthesis, Coll. Vol. V, 1 (1973).

26. A. Yamamoto, C. Miyashita, and H. Tsukamoto, *Chem. Pharm. Bull.*, Tokyo, *13*, 1036 (1965).

27. C. H. Bolton, L. Hough, and M. Y. Khan, *Biochem. J. 101*, 184 (1966).

28. T. Usui, N. Yamaoka, K. Matsuda, K. Tuzimura, H. Sugihama, and S. Seto, *J. Chem. Soc., Perkin I*, 2425 (1973).

29. (a) B. K. Bhuyan, T. J. Fraser, H. H. Buskirk, and G. L. Neil, *Cancer Chemotherap. Rep.* (Part 1) *56*, 709 (1972). (b) B. Rudas, Arzneim, Forsch, *22*, 830 (1972).

30. T. P. Johnston, G. S. McCaleb, and J. A. Montgomery, *J. Med. Chem. 18*, 104 (1975).

31. (a) G. E. Foley, E. F. Barell, R. A. Adams, and H. Lazarus, *Exp. Cell. Res. 57*, 129 (1969). (b) K. A. Harrap and D. E. M. Speed, *Brit. J. Cancer Res. 18*, 809 (1964).

Index

G

J

K

L

S

W

X

Y

Z

A
B
C 8
D 9
E 0
F 1
G 2
H 3
I 4
J 5